Examining the Environmental Footprint of Artificial Intelligence

Momina Shaheen
https://orcid.org/0000-0001-9424-9787
University of Roehampton, UK

IGI Global
Scientific Publishing
Publishing Tomorrow's Research Today

Vice President of Editorial Melissa Wagner
Director of Acquisitions Mikaela Felty
Director of Book Development Jocelynn Hessler
Production Manager Mike Brehm
Cover Design Jose Rosado

Published in the United States of America by
IGI Global Scientific Publishing
701 East Chocolate Avenue
Hershey, PA, 17033, USA
Tel: 717-533-8845 | Fax: 717-533-7115
Website: https://www.igi-global.com E-mail: cust@igi-global.com

Library of Congress Cataloging-in-Publication Data

Names: Shaheen, Momina, 1993- editor
Title: Examining the environmental footprint of artificial intelligence / edited by Momina Shaheen.
Description: Hershey, PA : IGI Global Scientific Publishing, [2027] | Includes bibliographical references and index. | Summary: "This book analyzes AI's environmental impact on the world"-- Provided by publisher.
Identifiers: LCCN 2026004780 (print) | LCCN 2026004781 (ebook) | ISBN 9798337384900 hardcover | ISBN 9798337384917 paperback | ISBN 9798337384924 ebook
Subjects: LCSH: Data processing service centers--Energy consumption | Artificial intelligence--Environmental aspects | Smart cities
Classification: LCC TJ163.5.D38 E93 2027 (print) | LCC TJ163.5.D38 (ebook)
LC record available at https://lccn.loc.gov/2026004780
LC ebook record available at https://lccn.loc.gov/2026004781

British Cataloguing in Publication Data
A Cataloguing in Publication record for this book is available from the British Library.

Table of Contents

Detailed Table of Contents

The increased pace of Artificial Intelligence (AI) expansion had a significant impact on the modern economic system as well as fueling the debate on its environmental sustainability. In this chapter, the authors take a life-cycle approach to the assessment of AI, which covers the energy expenditure, carbon emissions, water use, material mining, and electronic waste considerations. The research highlights how the structural tension between ambitious climate pledges and the increasing material and energy requirements of large-scale AI infrastructure depicts the case studies of the most popular AI developers, Alphabet, Microsoft, Meta, OpenAI, NVIDIA and Salesforce finding that computational intensity and embodied supply-chain emissions confuse the sustainability narrative that focuses only on sourcing renewable energy or making AI infrastructure more efficient. The chapter also presents the new concepts, including Green AI and Responsible Innovation, and stresses the need to have criteria of environmental disclosure, and governance reforms.

Artificial intelligence (AI) is considered a key driver of sustainable transformation, yet its rapid development is leading to growing environmental impacts. We argue that the environmental footprint of AI cannot be understood as a technical side effect. Rather, it is a systemic outcome of intertwined technological, infrastructural, and economic dynamics. Drawing on a life-cycle perspective and ecological economics, as well as infrastructure research and technology governance, we identify three key drivers: scaling paradigms, material dependencies, and market-driven incentive structures. The analysis shows that efficiency gains are often offset or outweighed by rebound effects and structural growth dynamics. Building on this, we develop a multi-level governance framework that integrates lifecycle-based management, infrastructural embedding, and institutional coordination. Finally, we introduce 'net-positive AI,' a conceptual approach that shifts the focus from mere damage minimization to a positive net contribution to ecological and social systems.

Artificial intelligence (AI) is increasingly portrayed as an immaterial and efficiency-enhancing technology. We challenge this view by conceptualizing AI as a capital-intensive, infrastructure-driven system embedded in global material and energy flows. From an ecological-political-economic perspective, we argue that AI's ecological footprint is not merely a byproduct of technological progress but rather the result of financialization and scaling dynamics. Major investments in computing infrastructure turn AI into an asset class, thereby reinforcing incentives for continuous expansion. These dynamics lead to significant environmental impacts throughout the entire life cycle, from raw material extraction and hardware production to data center operations and electronic waste. At the same time, rebound effects and infrastructural lock-in mechanisms exacerbate resource consumption despite efficiency gains. We develop an integrated analytical framework that links financialization, scaling, and ecological externalities, underscoring the need for governance approaches.

Artificial intelligence has become central to modern marketing, enabling brands to position themselves as innovative and environmentally responsible. However, this positioning poses a fundamental tension: the infrastructural demands of AI, particularly the energy-intensive operations of large-scale data centres, contribute meaningfully to global greenhouse gas emissions. Despite this, brands routinely invoke AI as a symbol of sustainability, raising urgent questions about the integrity of such claims. This study examines how the environmental costs associated with AI are misrepresented within the sustainability narratives promoted by contemporary brands, a practice this research identifies and defines as 'AI greenwashing.' Using Keller's CBBE framework and green marketing theory, this interpretivist research builds a layered understanding of the phenomenon. The thematic patterns shed light on how brands construct narratives that separate technological advancement from ecological responsibility. This research offers timely insights for responsible AI-driven marketing.

This chapter examines critical minerals required for AI hardware, including semiconductor fabrication and data centre infrastructure. While sustainability discussions have focused on operational energy consumption, minerals enabling machine learning generate substantial environmental impacts warranting parallel examination. The contribution connects three largely isolated literatures: environmental assessments of mineral extraction, critical minerals policy frameworks, and AI's environmental footprint. Covering developments from 2020 to 2025 across the European Union, United States, United Kingdom, Australia, Canada, Japan, and South Korea, the analysis addresses gigawatt-scale 'AI Factory' facilities and strategic weaponisation of mineral supply chains. Supply concentration emerges as defining, with China controlling 98 per cent of gallium production and 94 per cent of heavy rare earth processing. The chapter identifies tensions between supply security and environmental sustainability, examining burden-shifting risks and 'clean premium' strategies.

The rapid ascent of the global technology sector has revolutionized economies, communication, and daily life, but i has also generated an equally unprecedented environmental challenge: electronic waste (e-waste). Often invisible to the average consumer, e-waste represents the physical byproduct of Big Tech's expansion, comprising discarded devices, obsolete components, and broken or outdated electronics. From smartphones, laptops, and tablets to servers, data centers, and IoT devices, the proliferation of electronic products has created towering e-waste "mountains" that strain natural resources, overwhelm recycling infrastructure, and generate toxic environmental hazards. While the economic and societal benefits of digital technology are widely celebrated, the environmental cost embedded in this growth highlights a critical paradox: the more digital innovation accelerates, the heavier the ecological footprint becomes. Understanding the scale, composition, and consequences of e-waste is crucial for framing sustainable practices, circular economy policies, and corporate responsibility.

Decentalized energy systems are a radical departure from the way we have traditionally produced, distributed and consumed electricity – relying upon centralized grids that depend on fossil fuels towards more sustainable, resilient and community-based models. These systems improve grid flexibility, save transmission costs and are fit for renewable power input like solar or wind, since prosumers can generate and distribute energy within the local area. Blockchain technology is going to be a key enabler of this transition, as it offers a reliable, transparent and decentralized platform for energy sharing and grid management. The blockchain technology utilizing the smart contracts and mutual system can provide reliable peer-to-peer power trading, accurate settlement, and less necessity of centralized agency. Its distributed ledger makes the equipment trustable for all participants, and meanwhile it realizes real-time transaction data sharing to optimize grid management like demand response of electricity and certificates tracing of green power production.

The notion of greening intelligent machines becomes prominent for promoting the sustainable development of intelligent environments. The integration of innovative

This study aims to evaluate the implementation of the Smart City Masterplan in Malang City as a transformation effort towards a sustainable smart city. Using qualitative descriptive methods, the evaluation focused on six main pillars: Smart Governance, Smart Economy, Smart Branding, Smart Living, Smart Society, and Smart Environment. The results show that Malang City has achieved a significant level of success with program realization reaching 96.22%. In terms of digital infrastructure readiness (E-Readiness), Malang City obtained a score of 58.3%, which has exceeded the minimum readiness standard of 50%. This success is driven by the Hexahelix collaboration model that involves synergy between government, academia, communities, the private sector, media, and banking. Despite showing positive progress, this study identified major challenges in the form of obstacles to interoperability of data integration between agencies, limited human resource capacity, and the need to strengthen regulations supporting innovation.

Artificial Intelligence (AI) is widely celebrated as a transformative force capable of solving some of humanity's most pressing challenges, from climate modelling to medical diagnostics. Yet this narrative of technological promise often obscures a less visible reality: AI systems are deeply material, dependent on vast quantities of energy, water, minerals, and hardware that carry significant ecological and social costs. This chapter critically examines the environmental footprint of AI across its full lifecycle, addressing five interconnected dimensions: escalating energy consumption, water usage in cooling systems, mineral extraction for hardware, e-waste generation, and carbon emissions. It argues that the structural forces driving AI's rapid expansion—

competitive scaling races, market incentives, corporate opacity, and regulatory gaps—are actively outpacing the development of sustainability safeguards.

Preface

INTRODUCTION

Artificial intelligence has become one of the defining technologies of the twenty-first century, transforming industries, reshaping scientific discovery, and accelerating innovation across nearly every sector of society. From healthcare and transportation to manufacturing, education, and environmental monitoring, AI has demonstrated extraordinary potential to solve complex challenges and improve decision-making. Yet, as the scale and sophistication of AI systems continue to expand, an equally important conversation has emerged regarding the environmental costs associated with their development, deployment, and long-term operation. This volume, *Examining the Environmental Footprint of Artificial Intelligence*, was conceived to address this critical and often overlooked dimension of AI research.

While artificial intelligence is frequently presented as a powerful tool for advancing sustainability through climate modeling, resource optimization, biodiversity conservation, and environmental monitoring, the infrastructure required to support modern AI systems demands substantial quantities of energy, water, raw materials, and specialized computing hardware. The rapid growth of large-scale data centers, increasing computational requirements for model training, and the environmental implications of hardware manufacturing and electronic waste have introduced new sustainability challenges that require rigorous scientific examination. Understanding these tradeoffs is essential if AI is to become a truly sustainable technology capable of supporting global environmental objectives rather than undermining them.

This edited reference book brings together a diverse collection of research that examines the environmental footprint of artificial intelligence throughout its entire lifecycle. Rather than focusing solely on AI's environmental applications, the contributing authors investigate the ecological consequences of AI itself, offering comprehensive perspectives grounded in environmental science, computer science, engineering, public policy, and sustainability studies. Through theoretical analyses, empirical research, practical case studies, and policy discussions, the chapters collectively present a balanced assessment of both the opportunities and environmental risks associated with the rapid expansion of AI technologies.

The volume explores numerous dimensions of sustainable AI, including the environmental impacts of computational scaling, energy-intensive model training, resource consumption, data center operations, hardware production, electronic waste management, life-cycle assessment methodologies, and emerging approaches for designing more energy-efficient algorithms and infrastructures. Equally important, contributors examine governance frameworks, regulatory considerations, and industry best practices that can guide the responsible development and deployment of AI systems. By combining technological innovation with environmental accountability, this collection encourages readers to consider sustainability as a fundamental design principle rather than an afterthought.

One of the distinguishing features of this publication is its interdisciplinary perspective. Addressing the environmental implications of artificial intelligence requires expertise that extends well beyond computer science alone. The chapters demonstrate how meaningful progress depends upon collaboration among AI researchers, environmental scientists, engineers, sustainability specialists, policymakers, and industry practitioners. By integrating these diverse viewpoints, the book provides readers with a comprehensive understanding of the complex interactions between intelligent technologies and the natural environment while identifying opportunities for innovation that reduce environmental impact without sacrificing technological advancement.

This reference is intended to serve a broad audience, including researchers and academics working in artificial intelligence, environmental science, sustainability, engineering, and public policy. Graduate students and educators will find valuable insights into emerging research directions and interdisciplinary methodologies, while engineers, technology developers, and industry professionals will benefit from practical discussions of sustainable system design and infrastructure optimization. Policymakers, environmental regulators, nonprofit organizations, entrepreneurs, and innovators will also find evidence-based guidance for developing responsible strategies that align technological progress with environmental stewardship.

As Editor, I extend my sincere appreciation to the distinguished chapter authors whose expertise, dedication, and scholarly contributions have made this volume possible. Their work reflects a shared commitment to advancing both technological innovation and environmental responsibility. I also express my gratitude to the anonymous reviewers whose thoughtful evaluations strengthened the quality and rigor of each chapter, as well as to the editorial and publishing team whose professionalism and support were instrumental throughout the development of this project.

It is my hope that *Examining the Environmental Footprint of Artificial Intelligence* will encourage meaningful dialogue, inspire future interdisciplinary research, and support informed decision-making regarding the sustainable evolution of artificial intelligence. As AI continues to shape the future of society, ensuring that its environmental impacts are understood, measured, and responsibly managed will be essential for achieving long-term sustainability goals. By recognizing environmental stewardship as an integral component of AI innovation, we can work toward developing intelligent technologies that benefit both humanity and the planet.

ORGANIZATION OF THE BOOK

Chapter 1, "AI at a Crossroads: Managing AI's Environmental Costs in the Age of Sustainability," establishes the foundation for this volume by examining the environmental consequences of artificial intelligence through a comprehensive life-cycle perspective. The chapter explores the substantial energy consumption, carbon emissions, water usage, material extraction, and electronic waste associated with modern AI systems, while highlighting the tension between ambitious corporate sustainability commitments and the rapidly expanding infrastructure required to support large-scale AI. Through case studies of leading technology companies, the authors challenge prevailing sustainability narratives and introduce emerging concepts such as Green AI and Responsible Innovation, emphasizing the importance of stronger environmental governance, transparency, and disclosure standards.

Chapter 2, "Scaling Intelligence, Scaling Impact: A Systemic Analysis of the Environmental Sustainability of Artificial Intelligence," expands the discussion by framing AI's environmental footprint as the result of interconnected technological, economic, and infrastructural systems rather than an isolated technical issue. Drawing upon ecological economics, life-cycle assessment, and governance theory, the chapter examines how scaling strategies, material dependencies, and market incentives collectively drive environmental impacts. It concludes by proposing a multi-level governance framework and introducing the concept of "net-positive AI," encouraging a transition from minimizing environmental harm toward creating measurable ecological and social benefits.

Chapter 3, "Financialization, Scaling Dynamics, and the Ecological Political Economy of AI," examines the economic forces that underpin the rapid expansion of artificial intelligence infrastructure. The chapter argues that AI should be understood as a resource-intensive industrial system shaped by financial investment, global supply chains, and continuous scaling pressures. By linking financialization with material extraction, energy consumption, data center expansion, and electronic waste generation, the authors present an integrated framework that demonstrates how economic incentives contribute to AI's growing environmental footprint while underscoring the need for governance mechanisms capable of balancing innovation with ecological responsibility.

Chapter 4, "Selling the Green Dream: How AI's Energy Appetite Shapes Sustainable Brand Narratives," investigates the growing disconnect between corporate sustainability messaging and the environmental realities of AI deployment. Focusing on AI-driven marketing strategies, the chapter introduces the concept of "AI greenwashing" to describe how organizations promote AI as an environmentally responsible innovation while overlooking the significant energy demands of the infrastructure supporting these technologies. Using established branding and green marketing frameworks, the authors provide valuable insight into the ethical challenges surrounding AI-centered sustainability claims and advocate for greater transparency in corporate environmental communication.

Chapter 5, "Critical Minerals for AI Hardware: Material Dependencies, Environmental Dimensions, and Industrial Policy Responses in OECD Economies," shifts attention to the often-overlooked material foundations of artificial intelligence. The chapter examines the critical minerals essential to semiconductor manufacturing and data center infrastructure while exploring the environmental consequences of resource extraction and the geopolitical complexities of global mineral supply chains. Through an analysis of policy developments across leading OECD economies, the authors highlight the tension between securing critical resources for technological advancement and maintaining environmental sustainability, offering important perspectives on industrial policy and responsible resource management.

Chapter 6, "The E-Waste Crisis Behind Big Tech: The Carbon Cost of Intelligence," explores one of the fastest-growing environmental consequences of digital transformation: electronic waste. The chapter examines how the rapid pace of technological innovation has generated increasing volumes of discarded devices and obsolete computing infrastructure, creating significant environmental and public health concerns. By discussing the composition, scale, and ecological impacts of e-waste, the author emphasizes the importance of circular economy strategies, responsible recycling systems, and stronger corporate accountability in reducing the long-term environmental costs of digital technologies.

Chapter 7, "Decentralized Energy Systems and Blockchain's Role in Sustainable Power Grids," examines how decentralized energy systems and blockchain technologies can support more sustainable electricity generation and distribution. The chapter explores the role of blockchain in enabling transparent peer-to-peer energy trading, improving grid flexibility, and facilitating greater integration of renewable energy sources. By demonstrating how decentralized digital infrastructures can strengthen energy efficiency and enhance trust among participants, the authors provide an important perspective on technological innovations that may help offset some of the environmental demands associated with expanding digital ecosystems.

Chapter 8, "Greening Intelligent Machines: Recent Innovations, Challenges and Future Directions," reviews recent advances in the development of environmentally sustainable artificial intelligence systems. The chapter discusses innovations such as low-power hardware, edge computing, model optimization techniques, sustainable data centers, and data-driven efficiency practices while acknowledging the technical, economic, and scalability challenges that continue to limit widespread adoption. By synthesizing current research and identifying future opportunities, the authors present a valuable roadmap for advancing Green AI through continued technological innovation and interdisciplinary collaboration.

Chapter 9, "Digital Pollution: Examining the Impact on Environmental Awareness," addresses the environmental implications of everyday digital behaviors that often go unnoticed. Focusing on unnecessary data storage, excessive digital communication, and inefficient technology use, the chapter explores how routine digital activities contribute to increased energy consumption and carbon emissions. Through an assessment of teacher candidates' awareness of digital pollution, the authors identify significant knowledge gaps while offering practical recommendations for improving environmental awareness and promoting more sustainable digital practices.

Chapter 10, "Evaluation of Smart City Policies on Effectiveness, Efficiency, and Socio-Economic Impacts: Smart City Analysis in Malang City," presents an evaluation of smart city implementation through the case of Malang City. Examining six key dimensions of smart city development, the chapter highlights the role of collaborative governance, digital infrastructure, and stakeholder partnerships in advancing sustainable urban transformation. While documenting significant implementation successes, the authors also identify continuing challenges related to interoperability, regulatory support, and institutional capacity, providing valuable lessons for future smart city initiatives.

Chapter 11, "When Innovation Outpaces Sustainability: The Environmental Risks of Rapid AI Expansion," concludes the volume by revisiting the central challenge explored throughout this book: the widening gap between the accelerating growth of artificial intelligence and the development of effective environmental safeguards. The chapter critically examines the interconnected impacts of energy consumption, water use, mineral extraction, electronic waste, and carbon emissions while arguing that competitive market dynamics and regulatory shortcomings continue to accelerate AI deployment beyond the pace of sustainable governance. In doing so, it reinforces the volume's overarching call for greater transparency, stronger policy frameworks, and environmentally responsible innovation to ensure that the future of artificial intelligence aligns with global sustainability objectives.

CONCLUDING REMARKS

Artificial intelligence is poised to become one of the most influential technologies of our time, driving innovation, accelerating scientific discovery, and reshaping industries on a global scale. Yet, as the chapters in this volume collectively demonstrate, the benefits of AI cannot be evaluated independently of the environmental systems that sustain it. The computational power that enables increasingly sophisticated AI models is supported by extensive physical infrastructure, vast quantities of energy and water, critical mineral extraction, and growing volumes of electronic waste. Recognizing these realities is essential if the next generation of AI technologies is to contribute meaningfully to sustainable development rather than unintentionally intensify existing environmental challenges.

Throughout this volume, contributors have explored the environmental footprint of artificial intelligence from multiple complementary perspectives. Together, the chapters reveal that AI's ecological impact extends far beyond operational energy consumption to encompass the full lifecycle of intelligent systems—from raw material extraction and semiconductor manufacturing to data center operations, infrastructure expansion, governance, policy development, and end-of-life disposal. At the same time, the collection illustrates that technological innovation and environmental responsibility need not exist in opposition. Advances in Green AI, energy-efficient computing, sustainable infrastructure design, lifecycle assessment, circular economy principles, and responsible governance provide encouraging pathways toward reducing AI's environmental burden while preserving its transformative potential.

A recurring theme throughout this book is the importance of interdisciplinary collaboration. Addressing the environmental consequences of artificial intelligence requires expertise that bridges computer science, environmental science, engineering, economics, public policy, business, and ethics. The complexity of these challenges demands integrated solutions that combine technological innovation with transparent governance, evidence-based policymaking, and responsible corporate leadership. As AI continues to evolve, sustainability must become a core design objective embedded throughout research, development, deployment, and regulation.

The discussions presented in this reference also underscore the need for greater accountability across governments, industry, academia, and society. Transparent environmental reporting, standardized sustainability metrics, responsible supply chain management, ethical marketing practices, and investments in greener infrastructure will all play essential roles in shaping a more sustainable AI ecosystem. Equally important is fostering greater public awareness of the environmental implications of digital technologies, enabling individuals and organizations alike to make more informed decisions regarding their development and use.

It is my hope that this volume serves not only as a valuable reference for researchers, educators, policymakers, industry professionals, and students, but also as a catalyst for continued dialogue and innovation. The questions explored throughout these chapters will remain highly relevant as artificial intelligence becomes increasingly integrated into every aspect of modern life. Future progress will depend not only on making AI more capable, but on ensuring that its advancement is aligned with the long-term health of our planet.

As Editor, I extend my sincere appreciation once again to all of the contributing authors, whose scholarship and diverse perspectives have made this volume possible, as well as to the reviewers and publishing team whose expertise and dedication have strengthened the quality of this work. Their collective efforts have produced a timely and comprehensive examination of one of the most important emerging issues at the intersection of technology and sustainability.

Ultimately, the future of artificial intelligence will be defined not solely by the intelligence of the systems we create, but by the wisdom with which we choose to develop, govern, and sustain them. By embracing environmental stewardship as a fundamental pillar of AI innovation, we can help ensure that technological progress contributes to a more resilient, equitable, and sustainable future for generations to come.

Chapter 1
AI at a Crossroads:
Managing AI's Environmental Costs in the Age of Sustainability

Priya Rani
http://orcid.org/0009-0001-5639-784X
Delhi Technological University, India

Mini Jain
http://orcid.org/0009-0001-0705-6395
Delhi Technological University, India

ABSTRACT

The increased pace of Artificial Intelligence (AI) expansion had a significant impact on the modern economic system as well as fueling the debate on its environmental sustainability. In this chapter, the authors take a life-cycle approach to the assessment of AI, which covers the energy expenditure, carbon emissions, water use, material mining, and electronic waste considerations. The research highlights how the structural tension between ambitious climate pledges and the increasing material and energy requirements of large-scale AI infrastructure depicts the case studies of the most popular AI developers, Alphabet, Microsoft, Meta, OpenAI, NVIDIA and Salesforce finding that computational intensity and embodied supply-chain emissions confuse the sustainability narrative that focuses only on sourcing renewable energy or making AI infrastructure more efficient. The chapter also presents the new concepts, including Green AI and Responsible Innovation, and stresses the need to have criteria of environmental disclosure, and governance reforms.

DOI: 10.4018/979-8-3373-8490-0.ch001

INTRODUCTION

Artificial Intelligence (AI) has become one of the most revolutionary technological impacts of the twenty-first century, reshaping industries, transforming business models, and changing the manner in which decisions are made by government, corporations, and financial systems. In capital-market algorithmic trading, healthcare predictive analytics, supply-chain optimization, urban infrastructural smartness, and many other new areas, AI-based systems are currently well integrated into the international socio-economic plumbing (Bughin et al., 2018; Makridakis, 2017). The high rate of AI-based technologies spread to industries has allowed companies to increase efficiency of their business processes and the accuracy of their forecasts, as well as automate decision-making and give rise to brand new data-driven innovations. With the fast-paced digital transformation that is sweeping across the world, AI is being incorporated into sustainability governance systems, climate-risk modelling systems, ESG disclosure systems, and green investment models.

With sustainable finance (specifically, textual sustainability reporting on Nifty-listed companies) in mind, AI is ceased to be a technological invention but is a structural background that determines sustainability dynamics in any company. AI-based analytics are currently being implemented to analyze climate risks, analyze environmental performance metrics, detect ESG-related scandals, and guide decisions made regarding investments in line with sustainable development objectives. Regulatory authorities and other institutional investors are moving towards AI-based software to work with voluminous amounts of data related to sustainability and to increase the transparency of the ESG reporting procedures (Kaack et al., 2022). As a result, AI is becoming a key facilitator of sustainable finance systems and climate governance systems.

Nevertheless, in addition to its potential to achieve gains in efficiency, climate modelling, emissions tracking, and sustainable finance analytics, AI also has a paradoxical side that has yet to gain scholarly and policy consideration, its fast-growing environmental footprint. The mass implementation of machine learning algorithms, specifically GenAI systems and large language models (LLMs), is based on computational learning models that require a lot of energy, and such processing is central to large-scale data centres. The infrastructures themselves also use high electricity in the training and inference of AI implementation, which, in turn, has a direct effect on greenhouse gas emissions depending on the energy mix of powering the facilities (IEA, 2023). Therefore, the environmental costs of AI go beyond the efficiencies of software to include the actual physical infrastructure that will underpin digital intelligence systems.

It has been estimated that in the recent past global data centres use about 11.5 per cent of all global electricity and generate almost 1 per cent of all energy-related carbon dioxide emissions in the world (IEA, 2023). This share is likely to grow exponentially during the next ten years as AI workloads will take over the computational demand in the future. Large-scale AI models demand high-performance computing clusters, which can consume megawatt-hours of electricity over long durations. This creates significant carbon emissions if they rely on fossil fuels for power (Strubell et al., 2019). Since the application of the GenAI is still growing in the financial, healthcare, educational, and logistics sectors, the energy needs corresponding to the applications are predicted to grow, which only exacerbates the issue of the environmental sustainability of digital-transformation directions.

In addition to electricity consumption, AI systems have significant environmental footprint in the form of substantial water use to cool down the high-performance computing infrastructure. The AI processors produce large thermal loads, and thus, the water-based cooling systems are required in an effort to maintain its operational efficiency and prevent hardware degradation. Evaporative cooling systems are also commonly used in hyperscale data centres, and they can consume large amounts of freshwater to cool the heat produced by computational workloads (Li et al., 2025). Empirical studies have shown that some data centres can use millions of litres of freshwater on a daily basis which puts extra strain on the local water supply and the local communities. In areas where the water shortage is already experienced, the spread of AI-based computing infrastructure has a potential to reduce available resources and increase socio-environmental tensions.

Sustainability of AI systems also poses a challenge due to the lifecycle effects. Environmental cost does not only depend on operational electricity consumption but also extends to the production of hardware, semiconductor production, and the construction of infrastructure and the supply chains logistics related to AI deployment (Gupta et al., 2020). The making of state-of-the-art processors such as graphics processing units (GPUs) and the tense processing units (TPUs) involves resource-intensive production processes that require rare-earth minerals, energy-intensive production plants, and the complex international supply chains. Lifecycle analyses suggest that the development phase of large language models can be a significant percentage of the total environmental impact associated with AI training processes. The production of complex AI models might both result in carbon emissions that could match the yearly usage of several household units, and it may additionally consume a large amount of freshwater during the process of computation (Strubell et al., 2019).

Furthermore, the release of both particulate matter and other air pollutants produced by the work of data centres has also been linked to a negative effect on the well-being of the population and is disproportionately relevant to economically disadvantaged communities located near large computing infrastructure centers (Kaack et al., 2022). Clustering of geographic centres of data often motivated by conducive regulatory environments or access to low-cost power raises doubt on the subject of environmental justice as well as equitable allocation of technological externalities. The people living near such facilities might take up the environmental implications of AI-led digitalisation without correspondingly enjoying the economic rewards.

Ironically, AI technologies are also being advocated as absolutely necessary to realize the goals of climate mitigation and sustainable development. Analytical systems powered by AI have shown the capacity to improve transportation networks in cities, achieve better energy efficiency in buildings, renewable energy forecasting, and intelligent resource management systems in smart cities (Kaack et al., 2022) As an example, AI-based predictive maintenance systems will be able to improve the efficiency of wind turbines and solar panels, which will subsequently result in making renewable energy systems more reliable. In the same way, smart grids can optimise power demand through machine learning algorithms and decrease energy wastage and enhance load balancing within power networks.

A developing amount of evidence indicates that AI-powered applications would play an important role in reducing emissions in industries like energy systems, agriculture, and industrial procedures in the event that they are implemented in a strategized manner. Optimisation methods that AI can help optimise the efficiency of manufacturing systems, aid accuracy cultivation practices (and, as a result, lessen fertilizer application) and create intelligent transportation systems (and, consequently, lessen fuel usage). The applications illustrate the possible role of AI to become a facilitator of sustainability transitions and climate resilience policies.

This two-sided nature of AI as a source of environmental degradation and a source of environmental sustainability solutions puts the AI at a pivotal juncture in the current climate governance conversations. On the one hand, AI provides the most significant opportunities in terms of surveillance of environmental threats, predicting scenarios of climate change, and optimizing distribution of resources in economic systems. Conversely, its increased dependence on computationally intensive infrastructure stands to pose a threat of increased consumption of energy and carbon emissions unless handled effectively. With the world economies seeking far-reaching net-zero goal under international agreements like the Paris Climate Accord, the reconciliation of digitalisation strategies of AI and sustainability demands have become an urgent governance question.

On financial grounds, which is especially pertinent to the ESG-integrated analysis of investment, these environmental externalities bring new levels of climate-related risks and governance complexity. This could affect the valuation of the firm and investor sentiment on capital markets because the carbon intensity of AI-driven operations could impact corporate environmental performance measures, sustainability ratings, and cost-of-capital dynamics (Bolton & Kacperczyk, 2021). Companies with heavy energy usage of AI infrastructure are more likely to be scrutinised by different regulatory authorities and incur more transition risks that accompany decarbonisation policies. With the growing integrated use of climate risk by institutional investors when allocating their portfolios the environmental impact of AI infrastructure can become a material consideration of investment strategies and stewardship policies. Figure 1 depicts the framework linking AI adoption with environmental externalities and ESG outcomes.

Figure 1. Conceptual framework linking AI adoption with environmental externalities and ESG outcomes

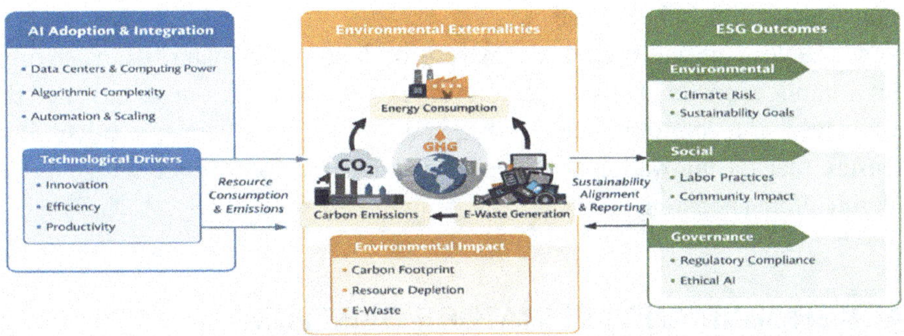

This figure illustrates the relationship between AI adoption, infrastructure expansion, and associated environmental externalities. It also highlights the mediating role of governance and sustainability strategies in shaping ESG performance outcomes.

The increase in the AI infrastructure has started influencing the sustainability paths of corporations in the technology sector. Companies that have made a significant investment in AI-powered services have reported an increase in greenhouse-gas emissions that can be attributed to an increase in electricity demand due to the operation of data-centres and the supply-chain. Such advances can make corporate net-zero goals more difficult to achieve, especially in areas where fossil-fuel technology is still the leading electricity generation method. As a result, policy-makers, regulators and investors are now faced with the challenge of ensuring the alignment of digital-innovation policies with the climate pledges expressed in the national sustainability policies.

With the pace of AI implementation in the economic sector, the cost of this AI in the environment has become an urgent governance agenda in the current sustainability age. The key issue is not to stop the development of technologies but to make sure that the new trends of AI development are based on the principles of resource efficiency, integration with renewable energy, and circular-economy adoption. The potential solutions to the ecological footprint of AI include emerging technological solutions, such as energy-efficient algorithmic architecture, low-carbon data-centre location planning, waste-heat recovery and new cooling technologies (IEA, 2023). However, the properly functioning of these measures requires the synchronous policy responses, clear environmental-reporting frameworks and the cooperation between the stakeholders within the industry, universities, and regulatory bodies.

AI is at a cross road in this regard. Although the possibility of its further growth has an enormous potential of fostering sustainability goals, uncontrolled growth is a threat of reversing the global efforts to reduce climate-emissions by increasing the energy and resource consumption. In this regard, it will be necessary to comprehend and handle the trade-offs that AI implementations pose to the environment to create a more sustainable digital future.

The chapter will thus provide an analysis of the dynamic nature of the environmental implications of AI systems, to evaluate the challenges of governance that are involved in responsible deployment, and to outline the strategic possibilities to ensure that AI innovation is aligned with the sustainability requirements in a changing global economy.

BACKGROUND AND LITERATURE OVERVIEW

The Rise and Diffusion of AI

AI has evolved into a highly niche academic field into a widespread general-purpose technology which is becoming the foundation of economic productivity, digital governance, and organizational decision-making, in a variety of fields (Figure 2). The most recent developments in deep learning and GenAI models have increased the openness of AI technologies though the intellectual heritage of AI can be traced to the mid-twentieth century when the pioneers of computationalism tried to recreate human cognition using programmable machines. Since its advent, AI has gone through a series of paradigm shifts where symbolic reasoning systems and rule-based expert systems of the 1960s and 1970s, were replaced by statistical machine learning models of the 1990s and more recently, deep neural network architectures and massive generative systems (Makridakis, 2017).

The history of AI development has been characterized by various phases of technological excitement and stasis, also known as AI springs and AI winters. The limits to the practical application of AI systems outside the laboratory environment at its initial stages were the limitations of computational power, data infrastructure, and scalability of the algorithms applied. Over the past few years, AI innovation has seen another resurgence due to the connectivity of three key technological enablers: high-performance computing infrastructure, large and digital datasets, and cloud storage resources. New AI models today have become capable of naturally-language processing, predictive analytics, pattern recognition, and autonomous decision-making faster and more accurately than ever before.

In economic terms, AI is being imagined as a general-purpose technology that can produce productivity spillovers and transform the structure over the long term in a variety of industries. The key feature of general-purpose technologies is that the technology can be spread across various industrial sectors, catalyse adjacent innovations, and transform productions systems due to network effect and technological complementarities. There is empirical data which shows that AI is not confined to the field of computer science, but is present in such areas as finance, healthcare, manufacturing, logistics, transportation, agriculture and environmental management (Bughin et al., 2018). This interdisciplinary dissemination characterizes the increased acceptance of AI as a facilitating technology capable of complementing human decision-making patterns and streamlining resource allocation across the complicated organizational settings.

The spread of AI in the economic sector has become even more significant over the last several years, with the main driving force being the development of computational capabilities, data analytics solutions, and continuous algorithm creation. In reaction, companies are gradually deploying AI-based applications into the business processes to accelerate predictive accuracy, automate routine tasks, and enhance strategic decision-making. According to surveys provided by McKinsey & Company (2023), an increasing percentage of companies now utilize AI technologies in one of their business functions, which speaks of the shift between experimental pilot projects to the implementation on the enterprise level. On the same note, the Stanford AI Index Report (Stanford, 2025) reveals the growing number of applications in the industries which are being automated using AI, such as marketing analytics, customer service automation, financial modelling, supply-chain optimisation, and sustainability risk assessment.

The advent of GenAI technologies has only made this process of diffusion even more widespread with the increase of the range of AI applications beyond the classic data analytics and automation work. This means that GenAI systems can create textual outputs, predictive simulations and decision-support insights, thus enabling high-level analytical processes in areas like financial services, healthcare, city planning and climate governance. In the financial market, AI-based predictive models are also currently being used to evaluate risk of investments, fraudulent transactions, and optimisation of portfolio strategies. Similarly, AI-driven models of climate modelling are utilized to assess environmental risks and make investment choices with sustainability in mind, thus being a part of the digital innovation of the ESG-oriented financial system.

Figure 2. Evolution of AI

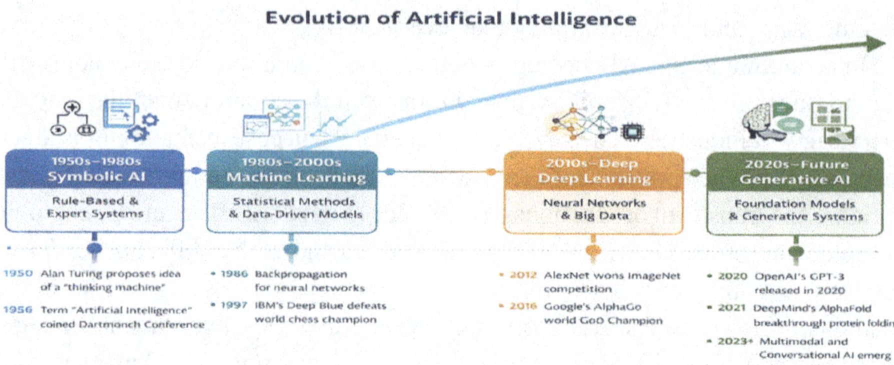

This figure depicts the chronological evolution of AI from rule-based systems and machine learning to deep learning and large-scale generative models. It highlights the increasing computational intensity and infrastructural scaling associated with each developmental phase.

The theoretical perspective on the organisational adoption of AI technologies can be framed with the help of the diffusion of innovation theory that explains the process of the technological innovation diffusion to the mainstream adoption by the economic systems. During the initial phase of adoption, AI applications were limited mostly to research institutions and technological companies that had access to specialised computational resources. Nevertheless, with the reduction in the cost of computing and an increase in the power of algorithms, AI technologies started to gain access to a wider range of industry. The current adoption trends indicate that AI has already moved into the accelerated diffusion stage leading to mass experimentation, incremental scaling, and integration of AI at the enterprise-wide level into organisational processes.

The trends of investment are also another testimony to the accelerating spread of AI. The level of global private investment in AI technologies has also grown significantly over the last ten years as it is becoming more convinced in its transformative power. According to market forecasts, the world AI market is likely to grow considerably in the next few years, as increasing demand in the automation, predictive analytics, and digital decision-support systems segments drives the market. This trend puts AI among the most rapidly growing technological areas in the history of modern economics, and productivity increases and changes in labour market relationships are the possible outcomes.

A series of facilitating factors, such as the development of cloud computing structure, algorithmic innovation, the development of digital skills, and enabling regulatory frameworks, have enabled the spread of AI. Cloud-based services enable companies to implement AI-based applications without incurring massive investments seen with the high-performance computer systems. As a result, small and medium-sized enterprises can currently incorporate AI-enhanced analytics into their workflows and, therefore, become more democratic in obtaining access to advanced computational forces. Besides, the emergence of open-source machine-learning systems has reduced obstacles to entry by companies interested in implementing AI technology in their business operations.

Although the adoption of AI technologies has largely applied to all geographic regions as well as industrial sectors, geographic, and industrial spread has not been even. Established economies, which have strong technological ecosystems, have recorded relatively high levels of AI adoption in various industrial sectors but emerging economies continue to experience structural barriers in the form of lack of sufficient computational infrastructure, skills, and regulatory ambiguities. However, recent developments suggest that countries like India are becoming major AI markets within a very short period of time, owing to their growing digital infrastructures and rising investment in technological-based innovation by their respective government.

Moreover, the spread of AI technologies opens important consequences in the field of labour markets and workforce relations. The use of AI is expected to transform job designs by enhancing human capabilities, automating low-order tasks, and generating new skills that are related to data science, machine-learning engineering, and algorithmic governance. Therefore, companies that implement AI-based systems might require conducting significant workforce reskilling programs to support the technological transformation and to make sure that everyone can participate in the digital transformation process.

Overall, the emergence and popularization of AI technologies is a complex process that is conditioned by technological innovation, economic factors, institutional relations, and the work of socio-cultural processes. With the ever-increasing penetration of AI into organizational operations and governance, a subtle insight into the patterns of its diffusion is needed to evaluate the economic value and the environmental impact of the technology.

The growing use of AI is a reminder, in the light of sustainability transitions, and especially in the ESG-integrated financial systems, of the need to consider its environmental impact, as well as to find a way to govern this process to ensure the convergence of technological progress and environmental goals.

Economic and Social Benefits of AI

The concept of AI has become an influential trend that defines the progress of the economy, social well-being, and efficiency within organizations in developed and underdeveloped economies. Being a general-purpose technology, AI provides the potential to affect productivity, labour-market dynamics, industrial competitiveness, and service provision by the government due to its capacity to process large amounts of data and give predictive insights. Economic and social benefits of AI are not only limited to single companies or industries but extend across the whole value chain to bring structural transformation to modern economies (Figure 3).

Figure 3. Benefits of AI

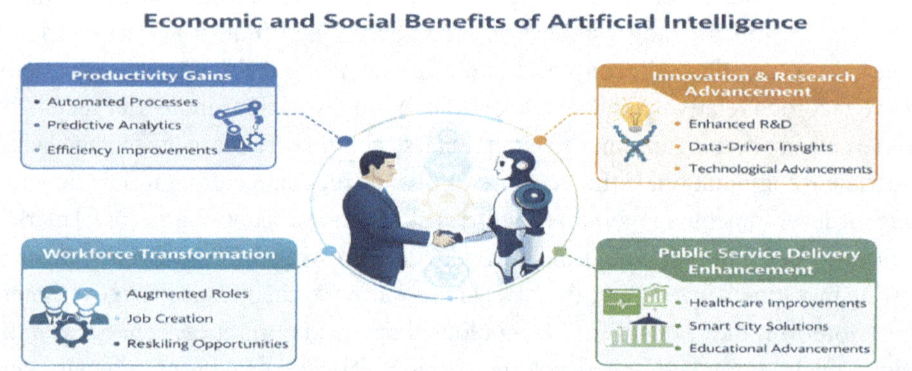

This figure outlines the key economic and social benefits associated with AI adoption, including productivity enhancement, innovation acceleration, cost efficiency, improved public services, and data driven decision making. It illustrates AI's role in supporting economic growth and societal advancements across sectors.

Economically, AI has been generally known to enhance the levels of productivity and operational efficiency in the various organizational processes. The automation based on AI will help companies to automate their routine, optimise resource distribution, and enhance the accuracy of decision-making using predictive analytics and machine-learning algorithms. Indicatively, AI-based predictive-maintenance systems are used in manufacturing where the organisations can predict equipment failures to reduce downtimes and minimise production losses and operational expenses. Similarly, in financial markets, AI-driven algorithm-trading systems allow real-time analysis of market dynamics and allow more accurate and faster transactions to be made, which can lead to an increase in market efficiency (Bughin et al., 2018) .

AI and its integration into supply-chain management systems have also led to great achievements in supply-chain logistics and demand forecasting. Machine-learning models have the potential to examine historical consumption trends, weather history, and transportation framework to maximise inventory and reduce wastage. Such applications are especially relevant to the industry where the level of demand uncertainty is rather high, and forecasting tools based on AI can enhance the resistance to disruptions in the supply chain. In this regard, the adoption of AI can bring about cost savings, revenue increase, and competitiveness at both the firm and the industry level.

AI technologies are becoming increasingly useful in becoming innovative and technologically progressive in knowledge-intensive industries. Through facilitating organisations to examine complex datasets, as well as identify emergent patterns, AI systems support the research and development (R&D) initiatives in various sectors like pharmaceuticals, renewable energy, and even advanced materials science. In health services, in example, AI-based diagnostic solutions have proven their ability to improve the accuracy of the disease diagnosis, and assist in elaboration of individual treatment strategies. In the same way, predictive analytics guided by AI may also serve to enhance the performance of wind turbines and photovoltaic panels in the renewable energy sphere by optimising maintenance scheduling and reducing the forecast of energy production (Kaack et al., 2022).

In addition to short-term productivity improvements, the implementation of AI has serious consequences on the labour market and the employment dynamics. As the fear of job displacement because of automation continues to rise, empirical studies have shown that AI technologies can at once introduce new job opportunities by enhancing human potential and promoting the demand of specialised skills in the field of data science, machine-learning engineering, and algorithmic governance. The AI-based systems have the ability to automate most of the routine tasks, thus allowing human workers to focus on more valuable tasks, including strategic planning, creative problem-solving, and interpersonal communication. In turn, the introduction of AI can also lead to the rise of the hybrid work environment, which can be described as the cooperation of the human and machine intelligence.

Macroeconomically, the spread of AI technologies is expected to drive economic growth due to the ability to increase productivity in several different areas. According to the results of scholarly research, AI-driven automation might have a significant impact on the gross domestic product (GDP) growth of the globe in the next several decades due to the increase in labour productivity and the ability to innovate in production processes (Makridakis, 2017). Some emerging economies, including India, might benefit in specific ways due to AI applications because of the scalability of digital technology and the increasing availability of cloud-based infrastructure. Platforms that incorporate AI will be able to support financial inclusion efforts by enhancing credit risk analysis and providing financial services to underserved groups, thus resulting in inclusive economic growth.

Besides having economic advantages, AI has significant social value through its use in the system of public service delivery. AI-based diagnostic tools can help medical professionals recognize the presence of diseases at their earliest stages in the healthcare industry, which will enhance patient outcomes and will lower the overall healthcare expenses. The use of AI in telemedicine systems has also helped to improve remote access to healthcare in rural and underserved areas, which helps in alleviating healthcare infrastructure disparities. Similarly, intelligence in learning applications and intelligent tutoring systems have also shown potential in education through the ability to personalise the learning process, as well as improve student engagement.

There is also the deployment of AI technologies to develop environmental sustainability and climate resilience. Intelligent-analytics can facilitate the smart-city projects by optimising energy use, traffic control systems and waste disposal systems. Precision-farming methods based on AI can also be used in agriculture so that farmers can gather data about the state of the soil, forecast weather conditions, and optimise irrigation methods, leading to a decrease in resource wastage and an increase in crop production. These applications highlight the fact that AI is capable of helping to achieve sustainable development objectives through efficient use of resources and prevention of environmental degradation.

Moreover, AI-based decision-support systems can enhance the practice of governance and the execution of the public policy due to their capacity to facilitate evidence-based policymaking. The use of AI technologies in analysing demographic data, tracking the performance of infrastructures, and forecasting economic trends by governments is increasing, thus improving the effectiveness of policy interventions. With the disaster management, predictive models built by AI can help officials foresee natural hazards and organize the efforts of emergency reaction to limit socio-economic losses.

However, the achievement of these economic and social benefits will be conditional upon the development of appropriate institutional frameworks and reskilling strategies of the workforce that will deal with issues related to both technological substitution and digital inequality. Fair access to the AI-driven technologies is an essential concern of policy makers who intend to leverage the transformative potential of AI and at the same time reduce socio-economic inequalities.

To conclude, AI has a wide range of economic and social positive impacts, including an increase in productivity, support in innovation, labour supplement, and better supply of governmental services. With AI technologies actively entering various economic fields, the socio-economic impacts of this technology in general should be clarified to find the way of structuring the system of governance to make technological progress and the goals of inclusive growth and sustainable development mutually compatible.

AI and the Sustainability Debate

The rapid proliferation of AI technologies in various sectors of the economy has generated more academic and policy debates on its overall impact on environmental sustainability and long-term social-economic viability. Although AI has been heavily proclaimed as a groundbreaking tool that can help to enhance climate mitigation, resource optimization, and sustainable development, there have also been parallel issues associated with the environmental costs that are created when AI is implemented. The result of such dualism, i.e. AI as a facilitator of sustainability transitions and as a cause of ecological externalities, has put it at the center of the present sustainability literature.

Regarding sustainability, AI has important opportunities in both terms of adding resource productivity and in climate adaptation policies. Predictive analytics, especially AI-based ones, are increasingly used to optimize energy use patterns across industrial production networks, reduce waste in supply-chain, and optimize demand planning in smart grid networks. Machine learning algorithms can be used to make wind turbines and solar panels more efficient in renewable energy systems by predicting equipment failures and scheduling maintenance (Rolnick et al., 2019). Similarly, the AI-based predictive models can be more specific in predicting the demand of electricity, which in turn can enhance the load balancing and reduction of energy waste within the urban infrastructure systems.

Precision-farming methods driven by AI make farmers control soil status, predict weather patterns, and optimise irrigation activities in the agricultural sector. These applications can help to achieve less usage of water, increase in crop production as well as reduction of fertilisers thus making farming activities environmentally friendly. Environmental surveillance systems which are AI-based are also used to monitor deforestation trends, determine the loss of biodiversity and identify unlawful fishing in the oceans. The examples of applications demonstrate the ability of AI to aid in conservation of the environment and enhance climate resilience in areas at risk.

In addition to environmental uses, AI technologies have also shown enhanced social sustainability by enhancing the delivery system of the public services. Examples of AI-based healthcare diagnostics include improving the detection of diseases and assisting in the planning of treatment, thus improving patient outcomes and decreasing healthcare expenditures. The use of AI-powered adaptive learning in education can be used to customize learning materials to match the specific learning requirements of a learner, thereby facilitating the provision of equal access to education of high quality. These developments highlight how AI can be used to combat socio-economic inequality and improve human well-being among the various groups of people.

Despite all these advantages, despite the mentioned advantages, there is also a high environmental cost to AI technologies implementation that can negatively affect the sustainability goals. AI systems are highly dependent on computationally intensive infrastructures, such as data centres, high-performance processors and cloud-based storage infrastructure, which has large power requirements both in the training and inference steps. It has been demonstrated that energy requirements incurred during the training of large-scale machine learning models produce significant carbon emissions, especially in those areas where electricity generation depends on energy systems powered by fossil fuels (Strubell et al., 2019). Thus, AI resources can neutralise the sustainability benefits of the tools when it comes to the environmental footprint.

Besides the energy consumption, the environmental externalities of AI systems also apply to the water usage, which is necessary to cool the computational infrastructure. Rack Data centres often make use of water-based cooling to draw heat out of high-performance computing systems. The increase in AI-based data centres can thus place further strain on the freshwater resources, especially in areas that have known shortage of water (Li et al., 2025). In addition, the manufacturing of advanced semiconductor parts to be used in AI devices entails resource-heavy production processes and can be a source of land degradation, water contamination, and loss of biodiversity.

The environmental effects of the use of AI technologies do not just relate to energy use, but also to the creation of electronic waste (e-waste) that is based on obsolescent hardware devices. The fast advancement in technology in the field of AI systems often requires the update of the computational infrastructure, which contributes to the increase in the amount of discarded electronic devices. When e-waste is not handled properly, it can lead to soil pollution and negative health impacts on the populace with the worst impacts being recorded in the developing areas where there are no adequate e-waste recycling systems. Such observations support a strong call to have life-cycle evaluations of AI technologies that evaluate its long-term environmental impacts.

In a governance perspective, the discourse of sustainability relating to AI has been met with a concerted demand of regulatory frameworks meant to restrain environmental externalities and promote responsible innovation. The use of renewable energy in data-centers and the implementation of energy-efficient algorithmic structures are two areas of policymaking and industry stakeholders who are constantly urging the world to incorporate them. The potential to reduce the environmental impact of AI infrastructure has the potential to be achieved through green computing technologies, including low-power processors and waste-heat recovery (Gupta et al., 2020).

The financial markets and ESG-based investment strategies are also influenced by the sustainability implication of AI. With the institutional investors integrating climate-related risks in their portfolio allocation, the companies involved in energy-consuming AI activities face increased examination of these variables of environmental performance. The carbon footprint of AI-driven business processes may have an impact on the corporate sustainability score and capital-cost relationships, hence on the price of the company and investor interest (Bolton & Kacperczyk, 2021). As a result, the adoption of AI in the business activity requires a strict evaluation of its environmental trade-offs in ESG reporting domains.

Moreover, the sustainability discourse of AI is connected with the more general ones, which are digital equity and technological inclusivity. The unequal distribution of the computational resources and digital infrastructure to the geographic area may enhance socio-economic inequalities, limiting the availability of AI-based sustainability solutions to the developing economies. Balanced access to AI technologies thus constitutes one of the central issues of policymakers that are set to exploit the transformative potential of AI technologies and, at the same time, address the environmental and social risks.

Emerging Perspectives: Green AI and Responsible Innovation

The growing recognition of the environmental impact of AI has contributed to the development of new research paradigms based on the ideas of Green AI and Responsible Innovation. With the increased integration of AI into economic systems, the issue of energy consumption and carbon emissions, as well as the environmental impact of the lifecycle of the technologies, have become an increasing source of concern in scholarly and policy discussion around the question of whether the digital transformation is sustainable. To address these issues, the Green AI movement accepts the creation of computationally efficient algorithms and AI-infrastructure that are environmentally friendly so that ecological externalities are minimized without negatively affecting performance results.

Green AI was developed as an answer to the growing computational requirements of the training and deployment of large-scale machine learning models. Historically, the paradigms of AI development traditionally focused on performance optimization that is usually assessed by accuracy or predictive ability. But these performance-focused methods often fail to take into consideration the environmental cost associated with the use of computational resources. It can consume significant quantities of electricity to train state-of-the-art AI models due to the large amounts of electricity that are required, especially when this is done on energy-intensive machinery that is fed by non-renewable energy sources (Strubell et al., 2019). Green AI therefore aims to reflect the analysis of AI systems by integrating environmental efficiency measures such as energy consumption, carbon intensity and computational complexity in model analysis models.

Technologically, Green AI focuses on the need to create machine learning systems that are alleviated by the importance of algorithmic efficiency in minimizing the environmental impact of machine learning systems. The demonstration of new model compression methods, power-efficient neural network models, and low-power hardware displays can enable creation of AI applications using fewer computational resources. Also, there has been the incorporation of renewable energy sources in the data-centre processes as a major step in curbing the carbon emissions of AI infrastructure. Technological companies and cloud providers are adopting renewable energy development efforts at a rapid pace in order to facilitate sustainable computing practices, making the implementation of AI compatible with climate-saving goals.

Along with algorithmic efficiency, AI system lifecycle sustainability is now an issue of concern in the Green AI paradigm. The environmental effects of AI technologies are not limited to the electricity used in its operation but also the production of the hardware, semiconductor fabrication, and the production of electronic waste (e-waste). The manufacturing of high-performance computing devices are usually resource consuming operations that can cause land degradation and water pollution. Electronic waste is therefore a significant issue of responsible management to reduce the environmental externalities of AI implementation by recycling and practices of the circular economy (Gupta et al., 2020).

Adding to the Green AI movement, the notion of Responsible Innovation has been becoming more noticeable as a form of governance that is meant to provide the alignment of the advances in technology with the overall social and environmental priorities. Responsible Innovation emphasizes how a combination of ethical consideration, stakeholder participation, and sustainability concepts should be implemented in the design and delivery of emerging technologies. When applied in AI, this strategy aims to resolve the issues associated with the environmental sustainability, privacy of information, transparency of algorithms, and social fairness.

Accountable AI governance systems motivate the organizations to consider the environmental and societal impacts of AI implementation in their business processes. Companies that adopt AI-based systems are becoming obliged to reveal data on the energy usage patterns, carbon emissions, and sustainability indicators of their performance as one of the parts of ESG reporting. This kind of transparency can become more accountable and allow investors to evaluate environmental risks that may occur in business models focused on AI (Bolton & Kacperczyk, 2021).

In addition, Responsible Innovation notes that it is crucial to align the implementation of AI with sustainable development goals (SDGs) and climate-policy frameworks. Governments and regulatory authorities are increasingly recommending the implementation of sustainability standards in the planning of digital infrastructure, such as the need to have energy-efficient data centres and a responsible manufacturing approach of hardware that is sustainable towards the environment. Additional policy measures that can be used to encourage the use of green computing technologies can also help in the shift to sustainable AI systems.

Green AI and Responsible Innovation also intersect and bring significant issues of digital equity and technological inclusivity into focus. Fair distribution of AI-based sustainability solutions is one of the key priorities of policy-makers who want to embrace the potential of digital technologies as a way to transform the world with the least possible harm to socio-economic inequalities. The investments in digital infrastructures and workforce reskilling campaigns can potentially facilitate the inclusive engagement in the AI-enabled innovation processes.

The concepts of Green AI and Responsible Innovation are impacting ESG-driven investing approaches in financial markets. Climate-related risks of AI infrastructure in portfolio allocation are being included by institutional investors, which is influencing corporate motivation to adopt sustainable technology. Companies focusing on efficient use of AI can also enjoy improved sustainability scores and decreased risks of transition, related to climate-policy systems.

ENVIRONMENTAL IMPACT OF AI: ENERGY, WATER, CARBON, AND E-WASTE

AI exerts environmental impacts across its entire lifecycle, from data processing and model training to infrastructure operation and hardware disposal, as depicted in Figure 4. Understanding these effects requires a lifecycle perspective that situates AI within physical systems of energy, water, materials, and waste. Despite its digital character, AI relies on resource-intensive infrastructures whose ecological consequences are increasingly visible.

Figure 4. Environmental impact of AI

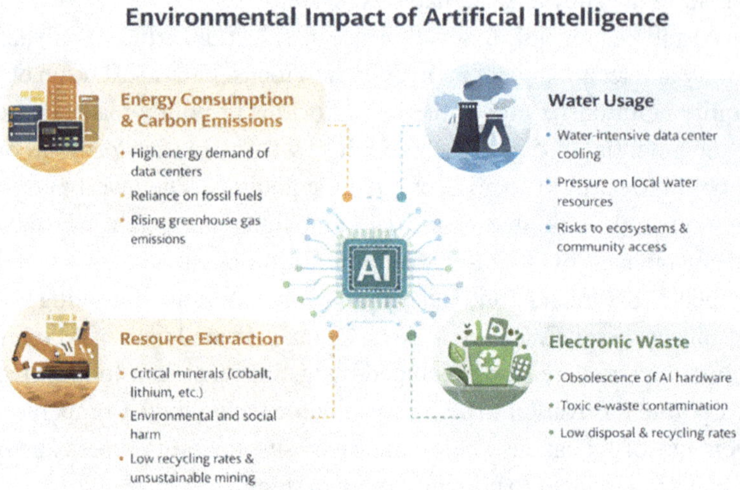

This figure illustrates the environmental impacts associated with AI systems, including energy consumption, carbon emissions, water use, hardware production, and electronic waste. It emphasizes the lifecycle footprint of AI infrastructure from semiconductor manufacturing to data center operations.

Energy Use and Carbon Emissions

The implementation and training of the large-scale AI models demand a lot of computing power hence boosting the increasing electricity usage in data centres all over the world. Deep neural networks and foundation models are machine-learning algorithms requiring large computational cycles both in their training and inference phases. These tasks involve repeated executions of data processing on the thousands of graphical processing units (GPUs) or tensors processing units (TPUs), which consume large amounts of electricity in the long run. As such, the spread of AI-based services has increased energy consumption in the digital infrastructural industry.

The data centres continue to use a significant percentage of electricity generated by energy systems based on fossil fuels particularly in areas where renewable energy facilities have not developed sufficiently. The high-energy use of the training processes involved in AI, in its turn, adds to the increase of greenhouse-gas emissions, which leaves climate-related risks that the digital transformation poses to the climate. The size of model and scale of deployment continue to increase, which places increasing energy demands, especially on foundation models that are the basis of real-time applications including natural language processing, recommendation systems, and autonomous decision-making platforms.

Despite the various technology companies committing to using renewable energy as a form of corporate social responsibility, AI growth often outpaces any efforts to implement clean energy infrastructure. This contradiction can provoke the question concerning the long-term carbon neutrality of digital ecosystems based on AI. On top of the operational carbon, the embodied carbon associated with the production of servers, GPUs as well as networking equipment adds to the overall carbon footprint of AI. The processes of semiconductor fabrication, an essential component of the manufacturing of high-performance computing products, are an energy-intensive process that could be a source of indirect carbon emission through upstream supply chains.

Water Consumption

In addition to relying on electricity, artificial-intelligence infrastructure serves as a significant burden to freshwater resources due to the use of cooling technologies that consume a lot of water. When performing computational processes, data centres generate a lot of heat which requires sophisticated cooling measures to ensure that equipment operations are maintained as well as to prevent hardware degradation. Most hyperscale facilities have adopted evaporative cooling that necessitates large amounts of water to cool down heat released by servers and networking devices.

The industrial demand of cooling water may further exert pressure in areas where there is water shortage on domestic and agricultural users. Such an increase in demand can impose pressure on the availability locally and upset the ecological balance of neighbouring population. Additionally, the environmental footprint that can be ascribed to water utilisation through AI infrastructure is not adequately disclosed in corporate sustainability reports, which creates the feeling of transparency and accountability.

The water consumption implications go beyond the direct depletion of the resources to the other wider environmental issues of justice. Populations that are located closer to data centre locations can experience reduced access to fresh water, particularly in the instances when governance structures are weak or not well implemented. Based on this, an AI infrastructure sustainability evaluation should also include the assessment of not only energy efficiency indicators but also the impact on the local aquatic environment and socioeconomic equity.

Resource Extraction and E-Waste

AI systems are also reliant on the minerals, such as cobalt, lithium, and rare earth so that they are vital components of semiconductor fabrication and battery technology utilized in computing hardware. The mining of these commodities is often linked with deterioration of the environment, which includes deforestation, soil pollution, and water pollution. Mining in developing areas is usually associated with significant social risks, including labour exploitation and displacement of communities, thus, raising ethical concerns about the sustainability of AI chain of supply.

Due to the increased rate of AI development, hardware specifications often become outdated in rather short terms. The ongoing need of advanced computational infrastructure has boosted the production of electronic waste (e -waste), which is currently one of the most rapidly growing waste categories all over the world. Poor recycling and ineffective disposal mechanisms allow them to seep into the soil and water resources carrying with them toxic chemical elements such as lead, mercury and cadmium, posing a risk to the integrity of the environment and human health.

The lack of appropriate recycling infrastructure in most the world economies complicates the lifecycle environmental effects of e-waste. The informal recycling processes subject the workers to toxic chemicals and cause pollution of the air and water. To address such challenges, the introduction of ideas of the circular economy will be required, including sustainable sourcing of materials, refurbishing of the components, and the disposal options.

CASE STUDIES

Alphabet (Google)

Alphabet Inc., the parent company of Google, is one of the largest developers and deployers of AI technologies, ranging from large language models to AI powered cloud services. Figure 5 depicts Alphabet's infrastructure demands provide insights into the environmental implications of large-scale AI integration.

Figure 5. Environmental impact of alphabet AI adoption

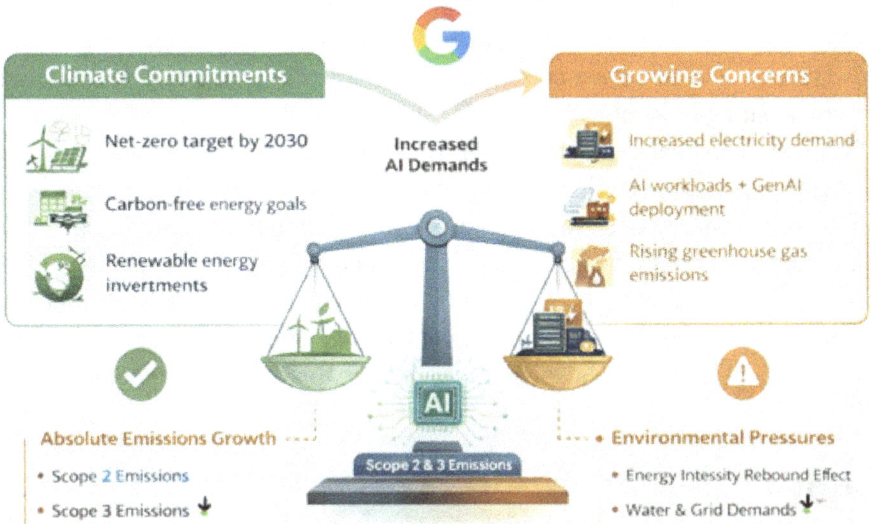

This figure presents the environmental implications of Alphabet's AI expansion, highlighting increased data center energy consumption, embodied carbon in TPU infrastructure, and water use for cooling. It situates these impacts within the company's renewable energy commitments and decarbonization strategy.

In recent years, it has publicly positioned itself as a climate leader within the technology sector. The company is committed to achieve net-zero emissions across its operations and value chain by 2030 and invested heavily in renewable energy procurement. It has also claimed progress in operating its data centers on carbon free energy through power purchase agreements and regional clean energy matching initiatives. These commitments situate Alphabet as a frontrunner in corporate climate ambition within the digital economy.

However, the rapid expansion of AI capabilities has turned this narrative. The recent reports indicate the surge in AI models training and deployment, particularly GenAI systems has significantly increased electricity consumption across Google's global data centers. The computational intensity of large-scale AI models requires substantial energy inputs, resulting in Scope 2 emissions (purchased electricity) and broader Scope 3 emissions linked to hardware manufacturing and supply chains. In some reports, Google's total greenhouse gas emissions have increased in recent years as well, raising concerns that AI-driven energy demand may outpace efficiency gains and renewable procurement efforts.

While the company has improved energy efficiency metrics, including power usage effectiveness, absolute electricity consumption continues to grow due to expanding

AI workloads. This reflects a broader phenomenon known as the 'rebound effect', wherein efficiency improvements reduce per unit energy intensity but stimulate greater overall demand. AI infrastructure requires high performance computing hardware and also extensive cooling systems, water usage, and grid integration, each contributing to localized environmental pressures.

From an analytical perspective, Alphabet illustrates the tension between technological innovation and climate responsibility. On one hand, the company invests in carbon-free energy and explores advanced grid solutions. On the other hand, the exponential growth of AI services risks undermining its emissions reduction trajectory. The discrepancy between relative efficiency improvements and rising absolute emissions highlights a structural challenge in assessing AI sustainability: progress cannot be measured solely through intensity metrics if total environment load increases.

Microsoft

Microsoft is a global leader in cloud computing and AI services through Azure and its integration of advanced AI models across enterprise and consumer platforms, Microsoft's environmental trajectory (Figure 6) offers insight into the material consequences of AI-driven digital transformation.

Figure 6. Environmental impact of Microsoft's AI growth

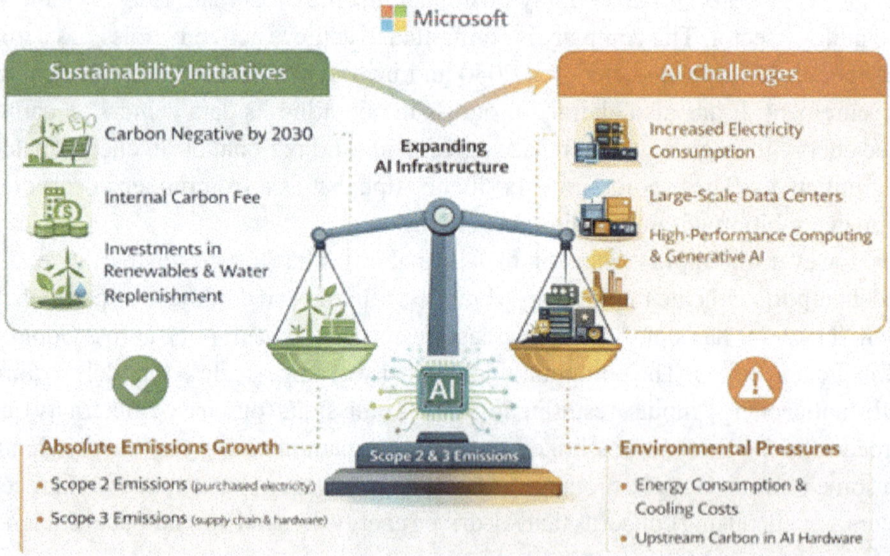

This figure illustrates the environmental consequences of Microsoft's AI-driven cloud expansion, including rising electricity demand, data center emissions, embodied carbon in hardware supply chains, and water use for cooling. It also reflects the interaction between AI scaling and Microsoft's carbon-negative and sustainability commitments.

Microsoft has positioned itself as a sustainability frontrunner within the technology sector. Notably, it committed to become carbon negative by 2030 and to removing all historical carbon emissions by 2050. The company has implemented an internal carbon fee, first introduced in 2012 and later expanded to cover Scope 3 emissions, thereby embedding carbon accountability into business operations. Its sustainability reports outline investments in renewable energy procurement, water replenishment initiatives, zero-waste goals, and the development of lower-carbon data center designs. These governance mechanisms suggest a structured and institutionalized approach to climate responsibility.

Simultaneously, Microsoft's rapid expansion of AI infrastructure presents environmental challenges. Announcements of multi-billion-dollar investments in AI-enabled data centers such as large-scale infrastructure projects supporting national AI ecosystems, indicate substantial growth in electricity demand. AI workloads, particularly those associated with generative models and high-performance computing, require significant computational capacity. This expansion increases energy consumption, require significant computational capacity. This expansion increases energy consumption, embodied emissions from hardware manufacturing, and upstream supply chain impacts. As a result, even with aggressive renewable energy procurement, absolute emissions may rise due to scaling effects.

Microsoft has also promoted the role of AI as a tool for climate mitigation, arguing that AI can enhance energy efficiency, optimize industrial processes, and contribute to emissions reduction across sectors. Previous research collaborations have suggested that AI-driven innovation may contribute to economic growth while supporting greenhouse gas reductions. However, this dual narrative of AI as both a climate solution and a source of emissions, highlighting a structural tension, central to the environmental footprint debate. The deployment of AI to accelerate decarbonization does not eliminate the energy and material intensity of the digital infrastructure itself. A critical dimension of Microsoft's environmental profile concerns Scope 3 emissions, which include supply chain activities and the production of data centre hardware. As AI systems require advanced semiconductors and specialized equipment, embodied carbon becomes a substantial component of total emissions. While the company's internal carbon pricing mechanism increases transparency and accountability, the globalized nature of hardware production complicates comprehensive decarbonization effects.

Meta

Meta, the parent company of Facebook, Instagram, WhatsApp, and a growing suite of AI-driven services, including GenAI assistants and recommendation systems. Meta has transitioned from a social media company to an AI-intensive infrastructure operator. This transformation has significantly expanded its computational requirements and consequently, its environmental footprint, shown in Figure 7. The company publicly advances ambitious climate commitments, including net-zero operational emissions and large-scale renewable energy procurement. Simultaneously, it is investing heavily in next-generation AI data centers and high-performance computing clusters to support GenAI and immersive technologies. This dual trajectory of climate leadership claims alongside rapid infrastructure growth, illustrates the structural tensions inherent in scaling AI within planetary constraints.

Figure 7. Environmental impact of Meta's AI growth

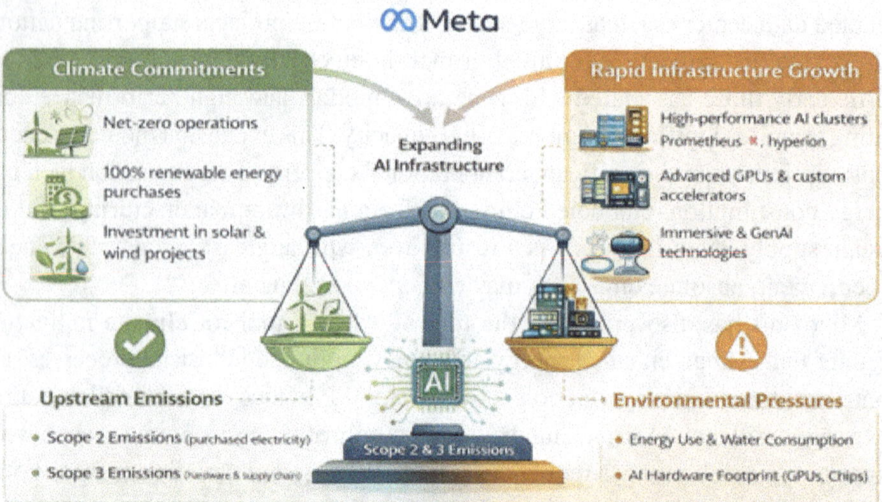

This figure depicts the environmental implications of Meta's AI expansion, including increased energy demand from hyperscale clusters (e.g., Prometheus and Hyperion), embodied carbon in semiconductor supply chains, and water use for data center cooling. It highlights the tension between infrastructure scaling and renewable energy commitments.

As Meta intensifies its focus on GenAI and large foundation models, it has invested heavily in next-generation computing clusters such as Prometheus and Hyperion. These systems represent more than technological upgrades, they signal a structural escalation in computational ambition, energy demand, and material throughput. When assessed through the lens of environmental sustainability, they illustrate the core dilemma between AI-driven innovation and ecological constraint. Prometheus was designed as a large-scale AI training cluster optimized for advanced generative and multimodal models. Hyperion represents an even more expansive trajectory intended to support future generations of increasingly complex AI systems. Both clusters rely on dense arrays of high-performance GPUs, custom accelerators, advanced networking infrastructure and sophisticated cooling architectures.

Beyond operational emissions, the lifecycle impacts of these AI clusters are substantial. The production of advanced GPUs and semiconductor components is highly energy and water intensive, relying on complex global supply chains. Fabrications processes require rare minerals, ultra-pure water, and fossil-fuel based electricity in many manufacturing regions. These upstream Scope 3 emissions may represent a significant share of the total environment footprint of hyperscale AI infrastructure, yet they are often less visible in corporate sustainability narratives focused on operational energy use.

Meta's sustainability reporting emphasizes progress in renewable energy procurement, often highlighting that its global operations are matched with 100% renewable energy through power purchase agreements. The company has invested heavily in solar and wind projects to offset electricity consumption associated with its data centers.

Meta's emphasis on renewable energy procurement highlights the distinction between carbon intensity reduction and absolute environmental impact. The case demonstrates that sustainability claims based on renewable matching must be evaluated alongside rising total electricity demand and embodied emissions. Also, Meta expansion underscores the lifecycle dimension of AI's environmental footprint, encompassing hardware manufacturing, water use, land transformation, and supply chain emission. The factors reinforce the need for comprehensive lifecycle assessment rather than narrow operational metrics.

OpenAI

OpenAI, the creator of widely used systems such as ChatGPT and other foundation models, it operates at the forefront of computationally intensive AI research and commercialization. The scale and frequency of model training and inference associated with these systems have placed the organization at the center of debates concerning AI's energy and water consumption (Figure 8).

Figure 8. Environmental impact of OpenAI's infrastructure development

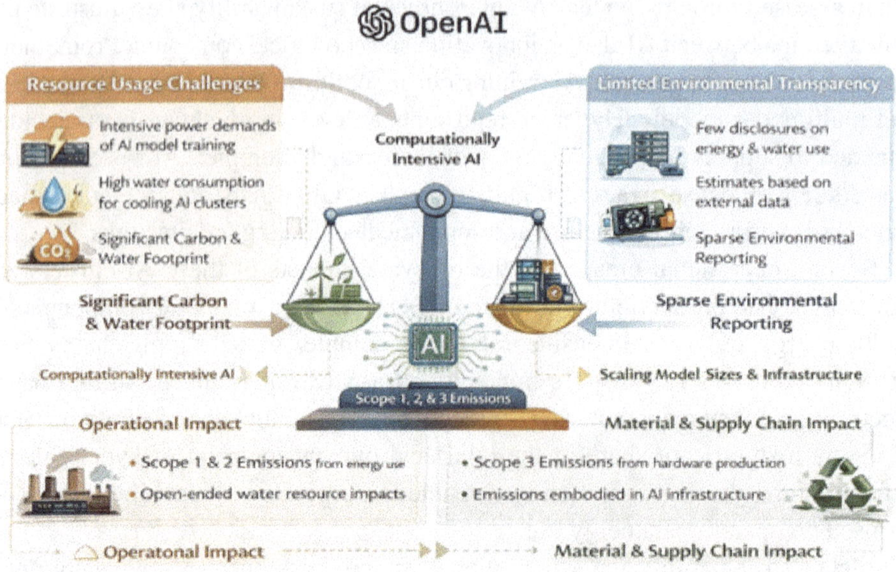

This figure illustrates the environmental implications of OpenAI's large-scale model training and deployment, including intensive electricity consumption, dependence on cloud-based data center infrastructure, and upstream semiconductor-related emissions. It highlights the growing resource demands associated with scaling frontier AI systems.

GenAI models require substantial computational resources during both training and deployment phases. Training large language models involves high-performance computing clusters powered by advanced graphics processing units (GPUs), often operating continuously for extended periods. This process entails significant electricity use and associated carbon emissions, depending on the energy mix of the data centers involved. In addition, cooling requirements for such infrastructure can demand considerable water resources, particularly in regions where evaporative cooling systems are employed. These environmental dimensions have intensified scrutiny of OpenAI's operational footprint.

Public discourse has been shaped in part by statement from OpenAI's leadership. Sam Altman, the company's CEO, has defended AI development against criticisms of excessive energy and water use, at times contextualizing AI's consumption relative to broader human or industrial energy demands. Such arguments emphasize the transformative economic and societal potential of AI, suggesting that its benefits may justify its resource intensity. However, these defenses have also reignited debate regarding proportionality, transparency, and the need for comprehensive environmental accounting frameworks.

Unlike some large technology corporations that publish extensive sustainability reports detailing emissions across Scopes 1, 2, and 3, OpenAI's public environmental disclosures remain comparatively limited. As a result, external assessments of its environmental footprint rely heavily on estimates derived from data center energy consumption, model size, and hardware requirements. This relative opacity complicates systematic evaluation and raises broader questions about governance, accountability, and industry standards for reporting AI-related environmental impacts.

The competitive dynamics of AI development incentivize increasingly larger models, greater computational power, and rapid deployment cycles. This scaling logic risks driving exponential growth in electricity demand and hardware production. Moreover, the environmental footprint extends beyond operational energy use to include embodied carbon in semiconductor manufacturing and server infrastructure.

As a flagship developer of GenAI, OpenAI embodies the material and energy intensity associated with cutting-edge AI systems. Also, the limited transparency in environmental reporting highlights a governance gap in the AI sector, underscoring the need for standardized disclosure practices.

NVIDIA

NVIDIA, the world's leading designer of graphics processing units (GPUs) and AI accelerators, NVIDIA provides the computational backbone for large-scale AI model training and deployment. While public debate often focuses on the energy consumption of data centers operated by technology firms, NVIDIA's role highlights the embodied and supply-chain emissions embedded in AI hardware production (Figure 9).

Figure 9. Environmental impact of NVIDIA's infrastructure development

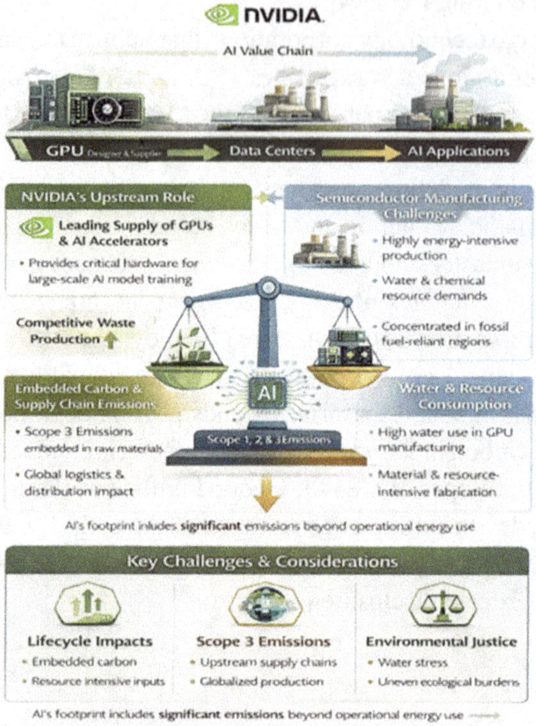

This figure depicts the environmental footprint associated with NVIDIA's expansion of AI chip manufacturing, including energy-intensive semiconductor fabrication, embodied carbon in hardware production, and upstream supply chain emissions. It highlights NVIDIA's central role in shaping the material and ecological intensity of global AI infrastructure.

The rapid expansion of GenAI and high-performance computing has significantly increased global demand for NVIDIA's chips. Advanced GPUs, such as those optimized for machine learning workloads, require complex semiconductor fabrication processes. These processes are energy-intensive and depend on highly specialized manufacturing facilities, often concentrated in regions where electricity grids remain partially reliant on fossil fuels. As a result, a substantial portion of AI's environmental footprint originates not only from operational electricity consumption but also from the carbon-intensive production of hardware components.

Beyond energy use, semiconductor manufacturing entails considerable water consumption and chemical inputs. Fabrication plants (fabs) rely on ultra-pure water and specialized materials, generating environmental pressures in regions already facing resource constraints. Some reports and environmental advocacy analysis have emphasized that communities located near semiconductor manufacturing facilities may experience disproportionate ecological burdens, including water stress and exposure to industrial pollutants. This introduces an environmental justice dimension to AI's hardware supply chain, raising questions about who bears the ecological cost of digital expansion.

NVIDIA's sustainability reporting indicates efforts to reduce operational emissions, increase renewable energy procurement, and improve supply-chain transparency. The company has outlined climate targets and emissions reduction strategies, including engagement with suppliers to decarbonize production processes. However, the scale of AI-driven demand presents structural challenges. As AI models grow larger and more computationally intensive, the volume of high-performance chips required increases correspondingly. This scaling dynamic amplifies embodied carbon emissions associated with raw material extraction, chip fabrication, assembly, and global logistics. It highlights Scope 3 emissions, particularly those embedded in supply chains, as a dominant yet often underemphasized component of AI-related environmental impact.

Salesforce

Salesforce integrates AI into customer relationship management (CRM) platforms and enterprise cloud services. Its approach, depicted by Figure 10, highlights the environmental implications of applied AI within business ecosystems rather than large-scale model training alone. Salesforce has publicly framed 'sustainable AI' as a strategic priority, emphasizing the need to align AI innovation with environmental stewardship. The company has articulated climate commitments, including net-zero targets and renewable energy procurement for its operations. It has also integrated sustainability considerations into its broader corporate governance framework, positioning environmental responsibility as a core organizational value. Through policy statements and sustainability outlook reports, Salesforce has acknowledged the growing energy intensity associated with AI and the importance of mitigating its environmental impact.

Figure 10. Environmental impact of Salesforce's infrastructure development

This figure illustrates the environmental implications of Salesforce's AI-enabled cloud expansion, including data center energy consumption, indirect Scope 3 emissions from cloud infrastructure providers, and resource intensity linked to enterprise AI deployment. It highlights the interaction between AI-driven service growth and corporate sustainability commitments.

A key feature of Salesforce's strategy is its governance-oriented approach to AI sustainability. The company has outlined a structured plan addressing environmental concerns, including energy-efficient model design, responsible data center selection, transparency in AI development, and collaboration with stakeholders to promote best practices. By emphasizing efficient model architectures and optimization techniques, Salesforce seeks to reduce computational intensity while maintaining performance. This reflects a broader industry shift toward 'efficiency-first' AI development as a means of limiting carbon growth.

However, the environmental footprint of Salesforce's AI offerings cannot be assessed solely through internal efficiency measures. As a cloud-based enterprise platform, Salesforce depends on extensive data center infrastructure, both owned and through third-party cloud providers. Consequently, its AI-related emissions include indirect Scope 2 emissions from electricity consumption and Scope 3 emissions embedded in supply chains and outsourced infrastructure. Even when renewable energy procurement offsets operational electricity use, the embodied carbon of hardware and the cumulative energy demand of expanding enterprise AI applications remain significant considerations.

Salesforce has also advanced the narrative that AI can contribute positively to climate action by enabling improved environmental data analysis, emissions tracking, and sustainability reporting for clients. Such applications suggest that AI may function as a multiplier for corporate climate management. Nevertheless, this dual role of AI as both an enabler of sustainability and a source of environmental burden reflects a recurring tension observed across the AI sector. Efficiency improvements at the model or application level may not fully counterbalance the systemic effects of scaling AI across millions of enterprise users. In sum, Salesforce represents a governance-centered model of sustainable AI. Its case highlights the importance of embedding environmental considerations into AI lifecycle management, enterprise integration, and stakeholder engagement.

DILEMMA: AI ADOPTION VS ENVIRONMENTAL CONCERNS

The rapid diffusion of AI across urban governance, industrial production, financial systems, and the built environment has created a structural dilemma at the heart of sustainable development. Policymakers and corporations increasingly portray AI as indispensable for achieving climate targets, optimizing energy systems, enabling sustainable engineering, and enhancing environmental monitoring. Simultaneously, the development, deployment, and scaling of AI systems require expansive digital and physical infrastructures that are energy-intensive, materially demanding, and environmentally consequential.

The dilemma, therefore, is not simply whether AI is environmentally beneficial or harmful. It rather concerns how to reconcile the infrastructural expansion necessary for AI adoption with the ecological constraints imposed by climate change, resource scarcity, and planetary boundaries. AI is both a potential instrument of sustainability and a driver of new environmental pressures. Understanding this contradiction requires a lifecycle, governance-oriented, and political-economic perspective on AI infrastructure.

AI adoption is contingent upon substantial infrastructural capacity. Smart cities require sensor networks, high-speed connectivity, distributed computing systems, and centralized data platforms. Industrial AI applications depend on integrated digital ecosystems and real-time analytics. GenAI systems necessitate high-performance computing clusters composed of advanced semiconductors and specialized accelerators. The existing empirical research on AI and IoT adoption consistently identifies lack of infrastructure, limited financial resources, cybersecurity vulnerabilities, and deficits in public trust as major barriers to implementation. These findings underscore a crucial point: meaningful AI deployment requires infrastructural scaling. However, infrastructure is not immaterial. It entails land use, resource extraction, energy consumption, and long-term environmental commitments.

The expansion of data centers illustrates this dynamic. AI-driven workloads demand continuous computational processing during both model training and inference. While training may be concentrated, inference generates persistent, distributed electricity demand as AI tools become embedded in everyday services. As AI applications proliferate across sectors, aggregate computational demand increases, placing sustained pressure on electricity grids and cooling systems. AI systems are embedded in material supply chains extending from mineral extraction to semiconductor fabrication, hardware assembly, and eventual disposal. Semiconductor manufacturing is highly energy and water intensive, relying on complex global production networks. Data center construction involves steel, concrete, rare earth elements, and advanced cooling technologies. Rapid hardware innovation cycles contribute to electronic waste and resource depletion.

Consequently, the environmental footprint of AI cannot be confined to operational electricity consumption. It encompasses embodied carbon in hardware production, water usage for cooling, land transformation, and waste management challenges. If infrastructural expansion outpaces renewable energy integration and circular economy practices, AI adoption may exacerbate carbon emissions and environmental degradation even as it is promoted as a sustainability tool.

AI as a Sustainability Enabler: Conditional Benefits and Systemic Uncertainty

AI's potential contribution to sustainability is substantial. Applications in energy grid optimization, climate modeling, predictive maintenance, sustainable construction, logistics efficiency, and environmental risk assessment offer pathways for reducing emissions and resource use. In the built environment, AI-powered systems can enhance lifecycle performance, optimize energy consumption, and support low-carbon design strategies. In industrial contexts, AI enables data-driven decision-making under uncertainty, aligning operational efficiency with Sustainable Development Goals (SDGs).

However, these benefits are contingent rather than automatic. They depend on reliable data, interoperable systems, cost-effective implementation, regulatory alignment, and organizational willingness to adopt AI-driven recommendations. The previous research consistently identifies gaps in data quality, standardization, and lifecycle integration, particularly in renovation and end-of-life phases of infrastructure systems. These limitations challenge the assumption that AI adoption inherently advances environmental sustainability.

Even when AI improves efficiency, systemic dynamics may undermine aggregate environmental gains. Efficiency improvements reduce per-unit resource intensity but can lower costs, stimulate demand, and expand usage, a phenomenon widely recognized as the rebound effect. As AI tools become cheaper and more accessible, adoption may accelerate, increasing total computational demand and energy consumption.

Moreover, AI development operates within competitive economic frameworks. Corporations and states pursue AI leadership to secure economic advantage and geopolitical influence. This growth-oriented paradigm incentivizes scaling larger models, more powerful chips, expanded data center capacity which often prioritizing performance and market share over ecological restraint. In such contexts, sustainability commitments may coexist with continued expansion in absolute resource use.

Ethical, Legal, and Justice Dimensions

The environmental consequences of AI infrastructure are unevenly distributed. Data centers and semiconductor facilities are frequently located in regions offering favorable regulatory conditions, low energy costs, or tax incentives. These siting decisions can intensify local water stress, strain electricity grids, and concentrate pollution burdens in specific communities. Furthermore, algorithmic biases or flawed decision-making systems may exacerbate environmental inequalities by misallocating resources or privileging certain populations over others. Thus, the dilemma of AI infrastructure extends beyond carbon accounting to include distributive justice and intergenerational ethics. Sustainable AI deployment requires a human-centric approach that safeguards equity, transparency, and accountability.

Existing regulatory frameworks primarily address safety, transparency, privacy, and fundamental rights. Environmental implications of AI systems often remain peripheral within legal regimes. Risk-based approaches may classify AI applications according to societal harm without explicitly integrating lifecycle carbon intensity or ecological impact into compliance standards.

Voluntary corporate sustainability commitments, though increasingly prominent, may not sufficiently constrain absolute growth in energy demand or embodied emissions. Carbon offsetting and renewable procurement mechanisms can mitigate operational emissions but may obscure supply-chain impacts and temporal mismatches between energy use and renewable generation. Effective governance must therefore integrate environmental metrics into AI regulation, including mandatory disclosure of energy consumption, lifecycle emissions, and resource intensity. Without such measures, AI infrastructure may expand in ways misaligned with climate targets.

Green AI Initiatives: Incremental Reform or Structural Transformation?

The rise of 'green AI' initiatives reflect growing awareness of environmental concerns. Efforts include model efficiency optimization, cloud resource management, carbon footprint measurement tools, and sustainability-oriented AI research communities. These initiatives seek to reduce computational intensity and promote environmentally conscious design.

Yet, empirical evaluation of their systemic impact remains limited. Many initiatives operate at the level of technical optimization rather than addressing macro-level growth dynamics. While they may reduce per-model energy consumption, they do not necessarily limit aggregate computational expansion. As a result, green AI efforts risk functioning as incremental mitigation strategies within an expansionary paradigm, rather than transforming the underlying logic of infrastructural growth.

The dilemma of building AI infrastructure or adopting AI while managing environmental concerns ultimately reflects a broader structural contradiction between expansionary digital systems and finite ecological limits. AI's transformative promise coexists with escalating material throughout and energy demand. Thus, resolving this dilemma requires more than technological refinement. It demands systemic coordination between AI innovation and energy transition strategies, integration of lifecycle assessment into policy and corporate governance, investment in renewable energy capacity proportional to computational growth, and the adoption of circular economy principles in hardware production and disposal.

The critical question is not whether AI should be developed, but under what ecological constraints and governance conditions. Sustainable AI adoption necessitates aligning infrastructure expansion with decarbonization trajectories, embedding environmental accountability into regulatory frameworks, and prioritizing long-term ecological resilience over short-term competitive advantage.

CONCLUSION

This chapter has examined the environmental footprint of AI through a multi-scalar analysis that integrates infrastructural materiality, lifecycle emissions, corporate strategy, and governance dynamics. Across the cases of Alphabet, Microsoft, Meta, OpenAI, NVIDIA, and Salesforce, a consistent pattern emerges: AI is simultaneously positioned as a sustainability enabler and embedded within an expansionary technological paradigm that intensifies energy demand, and global supply chain pressures. The central finding is not a simple contradiction but a structural tension between AI-driven growth and ecological constraint.

At the operational level, hyperscale data centers, advanced semiconductor manufacturing, and high-density AI clusters have significantly increased electricity consumption. While major firms have pursued renewable energy procurement, carbon accounting mechanisms, and efficiency improvements, these measures frequently reduce carbon intensity rather than absolute environmental impact. The distinction between relative decarbonization and absolute demand growth is analytically decisive. As computational workloads expand, particularly with the rise of GenAI and large foundation models baseline electricity consumption rises. Renewable matching strategies do not fully resolve regional environmental pressures.

At the lifecycle level, AI's environmental footprint extends far beyond operational energy use. Semiconductor fabrication, hardware production, cooling infrastructure, water consumption, and electronic waste generation constitute significant but often underemphasized dimensions of impact. The scaling logic that underpins frontier AI development, i.e. larger models, greater parameter counts, and intensified performance competition, magnifies upstream emissions embedded in global supply chains. NVIDIA's chip production, Meta's hyperscale clusters, Alphabet's TPU infrastructure, and Microsoft's cloud expansion collectively demonstrate that AI's environmental burden is distributed across extraction zones, manufacturing hubs, and energy systems worldwide. A comprehensive assessment must therefore incorporate Scope 3 emissions, embodied carbon, and material criticality.

At the political-economic level, competitive pressures among corporations and states incentivize rapid infrastructure expansion. AI capability has become synonymous with economic productivity, geopolitical leverage, and technological leadership. This growth-oriented paradigm complicates sustainability commitments. Efficiency gains at the model or hardware level can be offset by scale effects and rebound dynamics, wherein lower per-unit energy intensity stimulates broader adoption and increased aggregate consumption. Consequently, voluntary corporate sustainability initiatives operate within a systemic framework that prioritizes expansion.

AI can contribute meaningfully to climate mitigation, optimize resource allocation, and enhance environmental decision-making, yet its infrastructural foundations generate tangible environmental costs. Without deliberate alignment between AI scaling and ecological boundaries, the technology risks reinforcing the environmental crises it seeks to address.

Addressing this dilemma requires moving beyond narrow metrics of efficiency or carbon neutrality toward integrated governance architectures. First, lifecycle carbon accounting must become standardized, transparent, and mandatory, incorporating embodied emissions and supply-chain impacts. Second, infrastructure expansion should be explicitly linked to renewable energy additionality and grid decarbonization pathways, ensuring that AI-driven load growth does not outpace clean energy deployment. Third, sustainability regulation should extend beyond operational reporting to include material circularity, water stewardship, and responsible hardware end-of-life management. Finally, distributive justice considerations must inform infrastructure siting and supply chain governance to prevent the concentration of environmental burdens in vulnerable communities. Ultimately, the sustainability of AI will depend less on incremental technological optimization and more on whether its developmental trajectory can be reconciled with planetary boundaries. The question is not whether AI should be developed, but under what constraints, governance mechanisms, and normative commitments it should advance. A genuinely sustainable AI paradigm would align innovation incentives with decarbonization imperatives, embed ecological accountability into infrastructure planning, and evaluate success not solely in terms of computational performance but also environmental integrity.

In this sense, the environmental footprint of AI constitutes a defining test of 21st century technological governance. If AI expansion can be integrated within a framework of ecological limits, lifecycle responsibility, and equitable distribution, it may contribute to sustainable development. If not, it risks deepening material consumption patterns that undermine climate stability and environmental resilience. The future of AI, therefore, is inseparable from the future of environmental sustainability.

REFERENCES

At Meta. (2025). Meta 2025 sustainability report. https://sustainability.atmeta.com/wp-content/uploads/2025/08/Meta_2025-Sustainability-Report_.pdf

Bolton, P., & Kacperczyk, M. (2021). Do investors care about carbon risk? *Journal of Financial Economics*, *142*(2), 517–549. DOI: 10.1016/j.jfineco.2021.05.008

Bughin, J., Seong, J., Manyika, J., Chui, M., & Joshi, R. (2018). Notes from the AI frontier: Modeling the impact of AI on the world economy. *McKinsey Global Institute*, *4*(1), 2–61.

Green Digest. (n.d.). NVIDIA environmental and social impact. https://www.greendigest.co/p/nvidia-environmental-and-social-impact

Greenpeace International. (n.d.). The environmental cost behind NVIDIA's rise as the world's biggest AI chipmaker: Who is paying the price? https://www.greenpeace.org/international/story/79910/the-environmental-cost-behind-nvidias-rise-as-worlds-biggest-ai-chipmaker-who-is-paying-the-price/

Gupta, U., Kim, Y. G., Lee, S., Tse, J., Lee, H.-H. S., Wei, G.-Y., Brooks, D., & Wu, C.-J. (2020). *Chasing Carbon: The Elusive Environmental Footprint of Computing*. https://arxiv.org/abs/2011.02839

IEA. (2023). Data Centres and Data Transmission Networks. Https://Www.Iea.Org/Energy-System/Buildings/Data-Centres-and-Data-Transmission-Networks

Info, Q. (2025). Meta carbon AI. https://www.infoq.com/news/2025/10/meta-carbon-ai/

IRDO. (2025). Poster 63 [Conference poster]. https://www.irdo.si/irdo2025/posters/63.pdf

Kaack, L. H., Donti, P. L., Strubell, E., Kamiya, G., Creutzig, F., & Rolnick, D. (2022). Aligning AI with climate change mitigation. *Nature Climate Change*, *12*(6), 518–527. DOI: 10.1038/s41558-022-01377-7

Land, P. P. C. (n.d.). Meta confronts sustainability tensions amid massive AI infrastructure expansion. https://ppc.land/meta-confronts-sustainability-tensions-amid-massive-ai-infrastructure-expansion

Li, P., Yang, J., Islam, M. A., & Ren, S. (2025). *Making AI Less "Thirsty": Uncovering and Addressing the Secret Water Footprint of AI Models*. https://arxiv.org/abs/2304.03271

Makridakis, S. (2017). The forthcoming Artificial Intelligence (AI) revolution: Its impact on society and firms. *Futures*, *90*, 46–60. DOI: 10.1016/j.futures.2017.03.006

McKinsey & Company. (2023). *As organizations rapidly deploy GenAI tools, survey respondents expect significant effects on their industries and workforces The state of AI in 2023: GenAI's breakout year.*

Microsoft. (2019a, April 15). We're increasing our carbon fee as we double down on sustainability. https://blogs.microsoft.com/on-the-issues/2019/04/15/were-increasing-our-carbon-fee-as-we-double-down-on-sustainability/

Microsoft. (2025a, May 29). Environmental sustainability report. https://blogs.microsoft.com/on-the-issues/2025/05/29/environmental-sustainability-report/

Microsoft. (2025b, September 16). Microsoft invests $30 billion in UK AI future. https://blogs.microsoft.com/on-the-issues/2025/09/16/microsoft-30-billion-uk-ai-future/

MLQ.ai. (n.d.). Sam Altman counters AI environmental criticisms by comparing to human energy needs. https://mlq.ai/news/sam-altman-counters-ai-environmental-criticisms-by-comparing-to-human-energy-needs/

Rolnick, D., Donti, P. L., Kaack, L. H., Kochanski, K., Lacoste, A., Sankaran, K., Ross, A. S., Milojevic-Dupont, N., Jaques, N., Waldman-Brown, A., Luccioni, A., Maharaj, T., Sherwin, E. D., Mukkavilli, S. K., Kording, K. P., Gomes, C., Ng, A. Y., Hassabis, D., Platt, J. C., & Bengio, Y. (2019). *Tackling Climate Change with Machine Learning*. https://arxiv.org/abs/1906.05433

Salesforce. (n.d.-a). Sustainable AI. https://www.salesforce.com/news/stories/sustainable-ai/

Salesforce. (n.d.-b). Salesforce AI sustainability outlook. https://www.salesforce.com/en-us/wp-content/uploads/sites/4/documents/company/sustainability/salesforce-ai-sustainability-outlook.pdf

SalesforceBen. (n.d.). Salesforce tackles AI's environmental issues with 6-point plan. https://www.salesforceben.com/salesforce-tackle-ais-environmental-issues-with-6-point-plan/

Stanford. (2025). *AI Index Report 2025*.

Strubell, E., Ganesh, A., & McCallum, A. (2019). *Energy and Policy Considerations for Deep Learning in NLP*. https://arxiv.org/abs/1906.02243 DOI: 10.18653/v1/P19-1355

The Guardian. (2025, June 27). Google emissions and AI electricity demand derail green efforts. https://www.theguardian.com/technology/2025/jun/27/google-emissions-ai-electricity-demand-derail-efforts-green

Chapter 2
Scaling Intelligence, Scaling Impact:
A Systemic Analysis of the Environmental Sustainability of Artificial Intelligence

Silvio Andrae
http://orcid.org/0000-0002-8586-7812
Independent Researcher, Germany

ABSTRACT

Artificial intelligence (AI) is considered a key driver of sustainable transformation, yet its rapid development is leading to growing environmental impacts. We argue that the environmental footprint of AI cannot be understood as a technical side effect. Rather, it is a systemic outcome of intertwined technological, infrastructural, and economic dynamics. Drawing on a life-cycle perspective and ecological economics, as well as infrastructure research and technology governance, we identify three key drivers: scaling paradigms, material dependencies, and market-driven incentive structures. The analysis shows that efficiency gains are often offset or outweighed by rebound effects and structural growth dynamics. Building on this, we develop a multi-level governance framework that integrates lifecycle-based management, infrastructural embedding, and institutional coordination. Finally, we introduce 'net-positive AI,' a conceptual approach that shifts the focus from mere damage minimization to a positive net contribution to ecological and social systems.

DOI: 10.4018/979-8-3373-8490-0.ch002

INTRODUCTION: THE SUSTAINABILITY PARADOX OF AI

Artificial Intelligence (AI) is currently regarded as one of the key technologies for addressing global sustainability challenges. Applications in climate modeling, the optimization of energy systems, or the monitoring of biodiversity are frequently cited as evidence that AI not only enables economic efficiency gains but can also make substantial contributions to achieving ecological and societal goals (Rolnick et al., 2019; Vinuesa et al., 2020; Carmeno et al., 2026). From this perspective, AI appears as a technological enabler of sustainable transformation, making existing systems smarter, more resource-efficient, and more resilient.

At the same time, however, it is becoming increasingly clear that the rapid expansion of AI itself generates significant environmental impacts—a dynamic often referred to as the 'green paradox' (Chouksey et al., 2026). In particular, the scaling of data-intensive models, the exponential increase in computing capacity, and the global expansion of digital infrastructure are leading to rising energy and resource consumption that cannot be viewed as a mere side effect of technological progress (Strubell et al., 2019; Patterson et al., 2021; Andrae, 2025). Data centers are already among the most energy-intensive infrastructure in the digital economy, and the growing need for specialized hardware—such as GPUs and semiconductors—further increases demand for scarce mineral resources (IEA, 2023). This dual role positions AI not only as part of the solution but also as part of the problem, raising the question of whether current technological developments are compatible with environmental constraints. Understanding this paradox is central to assessing AI's sustainability and motivates the systemic perspective adopted in this chapter.

This tension points to a deeper sustainability paradox of AI. While much of the existing research focuses on efficiency gains—such as through more energy-efficient algorithms or optimized hardware architectures—the structural causes of environmental impact remain largely unaddressed. Efficiency gains are often offset by rising demand, a phenomenon described in environmental economics as the rebound effect or Jevons' paradox (Jevons, 1865; Polimeni et al., 2008). Consequently, technological progress does not necessarily lead to a reduction in absolute environmental impacts, but may even exacerbate them.

Against this backdrop, a purely technological perspective on sustainable AI falls short. Rather, it is necessary to understand AI as a socio-technical system whose ecological impacts arise from the interplay of technological architectures, economic incentive structures, regulatory frameworks, and global value chains (Geels, 2002; Hughes, 1987). The environmental impacts of AI are are distributed across the global life cycle, ranging from raw material extraction to hardware production and the operation of data centers, to the disposal of electronic waste. These processes are closely interlinked and lead to cumulative effects that unfold across different regions and sectors (Baccini & Brunner, 2012; Wiedmann et al., 2015).

Recent systematic reviews have advanced the assessment of the environmental footprint of artificial intelligence by adopting a life-cycle perspective that encompasses training, inference, and infrastructure processes (Oliveira et al., 2026). These studies show that environmental impacts are distributed across multiple phases and cannot be reduced to individual computational operations. By synthesizing a broad range of empirical findings, they provide a more comprehensive understanding of how energy consumption and CO_2 emissions accumulate across the entire AI lifecycle. At the same time, their results underscore that an exclusive focus on efficiency gains is insufficient to address the overall ecological footprint of AI systems.

Such a systemic perspective highlights the structural drivers of AI's environmental impacts. First, the current innovation dynamic is based on a scaling paradigm that views larger models, more extensive datasets, and increasing computational capacities as central determinants of technological performance (Kaplan et al., 2020). Second, the material basis of digital infrastructures is closely linked to global commodity markets, whose environmental and social costs are borne particularly in the Global South (Sovacool et al., 2020). Third, existing market and competitive structures favor the continuous expansion of AI systems, while regulatory frameworks have so far only partially internalized environmental externalities (Zuboff, 2019; Srnicek, 2017).

Against this backdrop, this chapter aims to analyze the environmental impacts of AI as structural characteristics of current innovation regimes. It integrates insights from environmental science, ecological economics, infrastructure studies, and technology governance to systematically examine the complex interactions between technological development, resource consumption, and institutional frameworks.

Four research questions guide this analysis:

(1) How do environmental impacts arise throughout the entire life cycle of AI systems, and how do these effects interact with one another?
(2) What technological, economic, and regulatory factors drive the growth of AI's ecological footprint?
(3) In what ways are these environmental impacts linked to global inequalities and issues of environmental justice?
(4) What governance approaches and systemic interventions are suitable for sustainably reducing the environmental impact of AI?

On this basis, the chapter argues that technical efficiency gains alone are insufficient to address AI's environmental challenges. Rather, the sustainable development of AI requires a fundamental realignment of incentive structures, infrastructure architectures, and governance regimes. Sustainability is thus understood as a constitutive principle for the future design of AI systems.

The remainder of the chapter follows a systematic, multi-stage analysis of the environmental impacts of AI. Section 2 first develops the conceptual foundations by situating AI as a socio-technical system within the context of environmental economics and infrastructure studies. Building on this, Section 3 analyzes the environmental impacts across the entire life cycle of AI systems and highlights their global interconnections. Section 4 identifies the key structural drivers of these dynamics, particularly the scaling paradigm, material infrastructure dependencies, and market- and governance-related incentive structures. Section 5 deepens the analysis through an environmental justice lens and demonstrates how ecological burdens are unevenly distributed. Finally, Section 6 discusses technical and systemic intervention options, before Section 7 develops a normative and strategic framework using the concept of 'net-positive AI.' The chapter concludes in Section 8 with a synthesis of the findings and a research and policy agenda.

CONCEPTUAL FOUNDATIONS: AI AS A SOCIO-TECHNICAL AND ECOLOGICAL SYSTEM

Analyzing the environmental impacts of artificial intelligence requires a perspective that goes beyond a purely technological view. AI should be understood as a complex socio-technical system whose functioning is inextricably linked to physical infrastructures, institutional arrangements, and global value chains. Early works in the sociology of technology and infrastructure research already emphasize that large-scale technical systems—such as electricity grids or communication infrastructures—are organized as 'system builders' (Hughes, 1987), whose development is shaped by the interplay of technical, economic, and political factors. Similarly, AI can be understood as part of a new generation of digital infrastructures that are deeply embedded in social and ecological systems.

AI as Material Infrastructure

Contrary to widespread assumptions, the digital economy is far from immaterial. Rather, it is based on a highly material infrastructure of data centers, networks, and hardware components, the operation of which requires significant amounts of energy, water, and raw materials (Kitchin, 2014; Starosielski, 2015). AI intensifies this materiality in a particular way, as data-intensive models and computationally demanding training processes require continuous scaling of physical resources. The growing importance of hyperscale data centers and specialized chips illustrates that algorithmic performance is closely linked to infrastructural capacities.

This perspective brings AI closer to classical infrastructure systems and suggests that its environmental impacts should be analyzed in the context of global infrastructure regimes. It becomes apparent that the expansion of AI is shaped by investment cycles, platform strategies, and geopolitical competitive dynamics. The material dimension of AI thus underpins its ecological effects and underscores the need to incorporate digital technologies into the analysis of resource systems.

Ecological Economics and External Effects

Ecological economics provides a central theoretical framework for analyzing these relationships. It views economic processes as embedded in biophysical systems and emphasizes the existence of planetary boundaries within which economic activity must operate (Daly, 1996; Rockström et al., 2009). From this perspective, environmental impacts should be as an integral part of economic processes.

The concept of externalities is particularly relevant for the analysis of AI. Energy consumption, water withdrawal, raw material extraction, and emissions are shifted onto the environment and society. This externalization enables the seemingly cost-effective scaling of digital technologies but obscures their actual ecological costs (Costanza et al., 1997). Added to this is the rebound effect, which describes how efficiency gains often lead to increased use and, in turn, to higher absolute environmental impacts (Polimeni et al., 2008). In the context of AI, this means that more efficient algorithms or hardware do not necessarily lead to a reduction in resource consumption but, on the contrary, generate additional demand through increased use.

These theoretical insights underscore that sustainability strategies for AI cannot focus solely on efficiency gains; they must also address the underlying incentive structures and growth dynamics.

Life Cycle Perspective and Systems Thinking

To adequately assess the environmental impacts of AI, a life-cycle perspective is also necessary. Life Cycle Assessment (LCA) approaches enable analysis of environmental impacts across the entire value chain—from raw material extraction through production and use to disposal. However, with regard to digital technologies, these approaches have so far been applied only in a fragmented manner, often focusing on individual components, such as data center energy consumption or the training of specific models.

A systemic expansion of this perspective makes it clear that the individual phases of the life cycle are closely interlinked. Decisions in model design, for example, influence hardware requirements, which in turn affect raw material demand and production processes. Similarly, market structures and regulatory frameworks impact the intensity of use and the lifespan of technologies. These interdependencies lead to cumulative effects that amplify across different sectors and regions (Wiedmann et al., 2015).

Systems thinking enables the visualization of these interactions and the understanding of AI as part of a global material and energy system. In this sense, AI is not merely a tool for analyzing ecological processes, but also an actor within them. Its development and application are inextricably linked to material resource flows, social structures, and political decision-making processes.

Against this conceptual backdrop, the following section systematically analyzes the environmental impact of AI throughout its entire life cycle to reveal the specific impact profiles and their global interconnections.

ENVIRONMENTAL IMPACTS THROUGHOUT THE LIFE CYCLE OF AI SYSTEMS

Recent research has begun to assess the environmental impacts of artificial intelligence more comprehensively, moving beyond isolated metrics to multidimensional assessments that incorporate energy consumption, infrastructure, and broader environmental impacts (Wenmackers, 2025; Jeanquartier et al., 2026). These approaches represent an important step toward capturing the systemic nature of AI's environmental footprint. However, they typically focus on emissions at the model level or during operation, thereby capturing only a portion of the broader environmental impacts identified in this chapter. In particular, upstream processes such as raw material extraction, hardware production, and infrastructure expansion—as well as systemic effects such as rebound effects—are still inadequately accounted for. Consequently, while CO_2 footprint assessments provide important insights, they offer only a partial perspective that must be embedded within a comprehensive life-cycle analysis. Although such assessments improve empirical understanding, they must be embedded within a broader analytical framework that accounts for the structural drivers of environmental impacts.

The environmental impacts of artificial intelligence cannot be adequately captured if individual system components are considered in isolation. Furthermore, life-cycle analyses of artificial intelligence show that environmental impacts arise not only at the levels of model training and operation, but also across interconnected infrastructure phases (Rohde et al., 2026). It is important to note that these impacts are shaped by underlying infrastructural conditions, including geographic location, energy mix, and supply chain structures. These effects arise along a globally distributed, materially embedded life cycle that ranges from raw material extraction through hardware production and the operation of digital infrastructures to the disposal of electronic waste. While traditional life cycle assessments (LCA) enable a structured capture of such effects, the context of AI reveals that the interactions between individual phases and their embedding in global value creation systems require an expansion toward a systemic perspective (Wiedmann et al., 2015). This supports the argument that the ecological footprint of AI must be analyzed as a distributed and networked life-cycle phenomenon rather than as a series of isolated technical processes.

In the following, the central phases of this life cycle are analyzed to highlight the specific environmental impacts and their interactions.

Raw Material Extraction and Material Base

The material foundation of modern AI systems relies on a variety of critical raw materials. In particular, lithium, cobalt, nickel, and rare earth elements are essential for semiconductors, battery systems, and electronic components. The extraction of these materials is associated with significant environmental impacts: Studies show that lithium mining in South America, for example, involves water consumption of several hundred thousand liters per ton of lithium, leading to significant conflicts over water use in arid regions (Sovacool et al., 2020).

At the same time, the extraction of these raw materials is highly concentrated geographically: around 70% of global cobalt production comes from the Democratic Republic of the Congo, often under conditions associated with significant environmental and social risks. This spatial concentration externalizes environmental costs onto the Global South and reinforces existing patterns of unequal resource use. The growing demand for AI hardware further exacerbates these dynamics, as rising investments in data centers and semiconductor production directly drive demand for raw materials.

Hardware Production and Manufacturing Processes

The subsequent processing of raw materials into high-performance hardware represents another major source of environmental impact. Semiconductor production is among the most energy- and resource-intensive industrial processes. The manufacture of a single high-performance chip requires large quantities of ultrapure water and energy-intensive lithography processes. Estimates suggest that the production of microelectronic components requires many times their own weight in resources and energy (Williams et al., 2002).

Furthermore, production chains are globally fragmented and highly specialized, which generates additional transport emissions and dependencies. The concentration of chip manufacturing in a few regions—particularly Taiwan, South Korea, and China—not only creates geopolitical risks but also spatially concentrates environmental impacts. This makes it clear that the environmental impacts of AI arise long before its actual use and are structurally embedded in global production systems.

Operation of Digital Infrastructures: Energy and Water Consumption

The operation of AI systems in data centers represents one of the most visible sources of environmental impact. Data centers are currently responsible for about 1–2% of global electricity consumption, and this figure is on the rise (IEA, 2023). Marinoni et al. (2025) investigate computational workloads under real-time grid conditions. Such approaches can reduce peak demand, improve load balancing, and increase energy efficiency within existing infrastructures. However, these strategies primarily operate at the operational level and do not fundamentally alter the underlying factors driving energy demand.

In particular, training large AI models can require significant amounts of energy. Estimates show that training a single large language model can consume several hundred megawatt-hours of energy, equivalent to the annual electricity consumption of dozens of households (Strubell et al., 2019; Patterson et al., 2021). For example, analyses of foundational models such as Moshi highlight the significant energy consumption and emissions associated with both training and inference processes (López-Rauhut et al., 2026). Jegham et al. (2025) show substantial variation across leading large language models, with energy requirements differing by more than an order of magnitude between model architectures and deployment environments. Their results suggest that infrastructure characteristics—including cooling systems, regional electricity mixes, and hardware configurations—are increasingly important determinants of AI's environmental footprint. Such studies provide valuable transparency by quantifying resource consumption in detail and illustrating the extent of environmental impacts. However, their focus on specific models also highlights a significant limitation: while they improve empirical understanding, they do not fully capture the broader systemic dynamics of scaling, infrastructure expansion, and cumulative environmental impacts that shape AI's overall footprint.

Recent review studies have systematically examined the ecological footprint of artificial intelligence in both the training and inference phases (Kim et al., 2026). These studies show that while training processes are often energy-intensive, inference workloads—due to their scale and persistence—can be the dominant factor in long-term environmental impacts. However, their focus remains largely on operational metrics, while broader infrastructural, material, and systemic dynamics are inadequately accounted for.

Recent analyses of AI-related data centers highlight the rising electricity demand associated with the training and inference of large-scale models, as well as their growing impact on regional power grids (Chen et al., 2025). These studies show that expanding AI infrastructure not only increases overall energy consumption but also affects load profiles, peak loads, and grid stability. In particular, the concentration of hyperscale data centers can put local power supply systems under pressure, requiring significant investments in grid capacity and energy supply. While such analyses significantly improve our understanding of operational energy consumption, they also underscore that AI's environmental footprint must be understood as part of a broader energy-infrastructure nexus. This highlights the need to incorporate infrastructure-level dynamics into life-cycle-based assessments of AI.

In addition to energy consumption, water consumption is becoming increasingly significant. Data centers require water for cooling, and individual training runs of large models can consume several hundred thousand liters, depending on location and cooling technology. In water-scarce regions, this can lead to conflicts over the use of digital infrastructure and local needs.

At the same time, a key systemic dynamic emerges: while efficiency gains in hardware or software reduce energy consumption per computational operation, they are more than offset by the exponential increase in the number of computational operations. This development corresponds to the classic rebound effect, in which efficiency gains lead to increased use.

Usage and Inference: Scaling in Everyday Life

While energy consumption during the training of large models is often the focus, a significant portion of the environmental impacts unfolds during the usage phase. The increasing integration of AI into everyday applications results in massive scaling of inference processes. Billions of queries to search engines, language models, or recommendation systems generate continuous computational loads that accumulate over time.

Recent review studies, such as that by Kim et al. (2025), on the ecological footprint of generative AI provide comprehensive summaries of energy consumption and CO_2 emissions across both the training and inference phases. Jegham et al. 2025suggest that inference, rather than training, may become the dominant contributor to the lifecycle energy demand of advanced AI systems, potentially accounting for up to 90% of total energy consumption over a model's operational lifetime. These studies highlight that while training processes are often highly energy-intensive, inference workloads—due to their scale and persistence—could become the dominant factor in long-term environmental impacts. By synthesizing the results of multiple studies, such review papers significantly improve the empirical basis for assessing the environmental footprint of AI. However, their focus remains largely on operational metrics such as energy consumption and emissions, while broader life-cycle dynamics, material dependencies, and systemic growth drivers are insufficiently accounted for. This underscores the need to complement the empirical synthesis with a structural analysis, as outlined in the following sections.

Current estimates suggest that, in widespread applications, the cumulative energy required for inference processes may exceed training costs in the long term. This shifts the environmental focus from one-off training events to ongoing, infrastructure-based impacts. This trend is reinforced by business models designed for maximum user interaction and continuous availability.

End-of-Life and Electronic Waste

At the end of the life cycle lie disposal and recycling. The global volume of e-waste recently exceeded 50 million tons per year and continues to grow (Forti et al., 2020). A significant portion of this waste is attributable to short innovation cycles and rapid hardware obsolescence, which also play a central role in the AI sector.

Recycling rates remain limited, particularly for complex electronic components. At the same time, large portions of e-waste are processed in informal sectors, often in the Global South, where environmental and health standards are inadequate. This creates additional burdens that conclude the life cycle of AI systems in ecologically and socially problematic ways.

Interdependencies and systemic effects

The individual phases of the life cycle should not be viewed in isolation but are closely interlinked. Decisions regarding model design, for example, influence the demand for computing power and, in turn, affect energy consumption and hardware demand. Similarly, market structures and regulatory frameworks impact the intensity of use and the lifespan of technologies.

These interdependencies lead to cumulative effects that amplify across different levels. For instance, an increase in model complexity can also boost demand for new generations of hardware, thereby intensifying pressure on raw material markets. At the same time, geopolitical tensions in supply chains can affect the availability of resources, thereby redirecting innovation pathways.

Figure 1. Lifecycle of AI systems and systemic environmental impacts

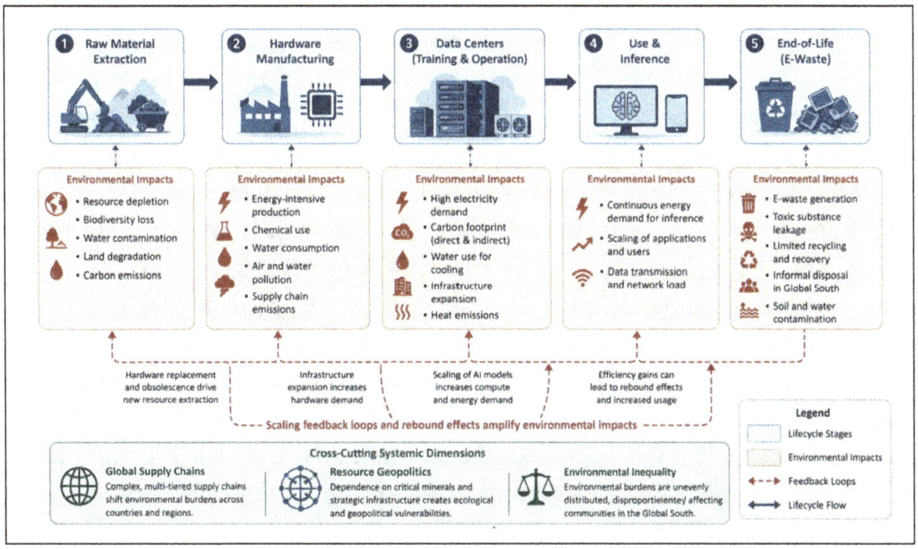

Source: Own representation.

Figure 1 presents a lifecycle perspective on artificial intelligence systems, linking five core phases—raw material extraction, hardware manufacturing, data center operation, usage and inference, and end-of-life disposal—to their associated environmental impacts. Each phase causes specific environmental impacts, including resource scarcity, energy consumption, water consumption, and waste generation. The model highlights systemic feedback loops, such as scaling dynamics and rebound effects, which amplify overall environmental impacts despite efficiency gains. At an overarching level, global supply chains, geopolitical resource dependencies, and ecological inequalities structure the distribution of these impacts.

The figure thus conceptualizes AI as an embedded socio-ecological system. Such a systemic perspective makes it clear that the environmental impacts of AI cannot be understood as the additive sum of individual effects. Rather, it is a dynamic, globally interconnected system in which technological, economic, and ecological processes are intertwined.

Recent research findings have made it clear that it is necessary to move beyond abstract computational metrics such as floating-point operations (FLOPs) and transition to a more comprehensive assessment of the material and ecological costs of artificial intelligence (Falk et al., 2025). While FLOPs are a useful indicator of computational intensity, they do not capture the full spectrum of environmental impacts associated with AI systems. By linking computational demands to energy consumption, hardware production, and infrastructure expansion, these approaches demonstrate that performance metrics are only an indirect indicator of the ecological footprint. This distinction highlights a fundamental limitation of model-centric assessments: computational effort cannot be linearly converted into environmental impact, as the latter is shaped by a broader range of material and systemic factors across the entire life cycle.

Importantly, life-cycle-oriented assessments also highlight the limitations of efficiency-oriented approaches. While technological improvements can reduce energy consumption per operation, they are often accompanied by increased model size, broader deployment, and expanded infrastructure (Oliveira et al., 2026). This suggests that environmental impacts are shaped not only by technological efficiency but also by structural dynamics such as scaling and demand growth. Therefore, addressing the ecological footprint of AI requires going beyond system-level optimization and analyzing the underlying factors that drive expansion and resource consumption. Against this backdrop, the question arises: which structural factors drive these dynamics, and why does the ecological footprint of AI continue to expand despite technical efficiency gains? These drivers are systematically analyzed in the following section.

STRUCTURAL DRIVERS OF AI'S ECOLOGICAL FOOTPRINT

An analysis of the life cycle of AI systems shows that their environmental impacts cannot be understood as isolated effects of individual phases. Rather, they are an expression of structural dynamics deeply embedded in current innovation regimes. These dynamics arise from the interplay of technological development logics, dependencies on physical infrastructure, and economic and institutional incentive structures. The environmental footprint of AI is thus not primarily the result of individual design decisions, but rather emerges from the systemic interplay of technological development, physical infrastructure, and economic incentives. Crucially, this interplay is reinforced by a scaling paradigm in which performance gains are closely tied to increasing computational intensity, thereby coupling technological progress with rising resource consumption. Efficiency improvements at the model or hardware level do not necessarily translate into reduced environmental impact; instead, they are frequently absorbed by expanding deployment, growing demand, and continuous system integration. In this sense, environmental pressures are not incidental side effects but an endogenous feature of contemporary AI development, reflecting a broader dynamic in which micro-level optimization drives macro-level expansion. Understanding these systemic interdependencies is therefore essential for moving beyond fragmented assessments toward a comprehensive analysis of AI's environmental footprint.

Figure 2. Structural drivers of AI's environmental footprint

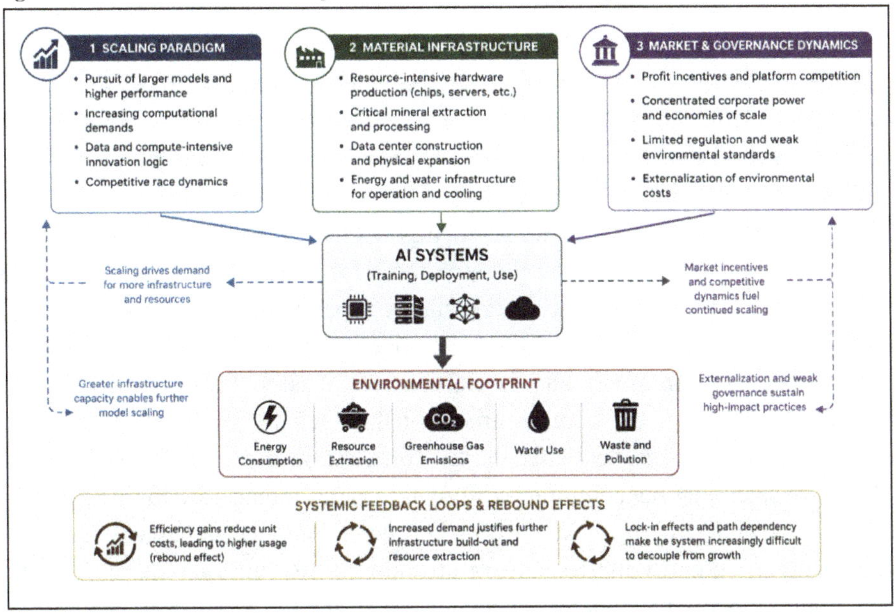

Source: Own representation.

Figure 2 illustrates the structural factors that shape the environmental footprint of artificial intelligence. Three central dynamics—scaling paradigms, dependencies on physical infrastructure, and market and governance structures—interact within AI systems, leading to rising energy demand, increased raw material extraction, and higher emissions. These factors reinforce one another through feedback loops, particularly via scaling dynamics and rebound effects.

The following analyzes three key drivers of this development: the scaling paradigm, the material nature of digital infrastructures, and the institutional and market-based dynamics that prioritize growth over sustainability.

The Scaling Paradigm: Performance Logic and Rebound Dynamics

A fundamental characteristic of current AI development is the so-called scaling paradigm. Empirical studies show that the performance of many AI models closely correlates with the size of the training dataset, the number of model parameters, and the computational resources used (Kaplan et al., 2020). This insight has led to an innovation logic in which progress is defined primarily through quantitative expansion: larger models, more data, and higher computing capacities are considered key determinants of technological performance. This logic generates a structural growth dynamic. Companies and research institutions invest in ever-larger computing clusters, specialized hardware, and global data infrastructures to gain competitive advantages. However, the associated resource consumption does not increase linearly; rather, it often increases exponentially. Training processes for large models can require several million kilowatt-hours of energy, while the scaling of inference applications further intensifies the demand for computing power.

The shift from computational metrics such as FLOPs to resource-based evaluations of AI systems also sheds light on the structural dynamics underlying the scaling paradigm. As model performance increasingly depends on higher computational intensity, performance gains are directly accompanied by rising energy demands, greater hardware requirements, and infrastructure expansion (Falk et al., 2025). This connection underscores the core argument of this chapter: scaling is a resource-intensive growth model embedded in market competition and innovation incentives.

A key mechanism underpinning this dynamic is the 'green paradox' in artificial intelligence, which highlights how efficiency gains can simultaneously stimulate new demand and thereby increase overall resource consumption (Chouksey et al., 2026). Improvements in model performance and computational efficiency do not necessarily reduce environmental impact; instead, they enable additional applications and intensify the use of existing systems. This dynamic is closely linked to the rebound effect described in ecological economics (Polimeni et al., 2008): declining costs per computing unit increase demand, offsetting or even exceeding efficiency gains. In the context of AI, this reveals a systemic pattern in which micro-level optimization drives macro-level expansion.

Beyond these efficiency dynamics, the scaling paradigm is also reinforced by competitive pressures within the AI industry. Recent research points to the emergence of a 'hidden AI race,' in which firms continuously expand model size, computational resources, and infrastructure capacities to secure or enhance their strategic position (Agarwal & Chakraborti, 2025). This competitive logic directly links technological progress to market leadership, thereby embedding scaling within the structure of innovation itself. Environmental costs are thus externalized as by-products of competition rather than acting as constraints on system development. In this context, efficiency gains are systematically reinvested in expansion, further amplifying resource consumption.

At a broader systemic level, the persistence of the scaling paradigm despite growing awareness of ecological impacts can be interpreted through Blühdorn's (2020) concept of the 'unsustainability of sustainability.' This perspective suggests that contemporary societies incorporate sustainability concerns into existing growth and innovation regimes without fundamentally transforming them. Applied to AI, this implies that efficiency improvements and emission reductions occur within—and ultimately reinforce—the dominant logic of scaling. Sustainability thus functions less as a transformative force than as a system-compatible adjustment, leaving the structural drivers of environmental impact largely intact.

The scaling paradigm is a development model embedded in economic and institutional frameworks. It links scientific performance indicators, market incentives, and investment logics into a system that favors continuous growth and systematically reproduces ecological constraints.

The Material Nature of Digital Infrastructures

The second key driver lies in the material basis of AI systems. Contrary to the notion of an 'immaterial' digital economy, AI is deeply embedded in physical infrastructures whose construction and operation have significant environmental impacts. Data centers, networks, and hardware components form a global infrastructure regime based on energy-intensive processes and resource-dependent supply chains (Kitchin, 2014).

These infrastructures exhibit several structural features. First, they require continuous investment and technological renewal, as rising performance demands shorten innovation cycles. Second, they are spatially concentrated, for example, in the form of hyperscale data centers or specialized manufacturing sites, which leads to local environmental impacts. Third, they are globally interconnected, thereby distributing environmental impacts along international supply chains.

Particularly relevant is the dependence on critical raw materials. The production of semiconductors and battery systems relies on materials such as lithium, cobalt, or rare earth elements, the extraction of which is associated with high environmental and social costs. These resources are geographically unevenly distributed and often linked to geopolitical tensions, which connect the environmental dimension of AI to issues of global power and resource sovereignty (Sovacool et al., 2020).

The material nature of AI thus leads to a structural coupling between digital innovation and the use of physical resources. This coupling is constitutive of the functioning of modern AI systems and amplifies their ecological impacts.

Market and Innovation Dynamics: Growth as a Guiding Principle

A third key driver lies in the economic and institutional conditions that shape AI development. The digital economy is characterized by platform markets, network effects, and significant economies of scale (Srnicek, 2017). In this environment, there is a strong incentive to expand, as larger systems generate competitive advantages and secure market dominance.

Companies, particularly large technology platforms and cloud providers, pursue strategies geared toward maximum scalability and availability. Investments in computing capacity, data infrastructure, and global networks are rather by growth and competitive imperatives. Startups, in turn, face pressure to scale rapidly to establish themselves in the market, leading to a prioritization of performance over resource efficiency.

Emerging empirical evidence suggests that these investment dynamics may themselves contribute to environmental pressures. Using panel data from nine advanced economies between 2012 and 2023, Konat et al. (2026) find that artificial intelligence-related venture capital investments are positively associated with carbon intensity in the long run. Their findings challenge the widespread assumption that digital innovation automatically promotes environmental sustainability and instead indicate that AI-driven investment flows can reinforce resource-intensive growth trajectories when embedded in carbon-dependent infrastructure systems. From the perspective of ecological political economy, this evidence highlights the importance of capital allocation as a mediating mechanism between technological innovation and environmental impact. Investments in AI do not merely finance technological progress; they also accelerate the expansion of computational infrastructures, data centers, and associated energy demand, thereby linking financialization directly to ecological outcomes.

These dynamics are exacerbated by regulatory gaps. To date, there are only limited binding standards for measuring and disclosing the energy, water, and life-cycle emissions of AI systems. Environmental costs are often externalized and not adequately factored into pricing structures or investment decisions. This creates an institutional environment in which ecological impacts are systematically underestimated or ignored.

The combination of market logic and regulatory shortcomings means that sustainability does not function as a key competitive factor. Instead, short-term efficiency and growth targets dominate, while long-term environmental costs take a back seat.

Interdependencies and Systemic Reinforcement Effects

The three drivers described do not operate in isolation but reinforce one another. The scaling paradigm increases demand for infrastructure, which in turn intensifies pressure on raw material markets and energy supply systems. At the same time, market incentives create the conditions for this expansion to accelerate further.

Material infrastructure acts as an intermediary through which technological and economic dynamics translate into concrete environmental impacts.

These interdependencies lead to systemic reinforcing effects. For example, an innovation in efficient hardware may reduce energy consumption in the short term, but in the long term, increased demand may lead to infrastructure expansion. Similarly, geopolitical tensions in commodity markets can influence the development of new technologies, thereby indirectly shifting environmental impacts.

From a systemic perspective, it becomes clear that the environmental challenges posed by AI cannot be solved through isolated interventions. They are the result of a complex interplay of technology, market, and governance that requires fundamental structural changes.

Against this backdrop, the question arises of how these structural dynamics are linked to global inequalities and environmental justice. Section 5, therefore, focuses on analyzing the social and geographical distribution of AI's environmental impacts.

ENVIRONMENTAL JUSTICE, THE GLOBAL SOUTH, AND THE POLITICAL ECONOMY OF AI

The analyses to date have shown that the environmental impacts of artificial intelligence arise along global value chains and are amplified by structural dynamics. However, these effects are not evenly distributed. Rather, they exhibit pronounced spatial and social asymmetry, raising central questions of environmental justice and global political economy. AI is thus also a mechanism that reproduces and, in some cases, exacerbates existing inequalities.

Rohde et al. (2026) emphasize that, with regard to AI's life cycles, the environmental impacts of artificial intelligence are inextricably linked to questions of justice and ethical responsibility. By focusing on the infrastructural prerequisites of AI—such as data centers, energy systems, and raw material extraction—these approaches demonstrate that environmental burdens are unevenly distributed across regions and communities (Carmeno et al., 2026). Crucially, those who benefit from AI-powered services often differ geographically and socioeconomically from those who bear the associated environmental costs. While AI-enabled applications can improve productivity, access to services, and environmental management, the associated environmental costs—such as raw material extraction, energy consumption, and infrastructure-related externalities—are often concentrated in specific geographic areas and disproportionately burden particularly vulnerable communities. This asymmetry makes environmental impacts a matter of distributive and procedural justice and underscores that AI must be understood as a socio-ecological system embedded in global inequalities. Incorporating these dimensions is essential to understanding the full impacts of AI as a socio-ecological system.

Digital Infrastructure and Environmental Impact

A central feature of AI's environmental impacts is its spatial decoupling from the sites of value creation. While a large portion of economic value creation, technological development, and platform control is concentrated in the Global North, the environmental costs are often borne by regions in the Global South. This dynamic corresponds to classic patterns of global production systems in which resource-intensive and environmentally damaging processes are spatially outsourced (Gereffi, 1994; Bair, 2005).

The extraction of raw materials for AI hardware illustrates this asymmetry particularly clearly. The mining of cobalt, lithium, or rare earth elements often occurs under conditions that entail significant environmental impacts and social risks. Local communities are affected by water scarcity, environmental pollution, and land-use conflicts, while the economic gains are captured further up the global supply chain. This form of externalization can be understood as an expression of a structural inequality in the distribution of environmental costs, which is exacerbated by the global organization of digital infrastructures.

E-Waste and Informal Waste Management Economies

The end-of-life phase of digital technologies represents another central dimension of global environmental inequality. A significant portion of the world's electronic waste is exported to countries in the Global South, where recycling and disposal often occur under informal, hazardous conditions (Forti et al., 2020). Workers in these sectors are exposed to toxic substances, while only a fraction of the valuable materials is efficiently recovered.

This dynamic illustrates that the environmental impacts of AI do not occur solely during the production and use phases but persist throughout the entire life cycle. The global distribution of these effects follows not primarily ecological but economic and institutional logics.

Digital Coloniality and Structural Dependencies

The patterns described can be interpreted in the context of the debate on 'digital colonialism.' This term describes asymmetrical power relations in which technological infrastructures, data flows, and platform economies are controlled by a few global actors, while other regions serve primarily as resource suppliers or markets (Couldry & Mejias, 2019; Andrae, 2026b).

In the context of AI, digital coloniality manifests not only in the control of data and platforms but also in the material dimension of digital technologies. Access to computing power, semiconductors, and infrastructure is highly concentrated, leaving many countries in the Global South in a position of dependency. At the same time, they bear a disproportionate share of the environmental burdens.

This constellation reinforces existing development disparities and hinders the independent shaping of technological transformation pathways. AI can thus function both as an instrument of development and as a mechanism of dependency.

Environmental Justice as an Analytical Framework

The perspective of environmental justice provides a suitable framework for systematically analyzing these interrelationships. It examines their distribution and the underlying power and decision-making structures (Schlosberg, 2007; Andrae, 2026a). With regard to AI, this means ecological effects must always be considered alongside access to resources, political participation, and institutional design.

From this perspective, it becomes clear that the sustainability issue of AI is not merely a technical or economic challenge, but also has a normative and political dimension. Who bears the costs of digital transformation? Who benefits from technological innovation? And which institutional mechanisms determine this distribution?

Ambivalence: Leapfrogging vs. Reproduction of Dependencies

However, the role of the Global South in the context of AI is not one-dimensional. On the one hand, AI opens up new opportunities for economic development, for example, through digital services, automation, or improved resource utilization. In this sense, AI can act as a tool for 'leapfrogging,' enabling countries to skip certain stages of development.

On the other hand, there is a risk that existing dependencies will be reinforced. If central infrastructures, data platforms, and technologies are controlled by external actors, local value creation and decision-making autonomy remain limited. At the same time, environmental burdens may continue to rise without corresponding economic benefits arising locally.

This ambivalence makes it clear that the development of AI does not automatically lead to sustainable or equitable outcomes. Rather, it depends on institutional frameworks, political strategies, and global power dynamics.

Against this backdrop, the question arises: which governance approaches and intervention strategies are suitable for addressing these inequalities and reducing AI's environmental footprint? Section 6, therefore, analyzes technical and systemic options for action, as well as their potential for designing more sustainable AI systems.

GOVERNANCE AND INTERVENTION PATHWAYS FOR SUSTAINABLE AI

The preceding analysis has shown that the environmental impacts of artificial intelligence do arise from structural dynamics embedded in innovation logics, infrastructures, and institutional frameworks. It follows that isolated efficiency gains, while necessary, are not sufficient to substantially reduce the environmental footprint of AI. The realization that environmental impacts occur across the entire life cycle of AI also has important implications for governance. Life-cycle-based analyses show that current regulatory approaches, which often focus on isolated components such as model transparency or energy reporting, fail to capture the systemic nature of environmental impacts (Oliveira et al., 2026). This underscores the need for governance frameworks that integrate life-cycle aspects and account for both operational and upstream impacts.

Figure 3. Governance architecture for sustainable AI

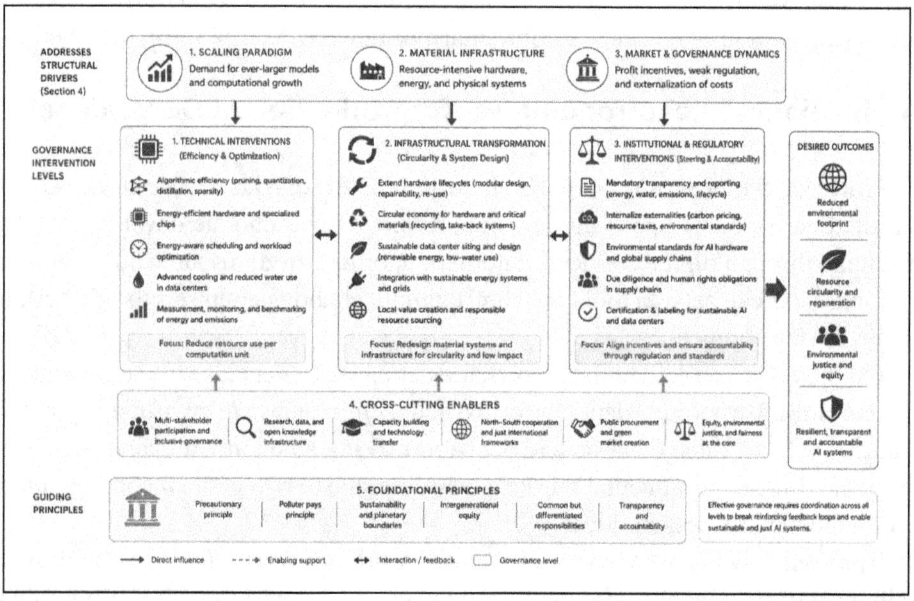

Source: Own representation.

Figure 3 illustrates a multi-level governance architecture that aims to systematically address the structural drivers of AI's ecological footprint. At its core are three complementary intervention dimensions. Technical interventions aim to increase efficiency and reduce resource consumption per computing unit. They thus directly address the impacts of the scaling paradigm but remain limited to this level, as they do not resolve the underlying growth dynamics. Infrastructure transformations go beyond this by redesigning the material basis of AI systems. Through principles of circularity, extended hardware lifecycles, and sustainable infrastructure planning, they directly intervene in the material and energy foundations of digital technologies. Finally, the third level of institutional and regulatory interventions aims to change incentive structures. Through transparency requirements, the internalization of external effects, and binding standards, market mechanisms are designed to make environmental costs visible and to ensure they are effectively taken into account.

In the following, a multi-level intervention model is developed that distinguishes between technical optimizations, infrastructural transformations, and institutional control mechanisms. The decisive factor here is the extent to which these interventions are suitable for addressing the structural drivers identified in Section 4.

Technical Interventions: Efficiency as a Necessary but Limited Strategy

Technical measures primarily aim to reduce resource consumption per computing unit. These include algorithmic optimizations such as pruning, quantization, and distillation, which enable models to be operated with fewer parameters and reduced computational effort. Similarly, energy-efficient hardware architectures and specialized chips improve energy efficiency (Patterson et al., 2021).

Another approach involves optimizing data center operations, for example, through energy-aware scheduling or the use of renewable energy sources. Advances in cooling technology, such as liquid cooling, can also reduce water and energy consumption.

Recent research findings on environmentally friendly AI architectures, such as federated learning approaches, highlight the potential of decentralized model training to reduce data transfer, improve energy efficiency, and strengthen data sovereignty (Sabella & Vitali, 2025). These approaches demonstrate how architectural design decisions can help reduce the environmental impact of AI systems. However, such technical solutions primarily focus on system-level efficiency and remain embedded in broader scaling dynamics. Therefore, their overall impact on resource consumption depends crucially on whether efficiency gains are offset by increased adoption and expanded use.

Despite these advances, the effectiveness of technical interventions remains structurally limited. Efficiency gains are often offset by rising demand, leading to continued growth in absolute resource consumption. Technical measures thus primarily address symptoms rather than the underlying growth dynamics. Their significance, therefore, lies less in solving the problem than in creating the conditions for more far-reaching transformations.

Figure 3 shows that technical efficiency gains materialize only when combined with appropriate infrastructure and regulatory frameworks. Conversely, regulatory measures remain limited in their impact without technological and infrastructural adjustments. The bidirectional arrows between the intervention dimensions highlight this systemic interdependence.

Infrastructural Transformation: Designing Physical Systems

A second level of interventions concerns the physical infrastructure of AI systems. This is not just about efficiency, but about the fundamental design of production and usage systems. A key approach is extending hardware lifecycles through modular designs, reuse, and improved repairability. Such strategies can help reduce demand for new raw materials and limit the generation of electronic waste.

Furthermore, the integration of circular economy principles is gaining importance. This encompasses both the recovery of materials from end-of-life devices and the development of closed-loop material cycles in hardware production. However, this requires significant investment in recycling technologies and regulatory frameworks that create appropriate incentives.

Another key lever lies in the spatial and energy design of data centers. Site selection, integration into local energy systems, and the use of renewable energy sources can significantly influence the environmental footprint. At the same time, the question arises as to what extent such measures contribute to reducing global emissions or merely shift environmental impacts.

Beyond global emissions, the environmental impacts of AI infrastructures are also evident at the local level. Recent research findings point to the emergence of a 'data heat island effect,' in which the concentration of energy-intensive data centers contributes to local temperature increases and microclimatic changes (Marinoni et al., 2026). These impacts are particularly relevant in regions already suffering from climate stress, where additional heat stress can exacerbate environmental pressures and adversely affect surrounding communities. Such findings underscore that the ecological footprint of AI is a spatially unevenly distributed phenomenon with local externalities that raise important questions of environmental justice and community-level impacts.

Institutional and Regulatory Interventions: Internalizing Externalities

The third and decisive level concerns institutional and regulatory measures aimed at internalizing environmental externalities and altering incentive structures. A key approach is to introduce mandatory transparency and reporting requirements for AI systems. Companies could be required to disclose the energy consumption, water consumption, and lifecycle emissions of their models. Such measures lay the groundwork for comparability and enable the integration of environmental criteria into investment and use decisions.

Current policy-oriented research increasingly emphasizes the need to account for the environmental impacts of AI within new regulatory frameworks, particularly in relation to data center infrastructure and the European AI Act (Ebert et al., 2026). Such approaches underscore the growing recognition that AI governance must go beyond algorithmic risk classification and also incorporate the material and energy dimensions of digital systems. However, current regulatory efforts continue to focus largely on transparency, compliance, and risk mitigation and only partially address the structural drivers of environmental impacts identified in this chapter. This suggests that while existing frameworks represent an important step toward integrating sustainability into AI governance, they remain limited in their ability to reshape the underlying dynamics of scaling and resource-intensive innovation.

Furthermore, regulatory instruments are needed that directly incorporate environmental costs into economic decision-making processes. This can be achieved, for example, through carbon pricing, resource taxes, or environmental standards specifically tailored to digital infrastructures. Certification systems for 'sustainable AI' could also play an important role by defining minimum standards and increasing market transparency.

Another key area is the regulation of global supply chains. International standards for responsible raw material sourcing, as well as binding due diligence obligations, can help reduce environmental and social risks along the value chain. However, this requires close international coordination, as national regulatory approaches alone are often insufficient to influence global production networks.

Market and Incentive Structures: Transforming Innovation Logics

In addition to formal regulations, market-based incentive structures also play a central role. Public procurement can serve as an important lever by specifically demanding sustainable AI solutions and thereby setting new market standards. Similarly, funding programs and investment incentives can be designed to support resource-efficient technologies and smaller, specialized models.

A particular challenge lies in questioning the prevailing scaling paradigm. As long as competitive advantages are defined primarily by size and performance, ecological criteria remain secondary. A transformation of the logic of innovation, therefore, requires new evaluation criteria that more strongly link efficiency, utility, and ecological impact.

In this context, alternative models of technology development are also gaining importance, such as open-source initiatives or cooperative infrastructure projects that are not exclusively geared toward competition but toward the shared use of resources. Such approaches can help reduce scaling constraints and enable new forms of sustainable innovation.

Limits and Conflicts of Interest: Efficiency, Growth, and Sustainability

Despite the multitude of possible interventions, fundamental conflicts of interest remain. Efficiency measures may clash with growth strategies, while stricter regulations could slow innovation. There is also a risk that environmental standards could lead to a shift of activities to less-regulated regions. These conflicting goals illustrate that sustainable AI cannot be achieved through individual measures but requires a coordinated transformation that integrates technological, economic, and political dimensions. This raises not only the question of how AI can be made more sustainable; it also the question of what role AI should play in an overall sustainable economic order.

The limited effectiveness of current governance approaches can also be understood in light of Blühdorn's (2024) argument that modern political and economic systems tend to absorb sustainability demands without fundamentally altering their underlying logic. Regulatory frameworks often aim to mitigate negative externalities while simultaneously preserving the conditions for continued technological expansion and economic growth. In the context of AI, this leads to governance strategies that emphasize transparency, efficiency, and incremental improvements, but rarely address the structural drivers of environmental impacts, such as scaling dynamics and market incentives. Consequently, regulation risks becoming part of a system-stabilizing mechanism in which environmental concerns are acknowledged and managed, but not disruptive enough to alter the course of resource-intensive innovations.

Against this backdrop, the question of whether and under what conditions AI can have also potentially positive ecological effects becomes significant. Section 7, therefore, develops the concept of 'net-positive AI' as a normative and strategic framework.

FROM EFFICIENCY TO REGENERATION: NET-POSITIVE AI AS A NORMATIVE AND SYSTEMIC FRAMEWORK

The analyses to date have shown that current approaches to the sustainable design of AI largely remain within an efficiency-driven framework. Technical innovations primarily aim to reduce resource consumption per unit of computation, while the underlying growth dynamics and structural incentive systems remain largely untouched. Against this backdrop, the question arises whether such a strategy is fundamentally sufficient to address AI's ecological challenges, or whether a paradigm shift in perspective is necessary.

This section develops such a shift in perspective through the concept of 'net-positive AI.' In contrast to existing approaches that aim for damage limitation or increased efficiency, net-positive AI conceives of artificial intelligence as a system whose overall ecological and social impacts must be positive. The central normative standard here is the achievement of a positive net contribution to ecological stability and social well-being.

Limitations of Existing Sustainability Paradigms

The debate surrounding 'Green AI' and 'Sustainable AI' has provided important impetus for reducing energy consumption and emissions. Nevertheless, these approaches often remain confined to a relative perspective: they measure progress by whether one system is more efficient than another, but not by whether it operates within ecological limits or makes a positive contribution.

This limitation is closely tied to the logic of incremental optimization. Efficiency gains reduce the environmental intensity of individual processes without questioning the system's overall development. As already shown in Section 4, such improvements can be neutralized or even overcompensated by rebound effects. In this context, sustainability becomes a relative category that loses significance as scaling progresses.

Furthermore, many existing approaches lack a systemic perspective on value creation. AI is often viewed as a tool embedded in existing economic structures rather than as a potential driver of structural transformation. This results in environmental and social impacts being treated as external factors that must be taken into account retrospectively.

Definition and Conceptual Classification of Net-Positive AI

Ethical analyses of the AI lifecycle show that sustainability cannot be meaningfully addressed without incorporating aspects of justice and responsibility (Rohde et al.,

2025). By drawing attention to the unequal distribution of environmental burdens and benefits, these perspectives highlight the limitations of approaches that focus exclusively on efficiency or emissions reduction. The concept of 'net-positive AI' proposed in this chapter builds on this insight by integrating environmental and social criteria into a unified evaluation framework. Net-positive AI can be defined as an approach that aligns the design of AI systems with the condition that their total life-cycle impacts have a net positive effect on environmental and social systems. This implies a shift in the evaluation metric from relative efficiency to absolute impact. Rather than treating environmental impacts and social justice as separate concerns, it brings them together as interdependent dimensions of a broader transformation toward more equitable and sustainable AI systems.

Conceptually, this approach can be situated within the tradition of regenerative approaches as discussed in ecological economics and sustainability research. Regenerative systems aim also to actively restore and stabilize ecological systems. Applied to AI, this means that technological systems must be designed to regenerate resources, reduce environmental impacts, and promote social resilience.

A central feature of net-positive AI is the integration of a life-cycle perspective. Assessment is conducted across the entire system, including upstream and downstream effects. This encompasses both direct emissions and indirect impacts on resource consumption, infrastructure, and social structures.

Conditions for a Net-Positive Impact

Implementing a net-positive approach requires several key prerequisites. First, robust measurability is necessary. Without suitable indicators for energy consumption, emissions, resource use, and ecological impacts, an assessment of net impacts is not possible. This requires standardized reporting and assessment procedures that are comparable across organizations.

Second, environmental costs must be systematically integrated into decision-making processes. This requires both regulatory instruments and changes in market mechanisms to ensure that environmental impacts are not externalized. Without such internalization, the incentive to scale up persists regardless of environmental consequences.

Third, a transformation of the logic of innovation is necessary. Instead of focusing on maximum performance and scaling, criteria such as resource efficiency, system impact, and long-term sustainability must take center stage. This can be achieved, for example, by promoting smaller, specialized models or through alternative development approaches that are less resource-intensive.

Fourth, institutional embedding plays a central role. Net-positive AI cannot be achieved through technical design alone but requires coordinated governance structures that involve various actors—companies, governments, and civil society.

Conflicting Goals and Limitations

The characterization of AI's environmental impacts as a 'green paradox' provides a valuable analysis of the tensions inherent in current development trajectories, highlighting the coexistence of positive and negative effects (Chouksey et al., 2026). However, this perspective remains largely descriptive, highlighting the contradiction without addressing the structural conditions that perpetuate it. The concept of 'net-positive AI' proposed in this chapter builds on this insight but goes further by redefining the basis for evaluation. Instead of weighing positive and negative impacts against one another, it asks under what conditions AI systems can make a net-positive contribution to ecological and social systems. In this sense, 'net-positive AI' can be understood as a response to the green paradox by shifting the focus from managing conflicting goals to transforming the underlying system dynamics that give rise to them.

However, implementing net-positive AI poses significant challenges. A central difficulty lies in quantifying net effects, particularly when indirect and long-term impacts are considered. Furthermore, conflicts of interest may arise between ecological, economic, and social goals. Measures that reduce ecological impacts may increase economic costs in the short term or call existing business models into question. Another problem lies in the global dimension of AI systems. Since value chains and environmental impacts are distributed globally, implementing net-positive approaches requires close international coordination. National measures alone are often insufficient to manage global effects. Finally, there is the question of the scalability of such an approach. While individual applications can achieve positive effects, it remains unclear whether and how these can be scaled up to the entire system without generating new rebound effects.

Net-Positive AI as a Transformation Framework

Despite these challenges, the concept of net-positive AI provides an important framework for the future development of digital technologies. It shifts the focus from how AI can be designed to be less harmful to what active contributions AI can make toward solving environmental and societal challenges. In this sense, net-positive AI functions as a heuristic tool that guides the development of new technologies, business models, and governance approaches. It calls for the systematic integration of ecological criteria into all phases of the innovation process and places the interaction between technological development and sustainable transformation at the center.

Blühdorn's (2024) analysis simultaneously highlights a central limitation of existing sustainability approaches: these tend to integrate ecological requirements with existing growth and innovation logics rather than fundamentally questioning them. Sustainability thus often appears as a system-compatible adaptation rather than a transformative force. Applied to AI, this implies that even approaches such as 'Green AI' operate primarily within the dominant scaling paradigm and leave its structural dynamics largely untouched.

The concept of 'net-positive AI' developed in this chapter addresses precisely this issue. It shifts the basis of evaluation from relative efficiency to absolute system impact and no longer asks about reducing negative effects, but rather about the conditions for a positive net contribution to ecological and social systems. In this sense, net-positive AI represents a conceptual break with system-compatible sustainability approaches and opens a perspective that explicitly aims to transform the structural framework conditions identified by Blühdorn.

Against this backdrop, we can finally assess the extent to which AI can serve as an instrument of sustainable transformation and the institutional, technological, and economic conditions that must be met for it to contribute to the stabilization of ecological systems. These questions are brought together in the following section.

CONCLUSION AND RESEARCH AGENDA: AI IN THE CONTEXT OF PLANETARY BOUNDARIES

Synthesis: AI as a Socio-Ecological System

This chapter has shown that the ecological impacts of artificial intelligence cannot be understood as isolated by products of technological innovation, but rather as structural properties of a globally networked socio-technical system. By integrating a life-cycle perspective (Section 3), an analysis of structural drivers (Section 4), a political-economic classification of global inequalities (Section 5), and a multi-level governance framework (Section 6), it becomes clear that AI is embedded in complex material and energy systems whose dynamics extend far beyond individual applications.

Central to this is the realization that the ecological impact of AI is caused by an interplay of scaling logics, infrastructural dependencies, and market-driven growth dynamics. These structural factors mean that efficiency gains are often offset by increasing usage, and absolute resource consumption continues to rise. AI is itself a driver of material expansion.

At the same time, it has been shown that these dynamics are closely linked to global inequalities. The spatial externalization of environmental costs along global supply chains results in environmental burdens falling disproportionately on the Global South, while value creation and control over technological infrastructures remain concentrated in the Global North. AI is thus part of a political economy in which technological innovation and environmental justice are inextricably linked.

Against this backdrop, the concept of net-positive AI (Section 7) was developed as a normative framework that goes beyond existing efficiency and sustainability approaches. This approach shifts the focus from reducing negative effects to actively generating positive ecological and social impacts and calls for a fundamental reorientation of the development and use of AI.

Theoretical and Conceptual Implications

The findings of this chapter contribute to several key debates. First, they expand the discussion on sustainable AI by introducing a systemic perspective on environmental impacts. Rather than viewing AI in isolation as a technical artifact, it is analyzed as part of a global socio-ecological system in which technological, economic, and institutional processes are intertwined.

Second, the chapter contributes to ecological economics by integrating digital technologies as relevant factors in material flows and energy consumption. This helps to critically question the frequently postulated 'dematerialization' of digital economies and replace it with a nuanced analysis of their material foundations.

Third, the chapter links approaches to technology governance with issues of environmental justice. It demonstrates that ecological sustainability and social distributional effects cannot be considered in isolation but are part of a shared analytical and policy framework. This integration is central, particularly for assessing global technological developments.

Finally, the concept of net-positive AI makes an independent theoretical contribution by introducing a new evaluation metric for technological systems. This approach shifts the perspective from relative efficiency to absolute impact, thereby opening up new possibilities for the analysis and design of sustainable technologies.

Policy and Practical Implications

The findings yield several key implications for policy and practice. First, it becomes clear that transparency is a fundamental prerequisite for sustainable AI. Without reliable data on energy consumption, emissions, and resource use, neither companies nor policymakers can make informed decisions. The introduction of standardized reporting requirements, therefore, represents a crucial first step.

Second, reducing environmental impacts requires greater internalization of external effects. This can be achieved through regulatory instruments such as CO_2 pricing, environmental standards, or mandatory sustainability criteria for digital infrastructures. At the same time, such measures must be coordinated internationally to avoid displacement effects.

Third, the design of innovation systems plays a crucial role. Funding programs, public procurement, and regulatory incentives can help strengthen alternative development approaches that are not exclusively focused on scaling and maximizing performance. In particular, promoting resource-efficient, specialized applications can make an important contribution.

Fourth, the findings underscore the importance of global cooperation. Since the environmental impacts of AI arise along international value chains, national measures alone are insufficient. International standards for raw material procurement, infrastructure development, and sustainability assessment are necessary to enable coherent governance

Research Agenda: Open Questions and Future Developments

Despite the insights gained, key questions remain that future research must address. An important first area concerns the measurability of environmental impacts. There is a significant need for standardized methods to track energy consumption, emissions, and resource use throughout the entire life cycle of AI systems. In particular, indirect and upstream effects have so far been inadequately quantified.

A second area of research concerns the dynamics of rebound effects in the context of digital technologies. While this phenomenon is well established in environmental economics, specific analyses for AI systems are lacking. A better understanding of these dynamics is crucial for realistically assessing the effectiveness of efficiency measures.

Third, further research is needed in the area of governance. What combination of regulatory, market-based, and cooperative instruments is suitable for effectively addressing the structural drivers of environmental impacts? What role do international institutions and transnational networks play in this context?

A fourth key area concerns the intersection of AI and environmental justice. Future research should focus more closely on how ecological and social impacts interact and which institutional mechanisms can contribute to a more equitable distribution of costs and benefits.

Finally, the concept of net-positive AI opens up new fields of research. These include developing suitable indicators to measure positive impacts, analyzing use cases with regenerative potential, and examining business models that combine ecological and economic goals.

CONCLUDING REMARKS

The growing importance of artificial intelligence presents societies with a fundamental challenge: integrating a highly dynamic, resource-intensive technology into a world of limited ecological carrying capacity. This chapter has shown that this challenge requires a comprehensive transformation of infrastructures, incentive systems, and governance structures.

The central question is therefore not whether AI can be made more sustainable, but under what conditions it will become part of a sustainable transformation. Net-positive AI offers a potential framework for this, redefining the relationship between technological development and planetary boundaries. Whether this approach can be implemented ultimately depends on political decisions, institutional innovations, and societal priorities.

REFERENCES

Agarwal, S., & Chakraborti, M. (2025). The hidden AI race: Tracking environmental costs of innovation. /arXiv.2511.22781DOI: 10.48550

Andrae, S. (2025). Energy-saving alternatives in Blockchain technology. In Gupta, B. B., Pramod, D., & Moslehpour, M. (Eds.), *Sustainable Information Security in the Age of AI and Green Computing* (pp. 229–246). IGI Global Publishing., DOI: 10.4018/979-8-3693-8034-5.ch012

Andrae, S. (2026a). Justice and climate change: An interdisciplinary perspective on global environmental limits. In Meçik, O. (Ed.), *Economic Growth Amidst Social and Environmental Uncertainty* (pp. 1–44). IGI Global Publishing., DOI: 10.4018/979-8-3373-1529-4.ch001

Andrae, S. (2026b). The role of local platforms and Non-Western perspectives on Platformization. In Çelik, N., Ateşoğlu, N., & Güner, S. S. (Eds.), *Sociological Dimensions of Platformization: Power, Culture, and Identity in the Digital Age* (pp. 31–70). IGI Global Publishing., DOI: 10.4018/979-8-3373-4267-2.ch002

Baccini, P., & Brunner, P. H. (2012). *Metabolism of the Anthroposphere: Analysis, Evaluation, Design* (2nd ed.). MIT Press. DOI: 10.7551/mitpress/8720.001.0001

Bair, J. (2005). Global capitalism and commodity chains: Looking back, going forward. *Competition & Change*, *9*(2), 153–180. DOI: 10.1179/102452905X45382

Blühdorn, I. (2024). *Unhaltbarkeit: Auf dem Weg in eine andere Moderne*. Suhrkamp.

Carmeno, N. L., Domingos, T., & O'Neill, D. W. (2026). The impacts of artificial intelligence on environmental sustainability and human well-being. /arXiv.2602.24091DOI: 10.48550

Chen, X., Wang, X., Colacelli, A., Lee, M., & Xie, L. (2025). Electricity demand and grid impacts of AI data centers: Challenges and prospects. /arXiv.2509.07218DOI: 10.48550

Chouksey, A., Rajan, A. K., Gurjar, V., Tiwari, R., & Mishra, P. K. (2026). The green paradox: The climate, environmental, and sustainability implications of artificial intelligence. *Global Environmental Change Advances*, *6*, 100029. DOI: 10.1016/j.gecadv.2025.100029

Costanza, R., d'Arge, R., de Groot, R., Farber, S., Grasso, M., Hannon, B., Limburg, K., Naeem, S., O'Neill, R. V., Paruelo, J., Raskin, R. G., Sutton, P., & van den Belt, M. (1997). The value of the world's ecosystem services and natural capital. *Nature*, *387*(6630), 253–260. DOI: 10.1038/387253a0

Couldry, N., & Mejias, U. A. (2019). *The Costs of Connection: How Data Is Colonizing Human Life and Appropriating It for Capitalism.* Stanford University Press.

Daly, H. E. (1996). *Beyond Growth: The Economics of Sustainable Development.* Beacon Press.

Ebert, K., Alder, N., Herbrich, R., & Hacker, P. (2026). AI, climate, and regulation: From data centers to the AI Act. *Computer Law & Security Review*, *61*, 106326. DOI: 10.1016/j.clsr.2026.106326

Falk, S., Kluge Corrêa, N., Luccioni, S., Biber-Freudenberger, L., & van Wynsberghe, A. (2025). From FLOPs to footprints: The resource cost of artificial intelligence. /arXiv.2512.04142DOI: 10.48550

Forti, V., Baldé, C. P., Kuehr, R., & Bel, G. (2020). The global e-waste monitor 2020: Quantities, flows and the circular economy potential. United Nations University. https://ewastemonitor.info/wp-content/uploads/2020/11/GEM_2020_def_july1_low.pdf

Geels, F. W. (2002). Technological transitions as evolutionary reconfiguration processes: A multi-level perspective and a case-study. *Research Policy*, *31*(8–9), 1257–1274. DOI: 10.1016/S0048-7333(02)00062-8

Gereffi, G. (1994). The organization of buyer-driven global commodity chains: How U.S. retailers shape overseas production networks. In Gereffi, G., & Korzeniewicz, M. (Eds.), *Commodity Chains and Global Capitalism* (pp. 95–122). Praeger.

Hughes, T. P. (1987). The evolution of large technological systems. In Bijker, W. E., Hughes, T. P., & Pinch, T. (Eds.), *The Social Construction of Technological Systems* (pp. 51–82). MIT Press.

International Energy Agency (IEA). (2023). Data Centres and Data Transmission Networks. https://www.iea.org/energy-system/buildings/data-centres-and-data-transmission-networks

Jeanquartier, F., Jean-Quartier, C., Rieder, P., Misirlić, V., Pasero, C., Hohensinner, R., Müller, H., & Holzinger, A. (2026). Assessing the carbon footprint of language models: Towards sustainability in AI. *Resources, Conservation and Recycling*, *226*, 108670. DOI: 10.1016/j.resconrec.2025.108670

Jegham, N., Abdelatti, M., Koh, C. Y., Elmoubarki, L., & Hendawi, A. (2025). How hungry is AI? Benchmarking energy, water, and carbon footprint of LLM inference. /arXiv.2505.09598DOI: 10.48550

Jevons, W. S. (1865). *The Coal Question: An Inquiry Concerning the Progress of the Nation, and the Probable Exhaustion of Our Coal Mines* (2nd ed.). Macmillan., https://oll.libertyfund.org/titles/jevons-the-coal-question

Kaplan, J., McCandlish, S., Henighan, T., Brown, T. B., Chess, B., & Child, R.. (2020). Scaling laws for neural language models. *arXiv*. https://arxiv.org/abs/2001.08361

Kim, M.-K., Yoo, T.-A., & Chung, J.-B. (2026). Toward sustainable Generative AI: A scoping review of carbon footprint and environmental impacts across training and inference stages. *IEEE Access : Practical Innovations, Open Solutions*, *14*, 18881–18896. Advance online publication. DOI: 10.1109/ACCESS.2026.3659894

Kitchin, R. (2014). The Data Revolution: Big Data, Open Data, Data Infrastructures and Their Consequences. *Sage (Atlanta, Ga.)*. Advance online publication. DOI: 10.4135/9781473909472

Konat, G., Salihoğlu, E., & Han, A. (2026). Financial investments in AI-based technologies and carbon footprint in selected advanced industrial economies. *Carbon Balance and Management*, *21*(1), 31. DOI: 10.1186/s13021-025-00383-4 PMID: 41483362

López-Rauhut, M., Landrieu, L., Aubry, M., & Ligozat, A. L. (2026). Environmental footprint of GenAI research: Insights from the Moshi foundation model. /arXiv.2604.11154DOI: 10.48550

Marinoni, A., Cambria, E., Lin, W., Mura, M. D., Chanussot, J., Ragusa, E., ... (2026). The data heat island effect: quantifying the impact of AI data centers in a warming world. /arXiv.2603.20897DOI: 10.48550

Marinoni, A., Shivareddy, S., Lio', P., Lin, W., Cambria, E., & Grey, C. (2025). Improving AI efficiency in data centres by power dynamic response. /arXiv.2510.11119DOI: 10.48550

Oliveira, A. P., Carraquico, T., & Martinez-Perez, C. (2026). Beyond efficiency: A systematic review of energy consumption and carbon footprint across the AI Lifecycle. *Sustainability (Basel)*, *18*(3), 1359. DOI: 10.3390/su18031359

Patterson, D., Gonzalez, J., Le, Q., Liang, C., Munguia, L., & Rothchild, D.. (2021). Carbon emissions and large neural network training. *arXiv*. https://arxiv.org/abs/2104.10350

Polimeni, J. M., Mayumi, K., Giampietro, M., & Alcott, B. (2008). *The Jevons Paradox and the Myth of Resource Efficiency Improvements*. Earthscan.

Rockström, J., Steffen, W., Noone, K., Persson, Å., Chapin, F. S.III, Lambin, E., Lenton, T. M., Scheffer, M., Folke, C., Schellnhuber, H. J., Nykvist, B., de Wit, C. A., Hughes, T., van der Leeuw, S., Rodhe, H., Sörlin, S., Snyder, P. K., Costanza, R., Svedin, U., & Foley, J. A. (2009). A safe operating space for humanity. *Nature*, *461*(7263), 472–475. DOI: 10.1038/461472a PMID: 19779433

Rohde, F., Nasruddin, Z., Kotova, E., & Ammon, S. (2026). Navigating justice in AI lifecycles: Ethical perspectives on infrastructural prerequisites and their environmental impacts. *AI and Ethics*, *6*(1), 156. DOI: 10.1007/s43681-026-01008-3

Rolnick, D., Donti, P. L., Kaack, L. H., Kochanski, K., Lacoste, A., & Sankaran, K.. (2019). Tackling climate change with machine learning. *arXiv*. https://arxiv.org/abs/1906.05433

Sabella, M., & Vitali, M. (2025). Eco-friendly AI: Unleashing data power for green federated learning. /arXiv.2507.17241DOI: 10.48550

Schlosberg, D. (2007). *Defining Environmental Justice: Theories, Movements, and Nature*. Oxford University Press. DOI: 10.1093/acprof:oso/9780199286294.001.0001

Sovacool, B. K., Ali, S. H., Bazilian, M., Radley, B., Nemery, B., Okatz, J., & Mulvaney, D. (2020). Sustainable minerals and metals for a low-carbon future. *Science*, *367*(6473), 30–33. DOI: 10.1126/science.aaz6003 PMID: 31896708

Srnicek, N. (2017). *Platform Capitalism*. Polity Press.

Strubell, E., Ganesh, A., & McCallum, A. (2019). Energy and policy considerations for deep learning in NLP. *Proceedings of the 57th Annual Meeting of the Association for Computational Linguistics*, 3645–3650. DOI: 10.18653/v1/P19-1355

Vinuesa, R., Azizpour, H., Leite, I., Balaam, M., Dignum, V., Domisch, S., Felländer, A., Langhans, S. D., Tegmark, M., & Fuso Nerini, F. (2020). The role of artificial intelligence in achieving the Sustainable Development Goals. *Nature Communications*, *11*(1), 233. DOI: 10.1038/s41467-019-14108-y PMID: 31932590

Wenmackers, S. (2025). Assessing the ecological impact of AI. /arXiv.2507.21102DOI: 10.48550

Wiedmann, T. O., Schandl, H., Lenzen, M., Moran, D., Suh, S., West, J., & Kanemoto, K. (2015). The material footprint of nations. *Proceedings of the National Academy of Sciences of the United States of America*, *112*(20), 6271–6276. DOI: 10.1073/pnas.1220362110 PMID: 24003158

Williams, E. D., Ayres, R. U., & Heller, M. (2002). The 1.7 kilogram microchip: Energy and material use in the production of semiconductor devices. *Environmental Science & Technology*, *36*(24), 5504–5510. DOI: 10.1021/es025643o PMID: 12521182

Zuboff, S. (2019). *The Age of Surveillance Capitalism*. PublicAffairs.

Chapter 3 Financialization, Scaling Dynamics, and the Ecological Political Economy of AI

Silvio Andrae
http://orcid.org/0000-0002-8586-7812
Independent Researcher, Germany

ABSTRACT

Artificial intelligence (AI) is increasingly portrayed as an immaterial and efficiency-enhancing technology. We challenge this view by conceptualizing AI as a capital-intensive, infrastructure-driven system embedded in global material and energy flows. From an ecological-political-economic perspective, we argue that AI's ecological footprint is not merely a byproduct of technological progress but rather the result of financialization and scaling dynamics. Major investments in computing infrastructure turn AI into an asset class, thereby reinforcing incentives for continuous expansion. These dynamics lead to significant environmental impacts throughout the entire life cycle, from raw material extraction and hardware production to data center operations and electronic waste. At the same time, rebound effects and infrastructural lock-in mechanisms exacerbate resource consumption despite efficiency gains. We develop an integrated analytical framework that links financialization, scaling, and ecological externalities, underscoring the need for governance approaches.

DOI: 10.4018/979-8-3373-8490-0.ch003

INTRODUCTION: FROM THE MYTH OF DEMATERIALIZATION TO THE MATERIAL REALITY OF AI

Artificial intelligence (AI) is often portrayed in public and academic discourse as an immaterial technology—a realm of algorithms, data, and 'cloud' infrastructures that seemingly operates detached from physical resources. This conception aligns with long-established narratives of the digital economy, in which information processing appears as a virtually cost-free, dematerialized activity (see Negroponte, 1995). However, the massive scaling of data- and compute-intensive models has made it clear that this perspective is analytically limited. AI is deeply embedded in energy, water, and raw material systems, as well as in global value chains and capital structures.

The growing importance of AI is accompanied by an exponential increase in computing demand, which directly translates into rising energy consumption, greater demand for high-performance hardware, and more intensive use of natural resources. Studies in computer science and environmental research have repeatedly shown that training large models in particular results in significant CO_2 emissions and requires energy-intensive infrastructure (see Strubell et al., 2019; Patterson et al., 2021). At the same time, the water consumption and cooling requirements of data centers are also increasingly coming into focus (see Li et al., 2023). These findings have sparked a growing debate about the ecological footprint of digital technologies.

Nevertheless, a significant part of this discussion remains limited to a technical-descriptive perspective. The focus is often on quantifying individual environmental indicators—energy consumption, emissions, water withdrawal—while the structural drivers of these developments are addressed only to a limited extent. In particular, there is a lack of systematic integration of AI expansion into the logics of capital markets, competitive dynamics, and infrastructural accumulation. This is precisely where we begin in this chapter.

Drawing on approaches from political economy and ecological economics, this paper argues that the ecological footprint of AI cannot be understood primarily as a byproduct of technological efficiency decisions. Rather, it is an expression of a specific socio-economic regime of scaling. This regime is shaped by the financialization of digital infrastructures (see Krippner, 2005; Epstein, 2005), by competitive dynamics in platform markets and by the inherent logic of growth-oriented innovation systems (see Perez, 2002).

Central to this is the thesis that computing power itself has become a strategic asset class. Companies are not only investing in AI to improve specific applications. They are doing so to build infrastructural capacities that secure long-term competitive advantages while simultaneously serving as a signal to capital markets. The size and performance of models increasingly serve as a proxy for innovation leadership, while the underlying infrastructure—data centers, energy contracts, semiconductor production—acts as a stabilizing foundation for this positioning. This development can be interpreted as a form of 'infrastructure capitalism,' in which digital infrastructures themselves become the central drivers of value creation and power (see Plantin et al., 2018).

After the quartet reported combined capital expenditures of $410 billion for 2025, statements from tech giants Amazon, Google, Microsoft, and Meta suggest that up to $725 billion in capital expenditures is expected for the current year. This would mean that more than 2.2% of the projected U.S. gross domestic product (GDP) would be accounted for by the capital expenditures of just four companies. The money is primarily intended to fund infrastructure expansion, such as data centers. Meta, the Facebook-owned company, and Google have once again raised their spending forecasts.

Figure 1. Capital expenditures of big tech firms (USD billions)

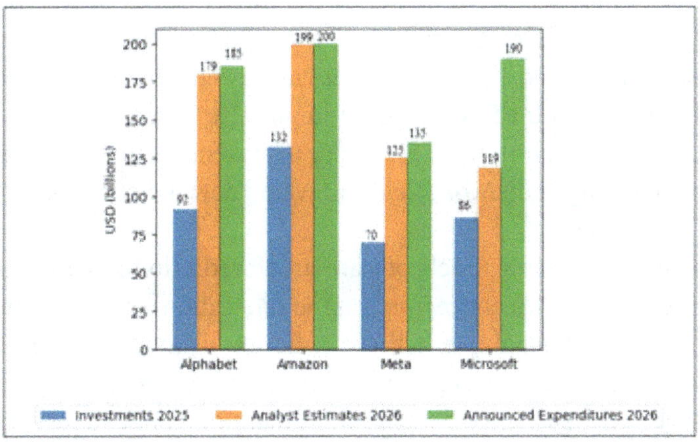

The environmental implications of these dynamics are far-reaching. First, the capital-intensive nature of AI infrastructure leads to long-term lock-in effects, as investments made once imply continued use in order to meet expected returns. Second, economies of scale systematically increase resource consumption, as efficiency gains can be offset or even more than offset by rebound effects (see Jevons, 1865; Polimeni et al., 2008). Third, the pace of innovation in the semiconductor sector accelerates hardware obsolescence, leading to growing volumes of electronic waste and generating additional environmental impacts along the entire value chain.

Against this backdrop, it is insufficient to analyze the ecological footprint of AI solely at the level of individual models or applications. Rather, a systemic perspective is needed that accounts for the entire life cycle—from raw material extraction through production and use to disposal. Life Cycle Assessment (LCA) approaches provide important methodological foundations for this, but fall short if they are not supplemented by an analysis of the underlying economic and institutional structures.

We therefore propose an integrative perspective that links three analytical levels: (1) the technological level of scaling and model architecture, (2) the economic level of financialization and infrastructure investments, and (3) the ecological level of material and energy impacts. This linkage reveals that environmental impacts result from complex interactions between technological innovations, market logics, and institutional frameworks.

Our central aim is to critically expand the debate on 'Green AI.' While many contributions focus on efficiency gains, energy-optimized algorithms, or the use of renewable energy (see Schwartz et al., 2020), the fundamental growth logic of AI development often remains untouched. We argue that sustainable AI requires a reflection on the underlying scaling paradigms. In this sense, sustainability is not a downstream correction. It must be understood as a structural design criterion.

Furthermore, this chapter contributes to the current governance debate. Regulatory approaches in the field of AI have so far focused primarily on issues of safety, transparency, and ethical responsibility. Environmental aspects, by contrast, play a subordinate role or are addressed only implicitly. Given the growing material impacts of AI, however, it appears necessary to systematically integrate ecological criteria into regulatory frameworks and to develop new instruments that go beyond voluntary commitments.

The remaining sections of the chapter are structured as follows: First, a theoretical framework is developed that conceptualizes the connection between financialization, scaling logics, and ecological externalities. Building on this, the material foundations of AI infrastructure and its energy and water requirements are analyzed. Another section addresses the challenges of emissions measurement and reporting. Finally, governance options are discussed that enable a more sustainable approach to AI development.

Overall, we aim to fundamentally shift the perspective on AI: away from the notion of an immaterial, virtually limitless technology toward an understanding of AI as a physically anchored, resource-intensive, and politically and economically shaped infrastructure. Only on this basis can ecological risks be adequately assessed and viable strategies for a sustainable transformation be developed.

THEORETICAL FRAMEWORK: AN ECOLOGICAL POLITICAL ECONOMY OF AI

Analyzing the ecological footprint of AI requires a theoretical framework that transcends disciplinary boundaries and integrates technological, economic, and ecological dynamics. While existing approaches often focus either on engineering-based efficiency questions or on isolated environmental impacts, the structural embedding of AI within capitalist regimes of growth and innovation remains largely underexplored. We therefore develop a conceptual perspective that understands AI as part of an ecological political economy of scaling.

At the core is the assumption that the environmental impacts of AI do not arise by chance or exogenously: they systematically emerge from the interaction of three central logics: (1) the financialization of digital infrastructures, (2) technological scaling dynamics, and (3) ecological externalities within planetary boundaries. These three dimensions are theoretically grounded below and translated into an integrated analytical model.

Financialization and Infrastructure Capitalism

The growing significance of AI is closely linked to processes of financialization. Following Krippner (2005) and Epstein (2005), financialization is understood as a transformation in which financial motives, markets, and actors play a dominant role in the allocation of resources. In the context of the digital economy, this dynamic manifests itself in the growing importance of intangible assets, platform ecosystems, and data-driven business models.

For AI, this means specifically that computing power and data infrastructure are becoming central assets, the development and control of which will decisively determine competitive advantage. This development can be interpreted as a specific form of infrastructure capitalism. Digital infrastructures are not merely means of production; they function as independent asset classes (see Plantin et al., 2018). Investments in data centers, semiconductor capacities, and energy contracts are therefore not driven solely by functional considerations; they also follow a logic of capital accumulation and market positioning.

A key feature of this regime is the close coupling of technological scaling and capital market expectations. The size of models, available computing power, and the speed of innovation cycles become signals for future growth. As a result, a structural incentive arises for the continuous expansion of infrastructure capacity—regardless of whether it is fully utilized in the short term. This dynamic fosters a form of overaccumulation that is problematic from an ecological perspective, as it implies long-term resource lock-in and increased environmental impacts.

Scaling Logics and Innovation-Economic Dynamics

Alongside financialization, the development of AI is shaped by specific scaling logics. Empirical studies have shown that the performance of large models systematically increases with the amount of training data, the model size, and the computational effort expended (see Kaplan et al., 2020). These so-called scaling laws have become a central paradigm of AI development.

From an innovation economics perspective, this dynamic can be interpreted as part of a technological paradigm in the sense of Dosi (1982), which is shaped by specific problem-solving heuristics and search directions. In the case of AI, this heuristic assumes that 'more'—more data, more parameters, more computing power—leads to better results. This logic is reinforced by intense competition, in which companies attempt to gain strategic advantages through larger models and more powerful infrastructures.

These competitive scaling dynamics generate several structural effects. First, they lead to a concentration of resources among a few actors who possess the necessary financial and infrastructural capacities. Second, they reinforce path dependencies, as existing investments in specific technologies and infrastructures determine future development paths (see Arthur, 1989; David, 1985). Third, it accelerates innovation cycles, which in turn leads to faster obsolescence of existing technologies.

From an ecological perspective, it is particularly relevant that this scaling logic is not neutral: it is systematically accompanied by rising resource demand. Improvements in model performance are thus directly linked to an intensification of material and energy inputs.

Ecological Economics: Rebound, Limits, and Externalities

The ecological dimension of AI development can be clarified using concepts from ecological economics. A central point of reference is the rebound effect, originally described by Jevons (1865), which states that efficiency gains do not necessarily lead to a reduction in resource consumption. Under certain conditions, they can even increase it. In modern applications, this concept was further developed by Polimeni et al. (2008).

In the context of AI, this means: Improvements in the efficiency of hardware or algorithms can lower the cost per unit of computation, thereby expanding usage. The total demand for computing power rises, so that absolute energy and resource consumption can increase despite relative efficiency gains. This mechanism is particularly relevant in an environment characterized by strong incentives for growth and competition.

Furthermore, ecological economics points to the existence of planetary boundaries (see Rockström et al., 2009) within which economic activities must take place. The growing energy and resource intensity of AI raises questions about whether this technology is compatible with long-term sustainability goals. In particular, dependence on non-renewable raw materials and rising emissions are at odds with global climate goals.

Another key aspect is externalities. The environmental costs of AI development—such as those resulting from raw material extraction, energy consumption, or waste—are often not fully internalized. Rather, they are shifted onto other actors or future generations. This externalization is a structural feature of capitalist modes of production (see Kapp, 1950) and is further amplified in the case of AI by global supply chains and complex value-added structures.

An Integrative Analytical Model

Based on the preceding considerations, an integrative analytical model can be developed that systematically captures the interactions between financialization, scaling logics, and ecological externalities.

At the core of this model lies a circular causal relationship:

- Financialization creates incentives for infrastructure accumulation and scaling.
- Scaling logics drive demand for computing power, data, and hardware.
- Ecological externalities manifest as energy consumption, emissions, and resource use.

However, these externalities are only partially internalized, leaving the underlying incentives for further expansion intact.

This cycle is continuously reinforced by competition and technological innovation, leading to a structural decoupling between efficiency gains and absolute environmental impacts.

The model makes it possible to view the environmental impacts of AI not as isolated effects of individual technologies, but rather as emergent properties of a complex socio-technical system. At the same time, it provides a basis for developing governance approaches that specifically address the structural drivers—for example, through the regulation of infrastructure investments, the internalization of external costs, or the limitation of scaling incentives.

The ecological political economy of AI makes it clear that environmental impacts are not determined solely by technological parameters. They result from the interplay of economic incentives, institutional frameworks, and material limits. By bringing financialization, scaling, and ecological externalities together within a common analytical framework, it creates the basis for a more comprehensive assessment of the sustainability of AI systems. At the same time, it opens up a perspective in which sustainability is understood as an integral part of technological and economic design.

Figure 2. Conceptual model of the ecological political economy of AI

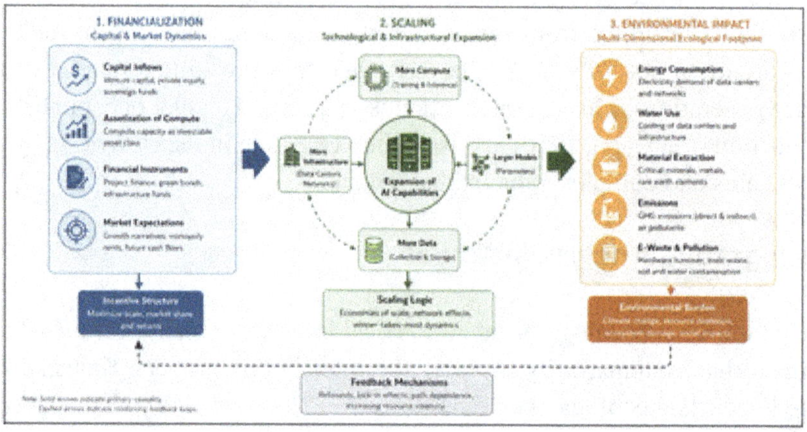

Figure 2 illustrates how financialization processes lead to the accumulation of infrastructure and computing capacity, which in turn drives scaling dynamics. These dynamics generate increasing energy, material, and environmental impacts. Feedback mechanisms such as rebound effects, path dependencies, and lock-in structures stabilize and reinforce the system.

COMPUTE AS AN ASSET CLASS: THE FINANCIALIZATION OF AI INFRASTRUCTURE

The current dynamics of AI development cannot be adequately understood if computing power ('compute') is viewed exclusively as a technical input. Rather, in recent years, computing has evolved into an independent economic category: a strategic resource that simultaneously serves as a factor of production, a barrier to entry, and a financial asset. This transformation reflects a deeper structural shift in which digital infrastructures are increasingly becoming central drivers of value creation and capital accumulation. The following section argues that the environmental impacts of AI are significantly shaped by this financialization of infrastructure.

The Rise of Computing as a Strategic Resource

With the transition from traditional machine learning approaches to large-scale neural networks, the demand for computing power has increased exponentially. Training and operating modern AI systems require highly specialized hardware (particularly GPUs and increasingly specialized AI accelerators), massive amounts of data, and complex data center architectures. This development has led to computing itself becoming a scarce and strategically controlled resource.

From an economic perspective, this represents a shift from variable to fixed cost structures. While early software development required comparatively low capital investment, the development and operation of modern AI systems involve significant upfront investments in infrastructure. This capital intensity leads to increasing returns to scale and favors players with access to substantial financial resources. Consequently, the competitive landscape is shifting toward oligopolies, with a few companies controlling a large share of global computing capacity.

At the same time, the function of computing is changing: it is a means of problem-solving and is itself becoming the object of strategic accumulation. Control over computing capacity enables companies to accelerate innovation cycles, aggregate data streams, and stabilize platform ecosystems. In this sense, computing functions as the infrastructural backbone of the digital economy. Recent research findings view computing power as a controllable resource due to its measurability, centralization, and strategic importance (Sastry et al., 2024). This leads to a redefinition of AI development, which views it as a capital-intensive infrastructure system in which control over computing power becomes a central lever for both economic power and environmental impact.

Capital Market Logic and Scaling Signals

The growing importance of computing is closely linked to the valuation mechanisms of modern capital markets. In technology-driven sectors, corporate valuation is increasingly based on expectations of future growth and technological leadership (see Corrado et al., 2009). In this context, investments in AI infrastructure serve as visible indicators of innovative capacity and scaling potential.

A central element of this logic is the symbolic function of model size and computing capacity. Large models—measured by parameters, training effort, or inference performance—are interpreted as an expression of technological superiority. This perception influences both competitive dynamics among companies and investor expectations. Infrastructure investments thus serve a dual function: they expand production capacity while simultaneously signaling in the capital market.

This signaling effect can reinforce investment dynamics that are not determined solely by real demand. Rather, a self-referential process emerges in which companies invest to meet expectations, which in turn are confirmed by these investments. In the literature on financialization, this phenomenon is described as a form of 'assetization,' in which assets are defined by their use, value, and tradability (see Birch & Muniesa, 2020).

In the case of AI, this means that computing functions as a valuable asset whose expansion itself becomes the goal of economic activity. This dynamic contributes to the perpetuation of high investment levels and reinforces the infrastructural dimension of AI growth.

Infrastructure Lock-In and Overaccumulation

The capital intensity of AI infrastructures has far-reaching implications for their use and environmental impact. Once data centers are built, energy contracts are signed, and supply chains are established, these commitments create long-term obligations that cannot be revised without high costs. This situation can be described using the concept of 'lock-in' (see Arthur, 1989): Past investment decisions determine future options for action and perpetuate existing development paths.

From a business perspective, there is a strong incentive to use existing capacity as fully as possible to amortize fixed costs. This can lead to the development of additional applications or the expansion of existing processes to increase infrastructure utilization. In this way, the demand for computing power is determined by external needs as well as by the logic of capital utilization.

This dynamic can lead to overaccumulation, in which more infrastructure is built than is necessary in the short term. While excess capacity is often considered inefficient in traditional industries, it can make strategic sense in the context of digital platform markets, for example, to secure market share or to deter potential competitors. From an ecological perspective, however, this overaccumulation is problematic because it is associated with increased resource consumption and long-term environmental impacts.

Another aspect of infrastructure lock-in concerns energy supply. Long-term power purchase agreements and the integration of data centers into existing power grids result in fixed energy flows for years to come. Even if technological alternatives or more efficient solutions emerge, existing structures can only be adapted slowly. This reinforces the system's inertia, making it difficult to rapidly reduce environmental impacts.

Ecological implications of infrastructure accumulation

The financialization of computing has direct and indirect ecological consequences. Directly, these manifest as energy consumption, water demand, and material use for the construction and operation of data centers. Indirectly, they operate through structural incentives that drive the continuous expansion of these infrastructures.

A key factor here is the temporal dimension. While individual models may have relatively short lifecycles, infrastructure is designed to last for decades. The ecological impact of AI is therefore a function of both current usage and future decisions regarding usage pathways. Infrastructure investments, in a sense, determine the 'ecological future' by dictating how much energy and resources will be tied up in the long term.

In addition, environmental impacts occur throughout the entire value chain. The construction of data centers requires large quantities of concrete, steel, and electronic components; the production of semiconductors is energy- and water-intensive; and the disposal of obsolete hardware generates significant amounts of electronic waste. These effects are spatially dispersed due to global supply chain fragmentation and are often externalized to regions with lower environmental standards.

Against this backdrop, the central thesis of this section can be clarified: The ecological footprint of AI is determined less by the characteristics of individual models than by the logic of infrastructural accumulation in which these models are embedded. While efficiency gains at the model level are certainly relevant, they cannot fully compensate for the structural effects of capital intensity and scaling dynamics.

Viewing compute as an asset class opens up a new perspective on the environmental impacts of AI. It makes it clear that the decisive drivers lie not solely in technological decisions but in how AI infrastructures are financed, valued, and used.

Financialization intensifies scaling incentives, long-term infrastructure lock-in, and the potential for overaccumulation of computing capacity. These processes are closely linked to environmental externalities, as they systematically increase resource consumption while simultaneously reducing the system's adaptability.

For further analysis, this means the ecological assessment of AI must not stop at considering individual applications. Rather, a systemic perspective is required that focuses on the interactions between capital markets, technological development, and material foundations. On this basis, the specific mechanisms of scaling dynamics and their ecological consequences can be analyzed in greater depth in the next section.

SCALING DYNAMICS AND SYSTEMIC EXTERNALITIES

The development of modern AI systems is significantly shaped by a specific scaling logic that both enables technological performance gains and generates substantial ecological externalities. While Section 3 demonstrated that the infrastructural accumulation of computing capacity is central to the environmental impacts of AI, this section shifts the focus to the dynamic mechanisms of AI's own scaling. It argues that the ecological effects arise from an interplay of technological laws, economic incentives, and systemic feedback loops.

Empirical Scaling Laws as Drivers of Expansion

A central feature of current AI development is the empirically well-documented relationship between model performance and the resources employed. Research on scaling laws shows that the performance of large neural networks systematically increases with the amount of training data, the number of model parameters, and the computational effort expended (see Kaplan et al., 2020). This insight has evolved into a dominant paradigm that shapes the strategic direction of many companies and research institutions.

From an analytical perspective, it is crucial that this scaling logic exhibits sublinear rather than linear return functions: while additional resource input leads to performance gains, these gains come with diminishing marginal returns. Nevertheless, the absolute improvement remains economically relevant for many applications, particularly in highly competitive markets where marginal differences in quality can have significant economic effects (e.g., in search engines, language models, or recommendation systems).

The consequence is a structural tendency toward further scaling, even as efficiency per unit declines. This dynamic is amplified by competition, as companies are forced to continuously improve their models to maintain their market position. Scaling thus becomes a systemic necessity.

Rebound Effects in the AI Economy

Environmental economics offers an important key to understanding the resulting environmental impacts through the concept of the rebound effect. As described by Jevons (1865) and elaborated by Polimeni et al. (2008), efficiency gains can paradoxically lead to an increase in total resource consumption.

In the context of AI, this effect manifests itself on several levels:

(1) Technological efficiency gains: Advances in hardware (e.g., more energy-efficient chips) or software (optimized training methods) reduce the cost per unit of computation.
(2) Expanding demand: Falling costs enable new applications and increase the use of existing systems.
(3) Market expansion: AI is being integrated into more and more sectors, thereby increasing the overall demand for computing power.

This cascade often leads to relative efficiency gains that are more than offset by absolute increases in consumption. The rebound effect can be particularly pronounced in the context of digital platforms that rely on scaling and network effects. The environmental benefits of more efficient technologies thus remain limited or are completely negated.

Infrastructure Lock-In and Path Dependence

Another key mechanism is the so-called infrastructure lock-in, introduced in the previous section and explored here in greater depth in its dynamic dimension. Lock-in describes the situation in which existing investments determine future development paths and crowd out alternative options (see Arthur, 1989). In the context of AI, this means:

- Once built, data centers create incentives for maximum utilization.
- Existing software and hardware architectures shape future innovation paths.
- Investments in specific technologies make it difficult to switch to alternative, potentially more sustainable solutions.

This path dependence is reinforced by institutional and organizational factors, such as existing business models, regulatory frameworks, or technical standards. The lock-in effect means that even when more efficient or resource-conserving technologies are available, their diffusion is delayed or prevented.

From an ecological perspective, it is particularly problematic that lock-in not only perpetuates the use of existing resources. It also ‘anticipates’ future environmental impacts. Infrastructure decisions thus function as long-term determinants of ecological impact trajectories.

Accelerated Obsolescence and E-Waste Dynamics

An often underestimated aspect of scaling dynamics is the accelerated obsolescence of hardware. Innovation cycles in semiconductors and AI accelerators are extremely short, leading to existing systems being rapidly replaced by more powerful generations. This process is driven by both technological and economic factors, as more powerful hardware promises competitive advantages.

The consequences are manifold:

- Shorter hardware lifespan: Devices are replaced before their physical lifespan is exhausted.
- Increase in electronic waste: Old hardware is disposed of or exported to less regulated markets.
- Resource inefficiency: The materials used in production are not fully utilized.

The literature on 'planned obsolescence' (see Packard, 1960) provides important points of reference here, although the dynamics in the AI context are shaped less by deliberate design than by technological competition. Nevertheless, a functionally similar effect arises: a systematic shortening of product life cycles with corresponding environmental consequences.

Systemic Feedback Effects

The mechanisms described—scaling laws, rebound effects, lock-in, and obsolescence—do not operate in isolation. They reinforce one another within a complex system of feedback loops.

A simplified model of these effects can be represented as follows:

(1) Improved models increase demand for AI applications.
(2) Rising demand justifies further investment in infrastructure.
(3) New infrastructure enables larger models.
(4) Larger models generate new applications and markets.

This cycle corresponds to a positive feedback loop that leads to continuous system expansion. At the same time, environmental costs are not fully internalized, so no natural braking mechanisms are at work. From a systems theory perspective (see Meadows, 2008), this is a growth-driven system without effective limiting mechanisms. Such systems tend to exceed ecological limits unless external interventions occur.

An analysis of scaling dynamics shows that the ecological footprint of AI is determined by a set of systemic mechanisms that favor expansion. Scaling laws provide the technical foundation, while rebound effects, lock-in structures, and accelerated obsolescence reinforce these dynamics economically and institutionally.

Taken together, these mechanisms lead to a structural decoupling between efficiency and sustainability: even if individual components become more efficient, the overall system can continue to grow, thereby expanding its environmental impacts.

For further analysis, this means that ecological interventions must not focus solely on efficiency gains. They must address the underlying scaling incentives. This opens up a view of the material foundations of AI infrastructure, which will be examined in detail in the next section.

Figure 3. Reinforcing feedback loop of AI scaling dynamics

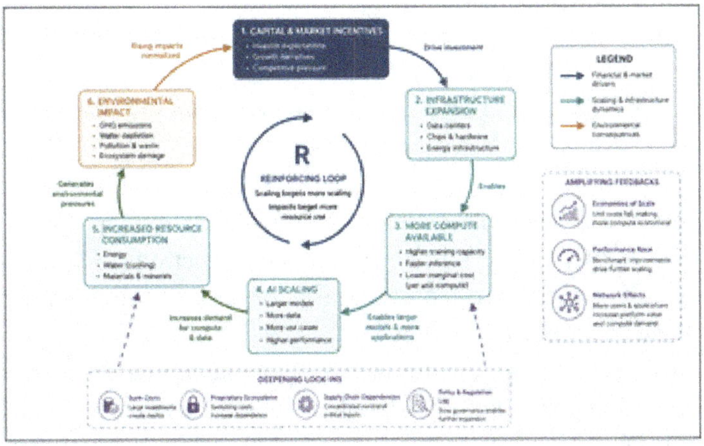

Figure 3 illustrates the circular interaction between capital incentives, infrastructure expansion, increasing computing capacity, and model scaling. These processes lead to growing resource consumption and rising environmental impacts, which are further amplified by mechanisms such as rebound effects, economies of scale, network effects, and lock-in structures. A self-reinforcing growth dynamic emerges as ecological intensity increases.

MATERIAL FOUNDATIONS: EXTRACTION, SUPPLY CHAINS, AND GEOPOLITICAL DIMENSIONS

The previous sections have shown that the environmental impacts of AI are significantly shaped by financialization and scaling dynamics. However, these dynamics do not materialize in the abstract. They are embedded in concrete physical processes, particularly in the extraction, processing, and circulation of raw materials. On the one hand, AI is a data- and energy-intensive technology. On the other hand, it is also deeply embedded in global extraction and production regimes. This section analyzes the material basis of AI infrastructure and demonstrates that its environmental and social impacts arise along complex global supply chains.

Critical Raw Materials as the Foundation of AI Infrastructure

The production of modern AI hardware—particularly high-performance chips and data center components—requires a variety of so-called critical raw materials. These include, among others:

- Rare earth elements (e.g., neodymium, dysprosium) for magnetic components,
- Copper for electrical wiring and infrastructure,
- lithium and cobalt for energy storage and power supply systems,
- High-purity silicon for semiconductor production.

The importance of these materials stems from their specific physical properties, which are essential for the performance of modern electronics. At the same time, many of these raw materials are geographically concentrated and subject to complex geopolitical and economic dependencies.

The rising demand for AI systems is increasing the pressure on these resources. While digitalization was long considered less material-intensive than traditional industries, it is increasingly clear that expanding digital infrastructure requires significant amounts of primary resources (see Bardi, 2014; IRP, 2019). AI is thus part of a broader trend in which digital technologies are not reducing global demand for raw materials, but transforming it.

Recent life cycle assessments of AI hardware show that environmental impacts are unevenly distributed across the value chain: While emissions during operation dominate climate-related impacts, manufacturing processes are responsible for the majority of toxicity and resource consumption impacts (Schneider et al., 2025; Luccioni et al., 2025). This distribution supports the view that AI is more of a material-intensive, resource-extracting system than a purely digital technology.

Environmental Impacts of Raw Material Extraction

The extraction and processing of raw materials required for AI are associated with significant environmental impacts. These can be categorized across several dimensions:

- Land-use changes and landscape destruction: Mining activities often lead to large-scale interventions in ecosystems, including deforestation, soil erosion, and loss of biodiversity.
- Water consumption and pollution: In particular, lithium mining and metal processing are water-intensive and can place a heavy burden on local water resources. At the same time, there is a risk of contamination from chemical residues.
- Emissions and energy consumption: The extraction and refining of raw materials are energy-intensive and contribute significantly to greenhouse gas emissions, particularly when fossil fuels are used.
- Toxic byproducts: Many extraction processes generate waste materials that pose risks to the environment and human health, such as heavy metals or acid residues.

These environmental impacts often occur in regions already characterized by ecological fragility or institutional weaknesses. As a result, part of the environmental costs of AI development is shifted to geographic areas that benefit only to a limited extent from the economic advantages.

Global supply chains and the externalization of environmental costs

The material basis of AI is embedded in highly fragmented and globally organized supply chains. Raw materials are mined in one country, processed in another, turned into components in a third, and finally deployed in data centers distributed around the world. This fragmentation makes it difficult to track environmental impacts and facilitates their externalization.

From a political economy perspective, this process can be interpreted as part of global value chains in which different regions assume specific functions (see Gereffi et al., 2005). High-value activities such as research, design, and platform operations are concentrated in industrialized countries, while resource-intensive, environmentally damaging processes often occur in the Global South.

This structure leads to an asymmetrical distribution of costs and benefits:

- Economic gains are primarily realized in the centers of the digital economy.
- Ecological and social costs often arise in peripheral regions.

To systematically capture these spatially distributed and functionally differentiated environmental impacts, a life-cycle-based perspective is required. As Figure 4 shows, the ecological footprint of AI unfolds along a multi-stage value chain, ranging from raw material extraction through production and use to disposal. It becomes clear that a significant portion of the environmental impacts occurs outside the actual operational phase of AI systems and is distributed along global supply chains.

Figure 4. Lifecycle footprint of AI

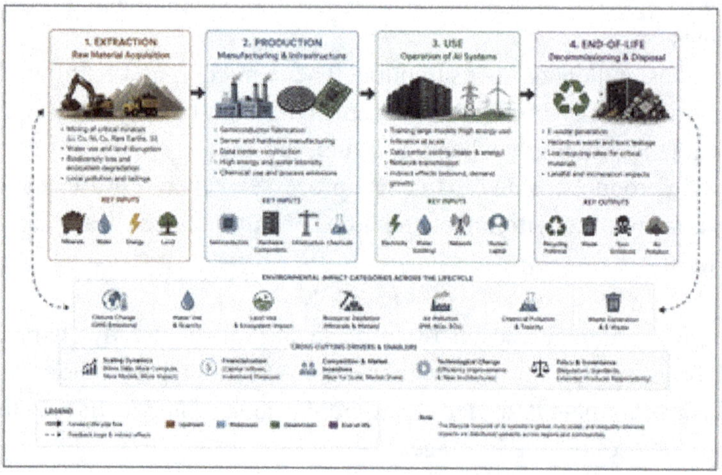

The concept of 'ecologically unequal exchange' (see Hornborg, 2011) offers an analytical framework for describing this asymmetry. In this sense, AI can be understood as part of a global system in which environmental impacts are systematically shifted across geographical boundaries.

Geopolitical Dimensions and Resource Conflicts

The growing importance of critical raw materials for digital technologies also has geopolitical implications. States and companies are increasingly competing for access to resources, production capacities, and key technological components. This is evident, for example, in:

- strategies to secure supply chains (e.g., diversification, 'friend-shoring'),
- government support programs for semiconductor production,
- export controls and technological restrictions.

These developments suggest that AI is also a geopolitical project. Control over resources and production capacities is becoming a central element of international power constellations (see Bridge, 2008).

At the same time, resource conflicts can arise at the local level, for example, between mining companies and local communities. Issues of land use, access to water, and the distribution of economic benefits can lead to social tensions. These conflicts are linked to global demand processes driven by the expansion of digital technologies—including AI.

AI as an Extractive Industry: A Conceptual Proposal

Against the backdrop of the developments described, it can be argued that AI is a form of digital extractivity. This term expands upon classical notions of extraction, which primarily refer to the extraction of raw materials, and integrates digital technologies into a more comprehensive analysis of resource use and value creation.

Digital extractivity encompasses several levels:

- Material extraction: extraction of raw materials for hardware,
- Energy extraction: Use of energy sources for operation and cooling,
- Data extraction: the collection and processing of large amounts of data as the basis for AI models.

This perspective enables systematic capture of the connection between digital and material processes. AI no longer appears as a contrast to the 'real' economy: it is an integral part of it, giving rise to new forms of resource use and control. At the same time, the concept opens up a critical perspective on the sustainability of digital transformations. If AI is understood as an extractive industry, the question arises of how far existing regulatory approaches, primarily geared toward industrial production, should also be applied to digital infrastructures.

An analysis of the material foundations of AI shows that its environmental impacts are deeply embedded in global extraction and production systems. The demand for computing power and digital infrastructure translates into rising demand for raw materials, the extraction of which entails high ecological and social costs.

These costs are spatially distributed along global supply chains and frequently externalized, giving rise to asymmetrical power and distribution structures. At the same time, geopolitical dynamics intensify competition for resources and exacerbate potential conflicts.

By interpreting AI as part of an extractive system, this section broadens the perspective on its environmental impacts and lays the groundwork for a more comprehensive analysis of the energy and infrastructural dimensions, which are the focus of the following section.

ENERGY AND WATER: PHYSICAL LIMITS OF AI EXPANSION

The preceding sections have shown that the ecological impact of AI is significantly shaped by financialization and scaling dynamics as well as by global material flows. However, these dynamics are particularly evident in the physical infrastructure of data centers, where energy and water flows converge. Section 6, therefore, analyzes the energy and hydrological foundations of AI and argues that these represent a central, hitherto underestimated limiting factor for further expansion.

Energy Requirements of AI Systems: Structure and Dynamics

The energy requirements of AI can be analytically divided into two main components: training and inference. While the training of large models is characterized by extremely high but time-limited energy consumption, inference—that is, the use of models during ongoing operation—generates a continuous base load. With the increasing integration of AI into everyday applications, the inference phase in particular is gaining significance for overall energy consumption.

Empirical studies show that the energy requirements of AI systems rise sharply with increasing model size and usage (see Patterson et al., 2021). At the same time, the structure of energy consumption is shifting: While individual training runs can reach spectacular peaks, it is the sustained scaling of applications—such as in search engines, voice assistants, or automated decision-making processes—that accounts for the largest share of energy consumption in the long term. Recent studies show that the ecological footprint of AI is increasingly determined by inference workloads rather than training alone, reflecting the integration of AI into infrastructures for ongoing operations (Kim et al., 2026). Through this shift, AI is transforming from a sequence of individual computational processes into a permanent source of resource consumption at the infrastructure level.

This development points to a central characteristic of the AI economy: energy consumption is an integral part of value creation. The ability to efficiently convert large amounts of energy into computing power becomes a competitive advantage in itself. This creates a direct link between economic expansion and energy resource consumption.

Data Centers as Energy Infrastructures

Data centers form the physical backbone of the AI economy. In particular, so-called hyperscale data centers operate on a scale that makes them significant players in the energy system. Their electricity demand sometimes reaches levels comparable to those of small cities and requires a stable, highly available energy supply.

From an infrastructure perspective, data centers can be understood as nodes in the energy system that fulfill several functions:

- Load centers: They generate a continuous demand for electrical energy.
- Grid integration points: They must be integrated into existing power grids and affect their stability.
- Flexibility options: In some cases, they can act as controllable loads, for example, by shifting computing processes.

The integration of these infrastructures places significant demands on energy supply systems. Especially in regions with limited grid capacity or high demand, data center expansion can create bottlenecks. At the same time, the concentration of large computing capacities in certain regions can contribute to spatial imbalances in energy consumption.

A central point of discussion is the role of renewable energy. Many companies emphasize that their data centers are increasingly powered by renewable energy. However, this statement must be viewed with nuance. On the one hand, the actual emissions balance depends on the temporal and spatial availability of renewable energy. On the other hand, the additional demand for electricity may indirectly necessitate the expansion of fossil fuel capacity if renewable sources are not sufficiently available (see Jones, 2018).

Water Consumption and Cooling Requirements

In addition to energy, water is another key resource for the operation of AI infrastructures. Server cooling is necessary to prevent overheating and ensure system performance. Various cooling technologies are used, ranging from air-based systems to water-intensive cooling methods.

Water-based cooling is particularly efficient but requires significant water consumption. This consumption includes both direct withdrawals (e.g., for evaporative cooling) and indirect effects, such as the water demand associated with power generation. Studies show that the water footprint of data centers can be significant in certain regions (see Li et al., 2023).

The ecological relevance of this water consumption depends heavily on the local context. In water-rich regions, the additional demand may be relatively unproblematic, whereas in water-scarce areas it can cause significant strain. The choice of data center location, therefore, becomes a decisive factor in their environmental impact.Furthermore, thermal effects can occur when heated water is discharged back into natural water bodies. This can affect local ecosystems and incur additional ecological costs.

Spatial Inequalities and Externalization

The physical infrastructure of AI is spatially unevenly distributed. Data centers are often built in regions that offer specific advantages, such as low energy costs, a favorable climate, or regulatory incentives. However, these location decisions have far-reaching implications for the distribution of environmental impacts.

A frequently observed pattern is the externalization of environmental costs. While the economic benefits of AI use are distributed globally or concentrated in specific hubs, the environmental burdens arise locally. This can lead to tensions between local communities and globally operating technology companies.

The literature on environmental justice (see Bullard, 1920) provides an important analytical framework here. It shows that environmental burdens are often distributed unequally and affect certain population groups more severely than others. In the context of AI, the question arises of how far the expansion of digital infrastructure reinforces existing inequalities or creates new ones.

Allocation Conflicts and Political Dimensions

As the importance of AI infrastructure grows, questions of resource allocation are coming into sharper focus. Energy and water are finite resources, and their use is subject to competition among different sectors. The expansion of AI applications can therefore lead to conflicting priorities, for example, between:

- industrial production,
- public infrastructure (e.g., households, healthcare),
- and digital services.

These conflicts cannot be resolved purely through technical means; they have a political dimension. They involve issues of prioritization, distributive justice, and long-term development strategies. In this sense, the expansion of AI becomes a subject of political negotiation processes in which different interests and values clash.

A particularly relevant question is whether and to what extent AI should be considered ‘critical infrastructure.’ Such a classification could have far-reaching consequences for the regulation and prioritization of resources.

The analysis of energy and water flows makes it clear that the expansion of AI is bound by physical limits. Data centers are deeply embedded in regional energy and water systems. Their operations require a continuous supply of resources and simultaneously generate environmental impacts.

The central finding of this section is that the environmental assessment of AI cannot proceed without an analysis of these physical foundations. Energy and water consumption are also expressions of structural dynamics driven by financialization and scaling.

This brings to the forefront the question of how these dynamics can be measured and managed. The following section, therefore, addresses the challenges of emissions measurement and the development of suitable assessment and reporting systems.

CARBON ACCOUNTING AND MEASUREMENT CHALLENGES: LIMITS OF QUANTIFICATION

Previous analyses have shown that the environmental impacts of AI are deeply embedded in infrastructural, material, and energy systems. Nevertheless, a central challenge remains: How can these impacts be adequately measured and compared? Quantification is a crucial lever for governance, regulation, and market steering. Without robust and standardized metrics, the environmental assessment of AI remains fragmented and difficult to address politically.

This section argues that existing approaches to carbon accounting for AI have systematic shortcomings and require further expansion. In particular, there is a lack of an integrated view across the entire life cycle and standardized, model-based metrics.

Current State of Emissions Reporting

Current emissions reporting practices in the technology sector are predominantly based on established frameworks such as the Greenhouse Gas Protocol, which categorizes emissions into Scope 1 (direct emissions), Scope 2 (indirect emissions from energy use), and Scope 3 (upstream and downstream emissions). Large technology companies regularly publish sustainability reports that detail their emissions across these categories.

However, several issues arise for AI-specific applications:

- Aggregation rather than differentiation: Emissions are reported at the corporate level, not at the level of individual models or applications.
- Lack of transparency in the data basis: Methodological assumptions and calculation bases are often not fully disclosed.

- Insufficient representation of dynamic usage: In particular, the inference phase, which is highly usage-dependent, is often captured only in a generalized manner.

Some studies have attempted to quantify the energy consumption and emissions of individual training runs (see Strubell et al., 2019; Patterson et al., 2021). These studies provide important insights but remain isolated and are difficult to compare, as they are based on different assumptions and system boundaries.

Systematic Measurement Gaps and Methodological Challenges

The shortcomings of existing approaches can be systematized across several dimensions:

- System boundaries (boundary problem): It is often unclear which processes are included in the emissions calculation. While some analyses focus on direct energy consumption, upstream processes such as hardware production or raw material extraction are not taken into account.
- Allocation problems: Data centers are used by multiple applications simultaneously. Assigning energy consumption to individual models or services is, therefore, methodologically challenging.
- Dynamics and Scalability: The energy consumption of AI systems varies depending on usage. In particular, the inference phase can account for the majority of emissions in widely used applications.
- Data availability: Much of the relevant data—such as information on hardware efficiency, data center utilization, or specific training processes—is proprietary and not publicly available.

These challenges mean that existing measurement approaches often capture only partial aspects, providing a fragmented picture of the actual environmental impacts.

Life Cycle Assessments (LCA) for AI Systems

Against this backdrop, the application of life cycle assessments (LCA) is gaining importance. LCA is an established environmental assessment tool that aims to capture the environmental impacts of a product or system throughout its entire life cycle (see ISO 14040).

For AI systems, this means extending the analysis to multiple phases:

- Upstream: Raw material extraction, semiconductor production, hardware manufacturing
- Midstream: Transport, installation, and construction of data centers
- Downstream: Operation (training and inference), maintenance, and disposal

Such an approach enables a more comprehensive assessment but places high demands on data availability and methodological consistency. In particular, integrating hardware and software components into a common assessment framework is complex.

Another problem is that LCA was traditionally developed for physical products, whereas AI systems are more difficult to define due to their dynamic use and continuous development . Models are updated, adapted, and deployed across different contexts, making it difficult to define clear system boundaries.

Need for Standardized Metrics

To improve the comparability and transparency of environmental assessments, there is a growing need for standardized metrics for AI systems. Possible approaches include:

- Energy consumption per training session (e.g., kWh per model training)
- Emissions per inference (e.g., CO_2e per request)
- Total life cycle footprint of a model or service

In addition, intensity-based metrics could be developed that relate environmental impacts to the performance delivered, such as emissions per increase in accuracy or per user interaction. Such metrics could help highlight efficiency gains without losing sight of absolute environmental impacts.

However, there is a risk that purely relative metrics may obscure the actual environmental impact if they are not supplemented by absolute values. A combination of absolute and relative metrics, therefore, appears necessary.

Transparency, Reporting, and Governance

Improving carbon accounting is also an institutional challenge. Without appropriate incentives and regulatory frameworks, companies are unlikely to voluntarily provide comprehensive and comparable data.

Possible starting points for further developing governance include:

- Mandatory disclosure requirements for large AI systems,

- Standardized reporting frameworks specifically tailored to digital technologies,
- Certification systems that incorporate environmental impacts as a quality criterion,
- Integration into existing regulatory regimes, such as in the context of digital or sustainability-related regulation.

A key issue here is granularity: While aggregated corporate data is relevant for macroeconomic analyses, regulatory and market-based control mechanisms require more detailed information at the level of individual applications or models.

Measuring the environmental footprint of AI is characterized by significant methodological and institutional challenges. Existing approaches to carbon accounting capture important aspects but remain fragmented and often insufficiently comparable.

A comprehensive assessment requires integrating life-cycle analyses, standardized metrics, and improved data access. At the same time, it is clear that measurement alone is not enough: without appropriate governance structures, quantifying environmental impacts remains largely inconsequential.

Thus, this section serves as a central bridge to further analysis: The question of how environmental impacts are measured is inextricably linked to the question of how they can be managed. The next section takes up this perspective by critically assessing AI as a potential tool for solving environmental problems.

AI AS AN ENVIRONMENTAL SOLUTION? A NET IMPACT ANALYSIS BEYOND THE DICHOTOMY

The previous sections have highlighted the material and energy costs of AI expansion. Nevertheless, AI is frequently presented in politics, business, and science as a key technology for ecological transformations—for instance, to optimize energy systems, monitor biodiversity, or model climate developments. This dual role of AI as both a driver of environmental impacts and a potential tool for their reduction creates an analytical tension that cannot be resolved through simple dichotomies ('problem vs. solution').

This section, therefore, develops a net impact approach that systematically assesses and contextualizes the ecological effects of AI. The aim is to identify the conditions under which AI can actually contribute to a reducing in environmental impacts—and to highlight the scenarios in which it exacerbates existing problems.

AI in Environmental and Climate Applications

In recent years, a wide range of application areas has emerged in which AI is used to support environmental goals. These can be broadly divided into four categories:

- Monitoring and data analysis: AI enables the analysis of large amounts of data from satellites, sensors, and other sources, for example, to monitor deforestation, air quality, or marine ecosystems (see Rolnick et al., 2019).
- Optimization of energy systems: Through predictive models and control algorithms, AI can contribute to increasing efficiency in power grids, integrating renewable energy, and reducing peak loads.
- Resource efficiency in industry and agriculture: Applications range from precision fertilization and smart irrigation to the optimization of production processes.
- Climate modeling and scenario analysis: AI-based models can simulate complex interactions within the climate system, thereby supporting decision-making processes.

These applications demonstrate that AI is, in principle, capable of reducing information asymmetries, optimizing processes, and utilizing resources more efficiently. In many cases, the potential environmental benefits are significant, particularly when AI improves systemic efficiency.

A Conceptual Net Impact Framework

To systematically assess the environmental impact of AI, a net impact approach is proposed: Net environmental impact = (environmental impacts avoided by AI) – (life-cycle-related environmental impacts of AI)

This approach integrates two dimensions:

(1) Avoided Impacts: Emissions reductions or resource savings made possible by the use of AI.
(2) Direct and Indirect Impacts: Environmental impacts resulting from the development, operation, and disposal of AI systems.

It is crucial that both dimensions are captured within consistent system boundaries. Otherwise, there is a risk that positive effects will be overestimated or negative effects underestimated. The net impact approach also makes it clear that the environmental assessment of AI is context-dependent. The same technology can yield very different outcomes across different applications.

Sectoral Differentiation and Impact Pathways

The net impact of AI varies greatly across sectors and use cases. Three ideal-typical scenarios can be distinguished:

- Net-positive applications: AI leads to significant efficiency gains or emission reductions that far exceed its own environmental costs. Examples include optimized power grids or improved forecasts for renewable energy.
- Ambivalent applications: The positive and negative effects are roughly balanced or difficult to quantify. This can be the case, for example, with applications that increase efficiency but simultaneously generate additional demand.
- Net-negative applications: The use of AI leads to additional environmental impacts without corresponding compensation through avoided emissions. This can occur particularly in computationally intensive applications with limited societal benefits.

This distinction underscores that a blanket assessment of AI as 'green' or 'not green' is not meaningful. Rather, a case-by-case analysis is required that considers both the specific application and the institutional context.

Rebound and Systemic Effects

Even when AI leads to efficiency gains, rebound effects can offset or even reverse the positive impacts. Efficiency gains can reduce costs and thereby expand usage, causing total resource consumption to rise (see Polimeni et al., 2008).

A classic example is the optimization of logistics systems: while individual transports become more efficient, the overall demand for transport services may rise, for instance, due to falling prices or new business models. Similar effects can also be observed in the energy sector or the digital economy.

Furthermore, systemic effects may arise that extend beyond the immediate scope of application. AI can create new markets, alter consumption patterns, or transform existing structures. These indirect effects are difficult to quantify but can have significant environmental impacts.

Greenwashing and Narrative Distortions

The portrayal of AI as an environmental solution is not free from strategic interests. Companies have incentives to highlight the positive effects of their technologies while making negative impacts less visible. This can lead to forms of 'greenwashing,' in which environmental benefits are overstated or insufficiently substantiated.

A central problem here is the asymmetry of measurability: While direct emissions are relatively easy to quantify, avoided emissions are often fraught with uncertainty and depend on assumptions about alternative scenarios. This uncertainty creates room for selective framing.

Furthermore, the focus on technological solutions can obscure structural problems. When AI is presented as a central lever for sustainability, there is a risk that fundamental questions of consumption, production, and distribution will be pushed into the background. In this sense, AI cannot merely be part of the solution. It is also part of the problem when it reinforces existing growth paradigms.

The analysis shows that AI's role in ecological transformation is ambivalent. Under certain conditions, AI can contribute to significant environmental improvements, particularly through efficiency gains and better information processing. At the same time, it is itself part of a resource-intensive system characterized by scaling and growth dynamics.

The proposed net impact approach enables systematic capture of this ambivalence and nuanced evaluation. It makes clear that the ecological impact of AI depends on technological characteristics. It manifests itself in its context of use, the underlying incentive structures, and the institutional framework.

This leads to a key implication for further analysis: whether AI contributes to solving environmental problems is closely linked to issues of governance and regulation. These are discussed in greater depth in the following section.

GOVERNANCE AND POLITICAL IMPLICATIONS: FROM EFFICIENCY TO LIMITATION

The preceding analyses have shown that the environmental impacts of AI result from the interplay among financialization, scaling dynamics, and physical infrastructures. This leads to a key conclusion: technical efficiency gains are insufficient to address the environmental challenges posed by AI expansion. Rather, a governance framework is needed that specifically targets the structural drivers.

This section develops such a perspective by critically reflecting on existing governance approaches and systematizing potential regulatory and economic instruments. At its core is the thesis that AI regulation must evolve from a primarily procedural, risk-based approach to a resource- and boundary-oriented framework.

Limits of Voluntary Commitments

The current governance architecture in AI is characterized by voluntary initiatives and commitments. Large technology companies publish sustainability reports, set climate targets, and invest in renewable energy. These measures undoubtedly raise awareness of environmental issues, but they have structural limitations.

First, voluntary measures are selective and not standardized. Companies can decide for themselves which data to publish and which methods to use. This makes comparisons difficult and reduces the binding nature of the commitments.

Second, such initiatives often address symptoms rather than causes. Efficiency gains or the use of renewable energy can reduce the environmental impacts of individual processes, but they do not alter the underlying growth logic of AI development.

Third, there is a conflict of objectives between ecological and economic interests. In a highly competitive environment, companies have incentives to prioritize scaling and market expansion. Without binding regulatory frameworks, it is therefore unlikely that ecological goals will be systematically taken into account.

These limitations are well documented in the literature on the regulation of environmental externalities (see Stavins, 2003) and can also be applied to the context of digital technologies.

Regulatory Instruments: Transparency, Standards, and Obligations

Against this backdrop, the question arises: which regulatory instruments are suitable for effectively addressing the environmental impacts of AI? A first step is to improve transparency and data availability.

(1) Disclosure requirements: Mandatory reporting on the energy consumption, emissions, and resource use of AI systems could lay the groundwork for further measures. It is crucial that the reporting is sufficiently granular to highlight differences between applications.

(2) Technical standards: Minimum requirements for energy efficiency, cooling technologies, or hardware lifespan could help limit particularly resource-intensive practices. Such standards are established in other industries and could be applied to digital infrastructures.

(3) Lifecycle-oriented regulation: Regulatory approaches should not only consider the operation of AI systems. The focus should also include upstream processes such as raw material extraction and hardware production. This would enable greater integration of environmental aspects throughout the entire value chain.

A key tension lies between promoting innovation and regulation. Excessively strict requirements could hinder technological development, while insufficient requirements could exacerbate environmental risks. A differentiated, risk-based regulatory approach, therefore, appears necessary.

Economic Policy Instruments

In addition to regulatory measures, economic instruments play an important role in internalizing environmental costs. These can help to change the incentive structures that currently drive the expansion of resource-intensive AI systems.

(1) CO_2 pricing: Including data centers and AI applications in existing emissions trading systems or CO_2 taxes could make energy consumption more expensive and create incentives for more efficient solutions.
(2) Differentiated energy prices: Time- and location-based electricity tariffs could help shift computing loads to times when renewable energy is highly available.
(3) Subsidies and support programs: Targeted support for 'Green AI'—such as energy-efficient algorithms or alternative hardware architectures—could steer technological innovation toward more sustainable directions.

These instruments address the cost structures of the AI economy and can help to internalize environmental externalities, at least in part. Their effectiveness, however, depends on their design and integration into broader regulatory frameworks.

Compute Governance and Ecological Limits

Beyond incremental measures, the question arises as to whether the scaling logic of AI itself should be regulated. In this context, the idea of compute governance is gaining significance.

One possible approach is to introduce 'compute budgets,' which link the use of computing power to environmental criteria. Similar to emissions budgets, such instruments could set upper limits on resource consumption, thereby limiting the expansion of particularly resource-intensive applications.

However, such an approach raises fundamental questions:

- How can appropriate thresholds be determined?
- Which applications should be prioritized?
- How can international coordination be achieved?

These questions highlight that compute governance is also a political project. It touches on fundamental issues of distribution, justice, and technological development.

Integration Into Existing Regulatory Frameworks

Another important aspect is integrating environmental considerations into existing regulatory frameworks for AI and digital technologies. Current regulatory initiatives focus primarily on issues such as security, data protection, and ethical responsibility.

The integration of environmental criteria could take place at various levels:

- Expanding existing AI regulations to include environmental requirements,
- Linking with sustainability regulations, for example, within the framework of ESG standards,
- coordination with energy and environmental policy, particularly with regard to infrastructure planning.

Such integration would help address environmental aspects within the context of comprehensive governance structures.

Governance of the environmental impacts of AI is still in its infancy. Existing approaches are fragmented and often focus on voluntary measures or technological efficiency gains. However, given the structural dynamics analyzed in the previous sections, a more comprehensive perspective is required.

This must encompass both regulatory and economic instruments and specifically target the drivers of AI expansion. In particular, the question of scaling limits comes to the fore. This shifts the focus from reactive regulation of individual risks to proactive shaping of technological development in harmony with ecological limits.

The concluding section takes up this perspective and discusses its implications for a fundamental rethinking of the development and use of AI.

DISCUSSION: RETHINKING AI WITHIN PLANETARY BOUNDARIES

The preceding sections have analyzed AI as an infrastructure-embedded, capital-intensive, and materially resource-dependent system. The discussion synthesizes these findings and derives implications for theory, empirical research, and governance. At the center is the question of how AI development can be reframed within planetary boundary constraints—beyond a paradigm focused exclusively on efficiency and scaling.

Synthesis: AI as an Infrastructural, Extractive System

First, it becomes clear that AI cannot be understood as 'lightweight' digital technology: at its core, it is heavyweight infrastructure involving long-term capital commitments. Data centers, energy contracts, and semiconductor capacities create path dependencies and perpetuate resource use. Second, AI is extractive: its value creation relies on the mobilization of energy, water, and critical minerals along global supply chains, with environmental costs distributed unevenly across regions. Third, its expansion is driven by financialization and competitive dynamics: computing power functions as an asset class, and scaling serves as a signal in the capital market.

These three dimensions—infrastructure, extraction, and financialization—form a system whose ecological effects emerge. Individual efficiency gains at the model or hardware level can only compensate for these systemic effects to a limited extent.

The Limits of the Scaling Paradigm

The dominant AI innovation paradigm is based on scaling: more data, more parameters, more computing power. Empirical scaling laws have supported this paradigm while, at the same time, revealing a sublinear return structure. As resource input increases, marginal utility decreases, while absolute environmental impacts rise.

Under conditions of planetary boundaries (e.g., climate targets, water availability, biodiversity), this logic becomes problematic. A dissonance arises between technological feasibility and ecological sustainability. The rebound effect amplifies this dissonance by inducing additional demand through efficiency gains. Lock-in structures also prevent rapid correction, even when more sustainable alternatives are available.

The central insight is therefore: scaling is a normatively charged development path with real ecological costs. Sustainable AI development consequently requires a reevaluation of this paradigm.

From 'More Compute' to 'Efficient and Bounded AI'

Against this backdrop, a paradigm shift is emerging: from a logic of unlimited growth ('more compute') toward an approach that combines efficiency with explicit limits ('efficient and bounded AI').

This approach encompasses several dimensions:

- Algorithmic efficiency: Development of resource-efficient models (e.g., smaller, specialized architectures, distillation, sparsity). The goal is relative efficiency, but also a reduction in absolute consumption.

- Systemic efficiency: Optimization at the system level (workload management, site selection, load shifting, waste heat utilization). Here, infrastructure design and operational strategies come to the fore.
- Constraints: Introduction of upper limits or budgets (e.g., energy, emissions, or compute budgets) that function as hard constraints and frame innovation pathways.
- Prioritization of applications: Differentiation based on societal benefit: Highly relevant applications (e.g., climate modeling, healthcare) are prioritized, while resource-intensive applications with low benefit are scrutinized.

The key difference from the previous approach is that constraints are understood as a design principle. Analogous to other infrastructure sectors (energy, transportation), sustainability becomes a constitutive condition of technological development.

Implications for Governance and Political Economy

The shift toward 'bounded AI' has far-reaching implications for governance theory. First, it requires realigning regulatory goals: in addition to safety, transparency, and fairness, ecological criteria must be systematically integrated. Second, the focus of the shifts from ex-post corrections (e.g., offsetting emissions) to ex-ante control (e.g., limiting resource inputs).

Third, questions of distribution and equity arise. Who gains access to limited computing resources? How are environmental costs distributed across global supply chains? These questions are political and normative. They touch on fundamental issues of the international order and development policy.

Fourth, the analysis implies an expansion of the concept of infrastructure regulation. If data centers and AI systems are understood as critical infrastructure, they must be regulated similarly to energy or transportation networks—including requirements for efficiency, resilience, and sustainability.

Research Agenda: Open Questions and Interdisciplinary Perspectives

The perspective developed here opens up several fields of research:

- Measurement and evaluation methods: Further development of LCA approaches for dynamic AI systems; integration of uncertainties and indirect effects.
- Economic modeling: Analysis of rebound and lock-in effects in digital markets; interactions between capital markets and infrastructure investments.

- Governance design: Development and evaluation of instruments such as compute budgets, differentiated energy prices, or mandatory disclosure standards.
- Socioeconomic impacts: Investigation of distributional effects along global supply chains; role of AI in North-South asymmetries.
- Technological alternatives: Research into new architectures and paradigms (e.g., neuromorphic systems, edge computing) that potentially exhibit lower resource intensity.

This agenda is necessarily interdisciplinary and requires integrating computer science, economics, environmental science, and political science.

The discussion makes it clear that the ecological challenge posed by AI cannot be solved through incremental efficiency gains. Rather, a fundamental shift in perspective is needed, one that places the material and political dimensions of the technology at the center.

AI must be understood as part of a comprehensive socio-technical system whose development can and must be actively shaped. Recognizing planetary boundaries does not imply the end of technological innovation, but rather its reorientation. This section thus prepares the transition to the conclusion: When AI is conceived within the constraints of ecological limits, its role in the economy and society also changes fundamentally.

CONCLUSION: SUSTAINABILITY AS A DESIGN PRINCIPLE OF AI

The preceding analyses have yielded a consistent picture: Artificial intelligence is not an immaterial technology decoupled from physical constraints. It is a resource-intensive, infrastructure-anchored, and politically and economically shaped system. Its ecological footprint emerges as an emergent property of the interplay between financialization, scaling logics, and global material and energy flows.

This conclusion has far-reaching implications for the evaluation and design of AI. It calls for a fundamental shift in perspective: away from a purely technology-centered view toward a systemic understanding that integrates ecological, economic, and institutional dimensions.

First, it is evident that the expansion of AI is largely shaped by a logic of scaling and capital accumulation. Computing power functions as a strategic asset class, whose expansion both enables technological progress and is driven by capital market expectations. This dynamic leads to a structural tendency toward infrastructure accumulation, which results in long-term resource lock-in and environmental impacts.

Second, it became clear that the environmental impacts of AI are not linear: they are amplified by systemic mechanisms. Rebound effects, lock-in structures, and accelerated obsolescence mean that efficiency gains are often offset by rising overall demand. The environmental footprint of AI is therefore the result of complex feedback processes.

Third, AI is deeply embedded in global extraction and production systems. The demand for computing power translates into rising demand for critical raw materials, the extraction of which entails high environmental and social costs. These costs are unevenly distributed along global supply chains and are often externalized.

Fourth, it has been shown that energy and water flows represent key physical constraints. Data centers are an integral part of regional infrastructures and compete with other sectors for scarce resources. The expansion of AI, therefore, inevitably raises questions of allocation and prioritization.

Fifth, there are significant shortcomings in the measurement and reporting of environmental impacts. Without standardized metrics and transparent data, the environmental assessment of AI remains fragmented and difficult to manage.

Finally, it became clear that AI's role in ecological transformation is ambivalent. While AI can contribute to increased efficiency and reduced emissions in certain applications, it is simultaneously part of a system characterized by growth-driven dynamics.

The analysis suggests that existing governance approaches are insufficient to address the environmental challenges of AI development. In particular, voluntary commitments and purely technological efficiency strategies fall short, as they do not alter the structural drivers of expansion.

What is needed instead is a multidimensional governance architecture that includes the following elements:

- Transparency and measurability through standardized disclosure requirements and life-cycle analyses,
- Regulatory minimum standards for energy efficiency, resource use, and supply chains,
- Economic incentives to internalize external costs, such as through CO_2 pricing,
- Limitation mechanisms that link the use of computing power to ecological criteria.

Particularly noteworthy is the concept of compute governance, which explicitly addresses AI's resource consumption. Such a perspective expands existing AI regulation to include a dimension that has been largely neglected to date and places the ecological sustainability of technological development at the center.

For scientific research, this requires a more robust analysis of AI as a socio-technical system. This requires interdisciplinary approaches that integrate computer science, economics, environmental sciences, and political science. In particular, the integration of life-cycle analyses, innovation economics models, and governance research appears central.

For practitioners—especially for companies and developers—this means that sustainability cannot be viewed as a downstream optimization task. Rather, it should be taken into account from the outset in the conception and design of AI systems. This includes:

- the selection of resource-efficient model architectures,
- the optimization of training and inference processes,
- considering location and infrastructure decisions,
- as well as the transparent communication of environmental impacts.

Our analysis can be placed within a broader context: the role of technological innovation in the Anthropocene. AI exemplifies a generation of technologies that, on the one hand, offer enormous potential for problem-solving, but on the other hand, are themselves part of the ecological challenge.

The central challenge lies in harnessing this ambivalence productively. This requires a shift in thinking that goes beyond the optimization of individual technologies and focuses on the fundamental structures of production, consumption, and innovation. In this sense, sustainability must be understood as a guiding principle. An AI that operates within planetary boundaries could be not only more environmentally sound but also more robust and sustainable in the long term.

We have shown that the ecological dimension of AI is neither marginal nor secondary. The integration of sustainability aspects into the design and regulation of AI is therefore not an optional add-on: it is a necessary prerequisite for its social acceptance and long-term legitimacy.

The challenge now lies in translating the concepts and approaches outlined here into concrete political, economic, and technological strategies. A decisive factor in this will be whether we succeed in reconciling the dynamics of scaling and growth with the requirements of ecological limits.

Only if this balance is achieved can AI realize its potential without undermining the ecological foundations upon which it is built.

REFERENCES

Arthur, W. B. (1989). Competing technologies, increasing returns, and lock-in by historical events. *Economic Journal (London)*, *99*(394), 116–131. DOI: 10.2307/2234208

Bardi, U. (2014). *Extracted: How the Quest for Mineral Wealth is Plundering the Planet*. Chelsea Green Publishing.

Birch, K., & Muniesa, F. (2020). *Assetization: Turning Things into Assets in Technoscientific Capitalism*. MIT Press. DOI: 10.7551/mitpress/12075.001.0001

Bridge, G. (2008). Global production networks and the extractive sector: Governing resource-based development. *Journal of Economic Geography*, *8*(3), 389–419. DOI: 10.1093/jeg/lbn009

Bullard, R. D. (1920). *Dumping in Dixie: Race, Class, and Environmental Quality* (3rd ed.). Westview Press.

Corrado, C., Hulten, C., & Sichel, D. (2009). Intangible capital and U.S. economic growth. *Review of Income and Wealth*, *55*(3), 661–685. DOI: 10.1111/j.1475-4991.2009.00343.x

David, P. A. (1985). Clio and the economics of QWERTY. *The American Economic Review*, *75*(2), 332–337.

Dosi, G. (1982). Technological paradigms and technological trajectories. *Research Policy*, *11*(3), 147–162. DOI: 10.1016/0048-7333(82)90016-6

Epstein, G. A. (2005). *Financialization and the World Economy*. Edward Elgar. DOI: 10.4337/9781781008263

Gereffi, G., Humphrey, J., & Sturgeon, T. (2005). The governance of global value chains. *Review of International Political Economy*, *12*(1), 78–104. DOI: 10.1080/09692290500049805

Hornborg, A. (2011). *Global Ecology and Unequal Exchange: Fetishism in a Zero-Sum World*. Routledge.

International Resource Panel (IRP). (2019). Global Resources Outlook 2019: Natural Resources for the Future We Want. United Nations Environment Programme. https://digitallibrary.un.org/record/4047319?v=pdf

Jevons, W. S. (1865). *The Coal Question: An Inquiry Concerning the Progress of the Nation, and the Probable Exhaustion of Our Coal Mines* (2nd ed.). Macmillan., https://oll.libertyfund.org/titles/jevons-the-coal-question

Jones, N. (2018). How to stop data centres from gobbling up the world's electricity. *Nature, 561*(7722), 163–166. DOI: 10.1038/d41586-018-06610-y PMID: 30209383

Kaplan, J., McCandlish, S., Henighan, T., Brown, T. B., Chess, B., & Child, R.. (2020). Scaling laws for neural language models. *arXiv*. https://arxiv.org/abs/2001.08361

Kapp, K. W. (1950). *The Social Costs of Private Enterprise*. Harvard University Press.

Kim, M.-K., Yoo, T.-A., & Chung, J.-B. (2026). Toward sustainable Generative AI: A scoping review of carbon footprint and environmental impacts across training and inference stages. *IEEE Access : Practical Innovations, Open Solutions, 14*, 18881–18896. Advance online publication. DOI: 10.1109/ACCESS.2026.3659894

Krippner, G. R. (2005). The financialization of the American economy. *Socio-economic Review, 3*(2), 173–208. DOI: 10.1093/SER/mwi008

Li, P., Yang, Y., Islam, S. M. R., & Ren, S. (2023). Making AI less "thirsty": Uncovering and addressing the secret water footprint of AI models. *arXiv*. https://doi.org//arXiv.2304.03271DOI: 10.48550

Luccioni, S., Gamazaychikov, B., Alves da Costa, T., & Strubell, E. (2025). Misinformation by omission: The need for more environmental transparency in AI. *arXiv*. https://doi.org//arXiv.2506.15572DOI: 10.48550

Meadows, D. H. (2008). *Thinking in Systems: A Primer*. Chelsea Green Publishing.

Negroponte, N. (1995). *Being Digital*. Knopf.

Packard, V. (1960). *The Waste Makers*. David McKay Company.

Patterson, D., Gonzalez, J., Le, Q., Liang, C., Munguia, L., & Rothchild, D.. (2021). Carbon emissions and large neural network training. *arXiv*. https://arxiv.org/abs/2104.10350

Perez, C. (2002). *Technological Revolutions and Financial Capital: The Dynamics of Bubbles and Golden Ages*. Edward Elgar. DOI: 10.4337/9781781005323

Plantin, J.-C., Lagoze, C., Edwards, P. N., & Sandvig, C. (2018). Infrastructure studies meet platform studies in the age of Google and Facebook. *New Media & Society, 20*(1), 293–310. DOI: 10.1177/1461444816661553

Polimeni, J. M., Mayumi, K., Giampietro, M., & Alcott, B. (2008). *The Jevons Paradox and the Myth of Resource Efficiency Improvements*. Earthscan.

Rockström, J., Steffen, W., Noone, K., Persson, Å., Chapin, F. S.III, Lambin, E., Lenton, T. M., Scheffer, M., Folke, C., Schellnhuber, H. J., Nykvist, B., de Wit, C. A., Hughes, T., van der Leeuw, S., Rodhe, H., Sörlin, S., Snyder, P. K., Costanza, R., Svedin, U., & Foley, J. A. (2009). A safe operating space for humanity. *Nature*, *461*(7263), 472–475. DOI: 10.1038/461472a PMID: 19779433

Rolnick, D., Donti, P. L., Kaack, L. H., Kochanski, K., Lacoste, A., Sankaran, K., ... (2019). Tackling climate change with machine learning. *arXiv*. https://doi.org//arXiv.1906.05433DOI: 10.48550

Sastry, G., Heim, L., Belfield, H., Anderljung, M., Brundage, M., Hazell, J., ... (2024). Computing power and the governance of artificial intelligence. *arXiv*. https://doi.org//arXiv.2402.08797DOI: 10.48550

Schneider, I., Xu, H., Benecke, S., Patterson, D., Huang, K., Ranganathan, P., & Elsworth, C. (2025). Life-cycle emissions of AI hardware: A cradle-to-grave approach and generational trends. *arXiv*. https://doi.org//arXiv.2502.01671DOI: 10.48550

Schwartz, R., Dodge, J., Smith, N. A., & Etzioni, O. (2020). Green AI. *Communications of the ACM*, *63*(12), 54–63. DOI: 10.1145/3381831

Stavins, R. N. (2003). Experience with market-based environmental policy instruments. In Mäler, K.-G., & Vincent, J. R. (Eds.), *Handbook of Environmental Economics* (pp. 355–435). Elsevier.

Strubell, E., Ganesh, A., & McCallum, A. (2019). Energy and policy considerations for deep learning in NLP. *Proceedings of the 57th Annual Meeting of the Association for Computational Linguistics*, 3645–3650. https://doi.org/DOI: 10.18653/v1/P19-1355

Chapter 4
Selling the Green Dream: How AI's Energy Appetite Shapes Sustainable Brand Narratives

Rhytheema Dulloo
http://orcid.org/0000-0002-8874-1270
Lovely Professional University, India

ABSTRACT

Artificial intelligence has become central to modern marketing, enabling brands to position themselves as innovative and environmentally responsible. However, this positioning poses a fundamental tension: the infrastructural demands of AI, particularly the energy-intensive operations of large-scale data centres, contribute meaningfully to global greenhouse gas emissions. Despite this, brands routinely invoke AI as a symbol of sustainability, raising urgent questions about the integrity of such claims. This study examines how the environmental costs associated with AI are misrepresented within the sustainability narratives promoted by contemporary brands, a practice this research identifies and defines as 'AI greenwashing.' Using Keller's CBBE framework and green marketing theory, this interpretivist research builds a layered understanding of the phenomenon. The thematic patterns shed light on how brands construct narratives that separate technological advancement from ecological responsibility. This research offers timely insights for responsible AI-driven marketing.

1. INTRODUCTION

The convergence of artificial intelligence (AI) and sustainability represents one of the most compelling, yet contested, intersections in contemporary marketing scholarship. AI has become a branding asset for corporate sustainability, companies

DOI: 10.4018/979-8-3373-8490-0.ch004

no longer just use it to cut costs; they showcase it as proof of environmental conscience, blurring the line between tech investment and green commitment. Phrases such as 'AI-powered sustainability,' 'intelligent resource optimisation,' and 'smart green operations' permeate corporate communications, annual sustainability reports, and digital marketing campaigns. Framing AI growth as environmentally conscious has proven to be a powerful move, one that shapes buying decisions and builds stakeholder confidence with striking efficiency (Dulloo et al., 2025a). Emerging scholarship has begun to formalise this dynamic, situating AI-powered communication within a strategic framework that links technological innovation directly to brand perception outcomes and sustainable marketing positioning, establishing that the communicative deployment of AI is not merely a tactical choice but a structural determinant of green brand identity (Dulloo et al., 2025b). Underneath that appealing surface, though, sits a contradiction significant enough to warrant far more scrutiny than it currently receives.

AI systems, particularly large language models (LLMs), generative AI platforms, and the expansive data centre infrastructure that supports them, are among the most energy-intensive technological systems in existence. Training a single LLM can emit carbon dioxide equivalent to the lifetime emissions of several petrol-powered automobiles (Strubell et al., 2019). At the global level, data centres processing AI workloads are estimated to account for roughly one to two per cent of worldwide electricity consumption, a share the IEA (2023) projects will climb substantially as uptake continues to expand. The irony is stark: the very technology brands invoke as a hallmark of their environmental credentials is simultaneously a significant contributor to the ecological crisis those brands claim to address. This study designates this phenomenon as 'AI greenwashing', the strategic deployment of AI-related rhetoric to construct or reinforce sustainable brand identities while the energy costs and emissions tied to that AI infrastructure remain unacknowledged or strategically obscured from public view.

Figure 1. The invisible environmental footprint of AI in sustainable branding

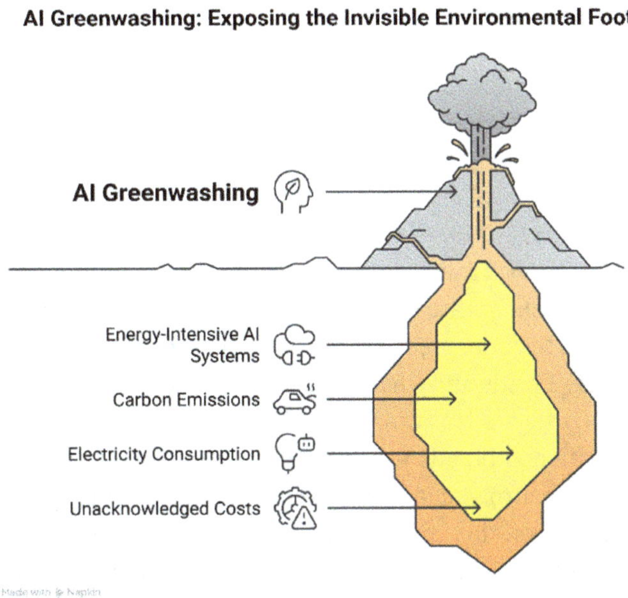

Greenwashing, broadly understood as the misleading communication of environmental claims, is not a new concern. Ever since the term entered academic discourse in the 1980s, researchers have traced its evolution across diverse industries and communication contexts (Lyon & Maxwell, 2011). However, the integration of AI into sustainable brand communication creates a qualitatively distinct challenge. Unlike traditional greenwashing, which typically involves exaggerated or unsubstantiated claims about product ingredients, manufacturing processes, or carbon offsets, AI greenwashing exploits the epistemic complexity of algorithmic systems. The inner workings of these technologies are rarely visible to the public, and the specialized knowledge required to interrogate claims about AI's environmental footprint remains out of reach for most consumers and even many practitioners within the marketing field. This asymmetry of understanding creates fertile ground for deliberate ambiguity, allowing organizations to frame AI-driven initiatives in sustainability terms that are difficult to meaningfully challenge or verify.

The significance of this inquiry is amplified by the regulatory landscape. In the European Union, the Green Claims Directive and the Corporate Sustainability Reporting Directive (CSRD) impose stringent obligations on firms to substantiate environmental claims with verifiable evidence (European Commission, 2023). In the United Kingdom, the Competition and Markets Authority (CMA) has issued explicit guidance on the substantiation of green marketing claims (CMA, 2021). In India, the Securities and Exchange Board of India (SEBI) has introduced Business Responsibility and Sustainability Reporting (BRSR) frameworks that require listed companies to disclose their environmental impact comprehensively (SEBI, 2021). Against this backdrop, brands that construct AI-enabled green narratives without corresponding transparency about energy consumption face significant reputational, legal, and ethical risks.

The theoretical framing of this study draws upon two interlocking scholarly traditions. First, Keller's (1993, 2001) Customer-Based Brand Equity (CBBE) model provides an established framework for understanding how brand associations, including sustainability claims are encoded in consumer memory and shape brand perception and loyalty. Second, green marketing theory, particularly the conceptualisations advanced by Peattie and Charter (2003), and more recently by Dangelico and Vocalelli (2017), offers a lens through which the communicative strategies of environmentally-positioned brands may be critically evaluated. By extending the CBBE model through green marketing theory and locating it within a critical discourse analysis (CDA) framework, this study offers a theoretically integrated perspective on how AI greenwashing is constructed and consumed in brand communication.

This research makes several distinct contributions to the literature. Conceptually, it introduces and theorises 'AI greenwashing' as a hitherto underexplored dimension of sustainable brand communication, distinguishing it from conventional greenwashing on the basis of its epistemic opacity and technological complexity. Empirically, it provides original qualitative evidence of the discursive mechanisms through which brands strategically decouple AI adoption from its ecological costs. Practically, it offers actionable guidance for marketing professionals seeking to build authentic AI-enabled sustainability narratives, and for regulators and standards bodies engaged in developing more robust frameworks for AI-related environmental disclosure. In so doing, it responds to calls in the marketing literature for deeper engagement with the ethical and ecological dimensions of AI-driven marketing practice (Rust, 2020).

2. REVIEW OF LITERATURE

2.1 The Evolution of Greenwashing in Marketing Communication

Greenwashing attracted scholarly attention following Terrence Holt's coinage of the term in 1986. Early research focused on product-level claims in consumer goods, establishing greenwashing as a manifestation of information asymmetry wherein firms exploit consumers' limited capacity to verify environmental claims (Carlson et al., 1993). Delmas and Burbano's (2011) typological framework distinguished between deep green firms, brownwashers, greenwashers, and pale green firms, while Lyon and Maxwell (2011) further distinguished between 'selective disclosure' and 'blue-sky' greenwashing, both directly relevant to AI greenwashing, where omission of energy consumption data is a characteristic strategy.

Subsequent scholarship demonstrated that greenwashing is most effective in low-scrutiny environments where consumers lack resources to verify claims (Liao and Li, 2026), and that nature-evoking visual imagery significantly enhances brand greenness perceptions independently of actual environmental performance (Parguel et al., 2011), a finding with direct implications for AI-themed sustainability communications.

2.2 AI in Marketing: Capabilities, Adoption, and Discursive Positioning

AI's marketing applications: personalisation, predictive analytics, programmatic advertising, and generative content have driven rapid adoption across the marketing function, with global expenditure projected to exceed USD 107 billion by 2028 (Statista, 2024). At the enterprise level, the strategic deployment of AI across diverse organisational environments intensifies pressure on brands to align technological capability with credible sustainability positioning, as AI-driven agility becomes both a competitive necessity and a reputational liability when its ecological dimensions are obscured (Dulloo, 2026a). Beyond functional utility, AI has acquired significant discursive power through what scholars term 'AI imaginaries', socially shared conceptions positioning AI as inherently progressive, efficient, and future-oriented, qualities that map readily onto green brand values (Zhong et al., 2025).

The discursive mobilisation of AI in sustainability marketing rests on what Brevini (2021) termed the 'myth of digital immateriality', the cultural assumption that digital technologies are ecologically neutral. This myth provides the scaffolding for AI greenwashing, compounded by the structural invisibility of AI's data extraction infrastructure in corporate communications (Zuboff, 2019) and the near-absence of environmental sustainability from AI ethics governance frameworks (Jobin et al., 2019).

2.3 The Environmental Footprint of Artificial Intelligence

Strubell et al.'s (2019) landmark study calculated that training a large neural network could generate approximately 626,000 pounds of CO_2 equivalent, roughly five times the lifetime carbon emissions of the average American car. Patterson et al. (2021) extended this analysis to distinguish training, inference, and data storage costs, demonstrating that aggregate AI environmental impact continues to grow despite algorithmic efficiency improvements. Crawford's (2021) Atlas of AI further encompassed rare earth extraction, data centre water consumption, and energy infrastructure, dimensions systematically absent from corporate sustainability narratives.

Data centres globally withdraw an estimated 1.7 billion cubic metres of water annually (Jones, 2018), further complicating the sustainability claims of brands deploying AI. These environmental dimensions are not confined to large-scale industrial applications: research applying the Technology Acceptance Model to voice-based AI systems demonstrates that sustainable adoption and continued consumer use are substantially modulated by perceived environmental impact, indicating that even consumer-facing AI products carry ecological trade-offs that seldom surface in brand-facing communications (Dulloo, 2026b).

2.4 Customer-Based Brand Equity and Green Brand Associations

Keller's (1993) CBBE framework grounds brand equity in the differential effect of brand knowledge comprising brand awareness and brand image on consumer response. Chen (2010) extended this to green brand equity, identifying green brand image, green satisfaction, green trust, and green brand associations as principal drivers, with confirmed positive relationships to willingness to pay premium prices (Hartmann & Apaolaza-Ibáñez, 2012), brand loyalty (Konuk et al., 2015), and advocacy (Kumar & Ghodeswar, 2015).

The vulnerability of green brand equity to greenwashing revelation is well-documented. Isac et al. (2025) showed that perceived greenwashing significantly reduces brand trust and purchase intentions. Crucially, negative information about environmental performance exerts disproportionately greater effects on trust than equivalent positive information (Yoon et al., 2006), an asymmetry compounded in AI contexts by the opacity of algorithmic systems.

2.5 Consumer Responses to Sustainability Claims and Greenwashing

The consumer-side literature on sustainability claims and greenwashing has produced a sophisticated but heterogeneous body of findings. Empirical research has consistently demonstrated that consumer awareness of greenwashing activates scepticism not only towards the implicated brand but also towards environmental claims more broadly a generalisation effect with significant market implications (Pham & Huynh, 2025). Consumer knowledge and environmental involvement moderate these responses. Newell et al. (1998) demonstrated that consumers with higher environmental knowledge are more effective at identifying misleading environmental claims, while Stadlthanner et al. (2022) showed that consumers with stronger environmental values exhibit greater scepticism toward corporate sustainability communications generally.

In the AI context, technical literacy constitutes an additional knowledge dimension that modulates consumers' capacity to evaluate AI-related sustainability claims. Kim et al. (2022) found that AI literacy, encompassing understanding of how AI systems function, their capabilities, and their limitations, significantly influences consumer responses to AI-powered services, though its specific relationship to AI greenwashing responses has not yet been examined. Extending this inquiry into the domain of AI-enabled sustainable marketing, Dulloo et al. (2026c) apply a Theory of Planned Behaviour framework to demonstrate that AI-driven green brand narratives in the Metaverse significantly shape consumer purchase intentions through attitude, subjective norms, and perceived behavioural control, underscoring the behavioural stakes of AI greenwashing in emergent digital environments. Related research on sustainable consumer behaviour in aviation contexts identifies an integrated theoretical model in which environmental values, technology acceptance, and contextual travel factors interact in shaping responses to sustainability communications delivered through technology-mediated service channels (Dulloo et al., 2026d), providing cross-sectoral evidence for the moderating dynamics central to understanding AI greenwashing reception.

2.6 Research Gap

The foregoing review reveals a substantial body of scholarship on greenwashing communication, AI in marketing, the environmental footprint of AI, green brand equity, and consumer responses to sustainability claims. However, critical examination of the literature reveals several significant lacunae that the present study is positioned to address.

(1) What discursive strategies do brands employ to construct AI-enabled sustainable brand narratives? (2) How do these strategies frame or obscure AI's energy footprint? (3) How do marketing professionals navigate the AI-capability/ecological-responsibility tension? (4) How do consumers decode AI-inflected green brand narratives? (5) What are the implications for brand equity, regulatory governance, and ethical marketing practice?

3. CONCEPTUAL MODEL AND THEORETICAL FRAMEWORK

3.1 Overview of the Integrated Framework

This study adopts an integrated theoretical framework that brings together three interlocking conceptual traditions: Keller's Customer-Based Brand Equity (CBBE) model (1993, 2001), green marketing theory as synthesised by Peattie and Charter (2003) and extended by Dangelico and Vocalelli (2017), and the critical discourse framework of Fairclough (1992, 2003). These three traditions are are brought into productive theoretical dialogue to enable a multi-level analysis of AI greenwashing spanning the production of brand communications (discursive construction), the brand equity consequences of AI sustainability claims (brand management), and the consumer-side reception and response to such claims (consumer behaviour). Figure 2 presents the integrated conceptual model.

Figure 2. Conceptual framework integrating CBBE, green marketing theory, and critical discourse analysis

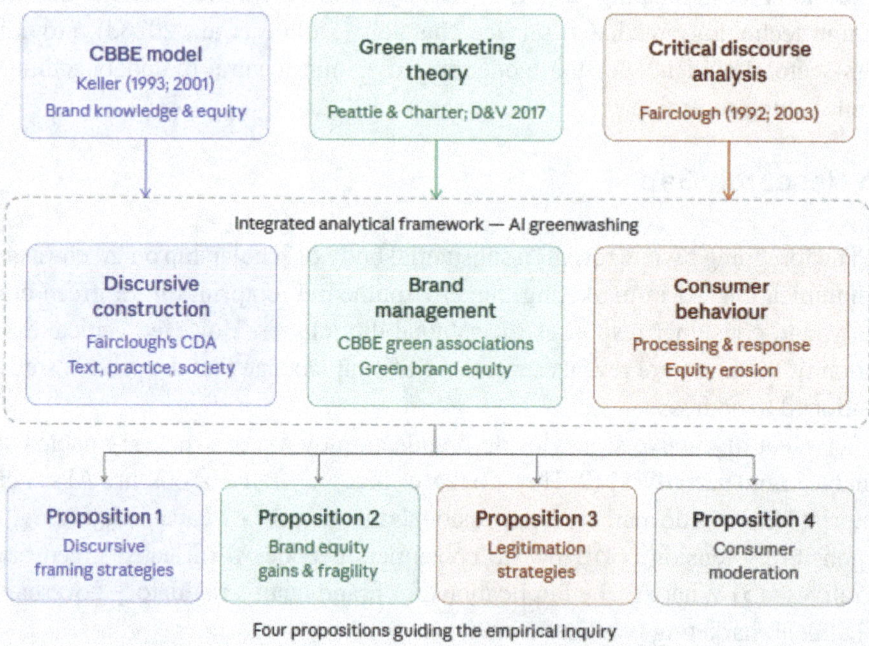

The rationale for this integration is grounded in the multi-dimensional nature of the AI greenwashing phenomenon. AI greenwashing is simultaneously a communicative practice (the production and dissemination of strategic discourse), a brand management strategy (the construction and management of green brand associations), and a consumer behaviour phenomenon (the processing and response to AI-inflected sustainability claims). No single theoretical tradition is sufficient to capture all three dimensions; the integrated framework enables a theoretically coherent account of the phenomenon across its full complexity.

3.2 Keller's Customer-Based Brand Equity Model

Keller's (1993) CBBE model conceptualises brand equity from the consumer's perspective, grounding it in the psychological effects of brand knowledge on consumer response to marketing activities. Brand knowledge is decomposed into two core components: brand awareness and brand image. Within this architecture, brand associations are further classified by their type, their favourability, strength, and uniqueness.

In the context of AI greenwashing, the CBBE framework enables analysis of how AI-related sustainability claims are encoded as brand associations. Specifically, the framework predicts that brands deploying AI as a sustainability signifier seek to establish: (a) strong awareness of their AI capabilities as environmentally relevant; (b) favourable associations between AI adoption and ecological responsibility; (c) unique positioning relative to competitors on the dimension of AI-enabled sustainability. These association-building strategies constitute the brand management dimension of AI greenwashing.

3.3 Green Marketing Theory

Green marketing theory encompasses a body of scholarship concerned with the marketing of environmentally responsible products, services, and practices, and with the management of environmental dimensions of the marketing mix (Peattie & Charter, 2003). Within this tradition, green brand management, the strategic deployment of environmental attributes to create brand differentiation and equity, has been a central research focus (Dangelico & Vocalelli, 2017). Chen's (2010) elaboration of green brand equity identifies four principal constructs: green brand image, green satisfaction, green trust, and green brand equity. This four-construct model provides the operationalisation framework for the green brand equity dimension of the integrated conceptual model.

Of particular relevance to AI greenwashing is the literature on the conditions under which green marketing claims achieve credibility and the mechanisms through which they are contested. Research has consistently identified transparency, specificity, and third-party verification as critical credibility-enhancing factors (Hartmann & Apaolaza-Ibáñez, 2012). Conversely, vagueness, future-orientation, and selectivity in environmental disclosure are identified as characteristic greenwashing markers. Applied to AI sustainability communications, this framework predicts that claims that specify concrete energy consumption metrics, reference independently verified data, and acknowledge trade-offs will achieve greater credibility than claims that rely on general appeals to AI's efficiency or innovation capacity.

3.4 Fairclough's Critical Discourse Framework

Fairclough's (1992, 2003) three-dimensional model of discourse provides the analytical architecture for the discursive dimension of the integrated framework. The model conceptualises discourse at three simultaneous levels: text, discursive practice and social practice. In the context of AI greenwashing, the textual level of analysis attends to the specific linguistic resources through which AI's environmental credentials are constructed including the lexical fields of innovation, efficiency, intelligence, and sustainability; the metaphorical framings of AI as a 'green' or 'clean' technology; and the rhetorical devices through which AI energy consumption is rendered invisible or marginal. The discursive practice level examines the institutional contexts shaping AI greenwashing communications including the marketing and corporate communications industries, the advertising ecosystem, and the regulatory frameworks governing environmental claims. The social practice level situates these communications within the political economy of AI development, the global sustainability governance regime, and the broader ideological formation of green capitalism.

3.5 Integration and Propositions

The integrated framework generates four conceptual propositions that organise the empirical inquiry:

Proposition 1: Brands deploy a repertoire of discursive strategies, including selective disclosure, technological complexity framing, and aspirational future-orientation, to construct AI-enabled sustainable brand narratives while marginalising evidence of AI's energy footprint. These strategies constitute AI greenwashing as a distinctive discursive formation.

Proposition 2: AI greenwashing strategies function to establish favourable, strong, and unique green brand associations within the CBBE architecture, temporarily enhancing green brand equity. However, these equity gains are structurally fragile, as they rest on brand associations that are vulnerable to disruption when AI energy consumption data enters consumer discourse.

Proposition 3: Marketing professionals navigate the tension between AI capability and ecological responsibility through a range of legitimation strategies, including the invocation of comparative efficiency gains, the deflection of systemic responsibility, and the deployment of future-oriented commitment narratives. These strategies reflect the institutional pressures and professional norms governing sustainability communications practice.

Proposition 4: Consumer responses to AI-inflected green brand narratives are moderated by AI literacy, environmental involvement, and prior brand trust. Consumers with higher AI literacy and environmental involvement exhibit greater scepticism toward AI sustainability claims and are more susceptible to negative brand equity effects when greenwashing is disclosed.

These four propositions collectively constitute a theoretically integrated account of AI greenwashing that spans the production, distribution, and consumption dimensions of sustainable brand communication. They are operationalised through the multi-method qualitative design described in the following section.

4. RESEARCH METHODOLOGY

4.1 Research Design

A qualitative research design is adopted, consistent with the interpretivist paradigm and the exploratory nature of the research questions. The design integrates three complementary qualitative methods: critical discourse analysis (CDA) of brand communications; semi-structured interviews with marketing professionals; and consumer focus groups. The three methods are analytically integrated: CDA findings inform the question schedules for interviews and focus groups, and vice versa, enabling a dialectical movement between textual analysis and participant perspectives.

4.2 Method 1: Critical Discourse Analysis of Brand Communications

A purposively constructed corpus comprises brand communications from companies operating in sectors with significant AI adoption (technology, energy, financial services, retail, consumer goods) that make environmental claims referencing AI capabilities. The corpus spans five text categories: corporate sustainability reports (2021–2024), advertising materials, website sustainability pages, press releases and executive speeches, and corporate social media drawn from Factiva, LexisNexis, the GRI database, and direct website harvesting. An initial pool of approximately 450 documents was refined to a final analytical corpus of 120 documents representing 45 companies across 12 countries. CDA proceeds through Fairclough's (2003) three-dimensional framework. Atlas.ti supports systematic coding. Inter-rater reliability is assessed via Cohen's kappa on 25% of the corpus, with a threshold of 0.75 adopted as the minimum for proceeding.

4.3 Method 2: Semi-Structured Interviews with Marketing Professionals

Participants are senior marketing professionals, marketing directors, sustainability communications managers, brand strategists, digital marketing leads with direct responsibility for AI-related or sustainability communications. Recruitment combines professional network outreach (Chartered Institute of Marketing; Global Association for Marketing Analytics), LinkedIn targeting, and snowball referrals. A target of 30–35 participants is guided by theoretical saturation (Fusch & Ness, 2015), with all participants required to have a minimum of five years marketing experience and sufficient seniority to speak to strategic communications decisions.

The interview protocol, covering six thematic areas (AI's role in sustainability communications; claim development processes; awareness of AI's environmental costs; management of capability-responsibility tensions; regulatory experience; ethical reflections), was developed through expert review and pilot testing. Interviews last approximately 60–90 minutes, are audio-recorded and professionally transcribed, and are subject to member checking. All participants are anonymised through pseudonymisation.

4.4 Method 3: Consumer Focus Groups

Eight focus groups of six to eight participants each (total 48–64) are stratified by AI literacy (Wang et al., 2022) and environmental involvement (Dunlap et al., 2000), producing a 2×2 design with two groups per cell. Participants are recruited through a market research panel provider; marketing and AI industry professionals are excluded. Groups are conducted in Delhi, Mumbai, and Bengaluru to ensure geographic diversity. Each session runs in three phases: baseline discussion of AI and sustainability understanding (~20 minutes); stimulus material presentation using real brand communications exemplifying CDA-identified strategies, with guided discussion (~40 minutes); and debriefing on factual AI energy consumption data with reflection on brand perception shifts (~20 minutes). Sessions are audio and video recorded, transcribed verbatim, and moderated using a standardised guide.

4.5 Data Analysis

Thematic analysis following Braun and Clarke (2022) is applied to interview and focus group data, proceeding through six phases iteratively with constant comparative refinement. Atlas.ti supports management, coding, and retrieval across all three methods. Integration across methods follows convergent mixed methods logic (Creswell & Plano Clark, 2018): findings are analysed separately, then synthesised through comparative meta-inference, with analytical decisions documented in audit trail memos.

5. FINDINGS AND DATA ANALYSIS

This section presents the integrated findings from the three analytical strands of the study. The findings are organised around the four conceptual propositions articulated in the theoretical framework, with each sub-section presenting convergent and divergent evidence across data sources. Participant quotations are presented verbatim, with identifying details replaced by pseudonyms; textual evidence is cited by document reference codes assigned during corpus management.

Theoretical saturation was achieved after 31 interviews, consistent with expectations for purposive qualitative samples addressing theoretically bounded phenomena. Focus group data reached thematic saturation within seven groups, with the eighth group confirming existing patterns without generating new conceptual material. Inter-rater reliability for the CDA coding scheme reached a Cohen's kappa of 0.81 (substantial agreement), validating the analytical consistency of the textual coding procedure.

5.1 Theme 1: The Architecture of AI Greenwashing-Discursive Strategies Identified

CDA of the 120-document corpus revealed a coherent repertoire of five primary discursive strategies through which brands construct AI-enabled sustainable brand narratives while marginalising or obscuring the attendant energy costs of AI systems. These strategies operate at multiple levels of the Fairclough (2003) framework, textual, discursive practice, and social practice and frequently co-occur within single communications, producing layered effects of environmental legitimacy that resist straightforward consumer scrutiny.

5.1.1 Strategy 1: Efficiency Substitution Framing

The most prevalent discursive strategy, identified in 89% of corpus documents, involves what this analysis terms efficiency substitution framing: the systematic replacement of absolute environmental impact data with relative efficiency metrics. Rather than disclosing the energy consumed by AI systems in absolute terms (kilowatt-hours, tonnes of CO_2), brands communicate comparative efficiency gains, the percentage reduction in energy per transaction, the proportional decrease in waste per unit of output, or the improvement in resource utilisation relative to a pre-AI baseline.

This strategy exploits the rhetorical structure of comparison to construct an impression of environmental improvement while rendering the baseline (and thus the absolute impact) analytically invisible. Textual analysis reveals characteristic lexical patterns including 'up to X% more efficient,' 'reducing energy per output by Y%,' and 'optimised resource deployment' formulations that grammatically foreground improvement while presupposing a reference point that is never specified. In the majority of cases, the baseline against which efficiency gains are measured is either undefined or refers to an outdated and unrepresentative pre-digital comparison, a pattern consistent with Lyon and Maxwell's (2011) 'selective disclosure' typology.

"Our AI-powered logistics platform has reduced per-delivery carbon emissions by 34% since 2019, demonstrating our commitment to sustainable operations at scale." [Document RC-07, Global Retail Corporation, Sustainability Report 2023]

The quotation above exemplifies efficiency substitution framing: a specific and credible-sounding percentage reduction is foregrounded, while the absolute carbon footprint of the AI platform itself including data centre energy consumption, hardware manufacture, and cooling water usage is entirely absent. This pattern of selective metric disclosure was confirmed by interview participants with direct experience of sustainability communications production:

"The brief is always about showing progress, showing improvement. Nobody asks us to put the total energy number in. The efficiency story is cleaner, it's positive, it's what the sustainability team can defend. The total number would just open up questions we can't answer." [Participant MP-04, Marketing Director, Technology Sector]

5.1.2 Strategy 2: Complexity Opacity Legitimation

The second strategy, present in 76% of corpus documents, exploits the epistemic complexity of AI systems to pre-empt scrutiny of environmental claims. Brands deploy highly technical language, references to algorithmic architecture, neural network optimisation, and computational efficiency improvements in ways that signal sophisticated environmental management while simultaneously rendering the claims practically unverifiable by non-specialist audiences.

Fairclough's (2003) concept of 'technologisation of discourse' is apposite here: the colonisation of sustainability communication by algorithmic and computational vocabulary constructs an authority effect that positions the brand as a credible environmental steward by virtue of its technical sophistication, independently of the substantive content of its environmental performance. This authority effect is reinforced by the deployment of data visualisations graphs of algorithm performance, heat maps of resource optimisation that carry a semiotic register of scientific rigour without necessarily conveying verifiable environmental data.

> *"Our proprietary transformer-based optimisation architecture, deployed across a distributed cloud infrastructure with dynamic load balancing and carbon-aware scheduling protocols, enables us to achieve Scope 2 emissions reductions that are fully integrated into our net zero pathway." [Document TC-19, Technology Company, ESG Report 2024]*

Complexity opacity legitimation was identified by marketing professionals as both a deliberate strategy and a consequence of genuine uncertainty about AI's environmental accounting:

> *"Honestly, half the time we don't fully understand the numbers ourselves. The AI team gives us efficiency metrics, we put them in the report. Nobody is deliberately lying, but nobody is asking the hard question about what's happening at the infrastructure level." [Participant MP-11, Sustainability Communications Manager, Financial Services]*

5.1.3 Strategy 3: Temporal Displacement

Temporal displacement the deferral of environmental accountability to a future state was identified in 71% of corpus documents. This strategy involves the displacement of present environmental costs onto a horizon of future improvement, typically framed around commitments to renewable energy procurement, AI efficiency roadmaps, or 'net zero by' target dates. The rhetorical effect is to reframe current environmental costs as transitional necessities on the pathway to a green technological future, borrowing credibility from the aspiration rather than from present performance.

Linguistically, temporal displacement is characterised by high concentrations of future-tense constructions ('will achieve,' 'is committed to,' 'plans to'), conditional formulations ('as our AI systems mature,' 'as renewable capacity expands'), and progressive aspect markers ('is working towards,' 'is developing') that signify ongoing positive trajectory without specifying current performance. These features are consistent with Szabo and Webster's (2021) identification of 'aspirational future-orientation' as a canonical greenwashing discursive mechanism.

5.1.4 Strategy 4: Proxy Naturalisation

Proxy naturalisation, present in 64% of corpus documents. Brands describe their AI systems as 'ecosystems,' 'growing intelligently,' 'naturally adaptive,' and 'as efficient as nature itself' metaphorical constructions that activate the cognitive associations between nature and sustainability established in the consumer psychology literature (Parguel et al., 2011) without making any substantive environmental claim. This strategy is particularly evident in visual communications, AI-themed advertisements that juxtapose server rack imagery with forest canopies, water droplets, and solar panels and in brand identity materials where AI-themed logos incorporate leaf motifs, circular economy symbols, and earth tones.

5.1.5 Strategy 5: Systemic Responsibility Deflection

The fifth strategy, identified in 58% of corpus documents, involves what this analysis terms systemic responsibility deflection: the attribution of AI's energy costs to the broader technological infrastructure rather than to the brand's specific AI deployment. Brands acknowledge the general challenge of AI's energy consumption at the industry level while framing their own AI operations as exceptional cases, citing cloud provider renewable energy commitments, hardware efficiency certifications, and third-party carbon offset programmes as evidence that their AI deployment is environmentally insulated from the sector-level problem.

This strategy operates through a classic deflection logic: the environmental cost is acknowledged in the abstract (demonstrating awareness and apparent transparency) but is simultaneously displaced onto cloud providers, hardware manufacturers, grid operators over whom the brand claims limited direct control. The effect is to preserve the brand's green credibility while distributing culpability diffusely across the supply chain. Interview participants confirmed this as a deliberate communications strategy:

> *"We can honestly say we use a data centre running on 100% renewable energy. That's true. What we don't say — what nobody asks — is how that data centre is actually powered on a Tuesday night in winter when the solar isn't producing. The annual renewable certificate story is much cleaner than the hourly grid reality." [Participant MP-22, Head of Digital Sustainability, Consumer Goods]*

Table 1. Summary of AI greenwashing discursive strategies identified in corpus analysis

Strategy	Prevalence in Corpus	Primary Mechanism	Fairclough Level
Efficiency Substitution Framing	89% (107/120 documents)	Relative metrics replace absolute impact data	Textual / Discursive Practice
Complexity Opacity Legitimation	76% (91/120 documents)	Technical language constructs unverifiable authority	Textual / Social Practice
Temporal Displacement	71% (85/120 documents)	Future commitments displace present accountability	Textual / Discursive Practice
Proxy Naturalisation	64% (77/120 documents)	Nature imagery constructs implicit green associations	Textual
Systemic Responsibility Deflection	58% (70/120 documents)	AI energy costs attributed to supply chain actors	Discursive Practice / Social Practice

5.2 Theme 2: Marketing Professionals' Navigation of the AI-Sustainability Tension

Thematic analysis of interview data revealed three principal orientations through which marketing professionals navigate the tension between AI's capabilities and its ecological costs. These orientations are not mutually exclusive and were frequently articulated in combination by individual participants; they represent ideal-typical positions along a continuum from strategic compliance to ethical discomfort.

Figure 3. Marketing professionals orientation in navigating the AI–sustainability tension

Navigation of AI-Sustainability Tension

Characteristic	Belief	Communication	Justification
Pragmatic Optimism	AI efficiency is a net environmental benefit	Selectively foreground efficiency gains	Overall positive contribution to sustainability
Strategic Compliance	Regulatory compliance is essential	Prepare for greater specificity in claims	Avoid disproportionate penalties for disclosure
Ethical Dissonance	AI greenwashing is reproduced institutionally	Institutional logics shape discursive practices	Individual actors resist, but greenwashing persists

5.2.1 Orientation 1: Pragmatic Optimism

The most prevalent orientation, expressed by 22 of 31 participants, involves what this analysis terms pragmatic optimism: the sincere belief that AI's efficiency gains represent a genuine net environmental benefit, combined with the pragmatic judgement that communicating this benefit selectively foregrounding efficiency gains, backgrounding energy costs is justified by the overall positive contribution of AI to sustainability goals. Participants holding this orientation did not regard their communications as misleading but as appropriately emphasising the most relevant and verifiable dimensions of AI's environmental impact.

> *"Look, the alternative is not running AI and going back to manual processes that are objectively less efficient. We're not hiding anything — we're leading with the story that's true and that consumers can act on. The energy story is complicated and it's improving every year." [Participant MP-08, Brand Strategy Director, Energy Sector]*

Pragmatic optimism reflects a genuine epistemic challenge in the AI sustainability domain: the absence of standardised methodologies for calculating and communicating the full lifecycle environmental cost of AI systems means that brands operate in a measurement vacuum that simultaneously enables and incentivises selective disclosure. Several participants with technical backgrounds acknowledged the limitation of current AI carbon accounting frameworks:

"There is no ISO standard for AI carbon footprint. There is no agreed methodology for allocating shared infrastructure costs to individual AI applications. We are making it up, and so is everyone else. The question is whether we're making it up in good faith." [Participant MP-17, Head of Sustainability, Technology Sector]

5.2.2 Orientation 2: Strategic Compliance

Twelve participants articulated what this study terms strategic compliance: a professionally managed approach to AI sustainability communications in which the primary driver is regulatory and reputational risk management rather than substantive commitment to environmental transparency. These participants demonstrated awareness of the gap between AI sustainability claims and AI energy realities, but framed their communications choices primarily in terms of what is legally defensible and reputationally safe rather than what is environmentally accurate.

"We have legal review everything now. Every claim has to be substantiable. But substantiable doesn't mean complete — it means we can defend what we say. We don't say anything that's wrong. We just don't say everything that's true." [Participant MP-26, Marketing Compliance Lead, Financial Services]

Strategic compliance participants were acutely aware of regulatory developments, particularly the EU Green Claims Directive, and several described active preparations for compliance that would require greater specificity in AI sustainability claims. However, they also expressed concern that regulatory convergence would disproportionately penalise companies that disclose more, creating what one participant termed 'a transparency paradox':

"If I disclose our AI's energy consumption honestly and my competitor doesn't, I look worse even though I'm more honest. The regulation needs to make full disclosure universal, not optional." [Participant MP-29, Director of ESG Communications, Retail]

5.2.3 Orientation 3: Ethical Dissonance

Nine participants, predominantly those with backgrounds in environmental science, sustainability consulting, or NGO communications who had transitioned into corporate marketing roles, expressed what this study terms ethical dissonance: a consciously experienced conflict between their professional responsibilities and their personal ethical judgements regarding the accuracy of AI sustainability communications. These participants described active internal advocacy for greater transparency, often encountering structural resistance from commercial and legal functions.

"I have raised the AI energy question in every sustainability report process for three years. Every year it gets noted and deferred. The commercial team doesn't want it in there. Legal says it creates liability. So I write the report that I'm told to write, and then I go home and I feel uncomfortable about it." [Participant MP-31, Sustainability Communications Specialist, Consumer Goods]

The ethical dissonance orientation points to the institutional mechanisms through which AI greenwashing is reproduced even in the presence of individual actors who recognise and resist it, highlighting the social practice dimension of Fairclough's framework — the ways in which commercial and legal institutional logics shape discursive practices independently of individual ethical commitments.

5.3 Theme 3: Consumer Decoding of AI-Inflected Green Brand Narratives

Focus group data revealed substantial variation in consumer responses to AI-inflected sustainable brand narratives, moderated by the two stratification variables of AI literacy and environmental involvement, consistent with the moderating dynamics anticipated in Proposition 4 of the conceptual framework. The analysis identifies four distinct consumer response patterns, designated Credulity, Critical Engagement, Resigned Acceptance, and Active Scepticism.

Figure 4. Consumer response patterns to AI-inflected sustainable brand narratives

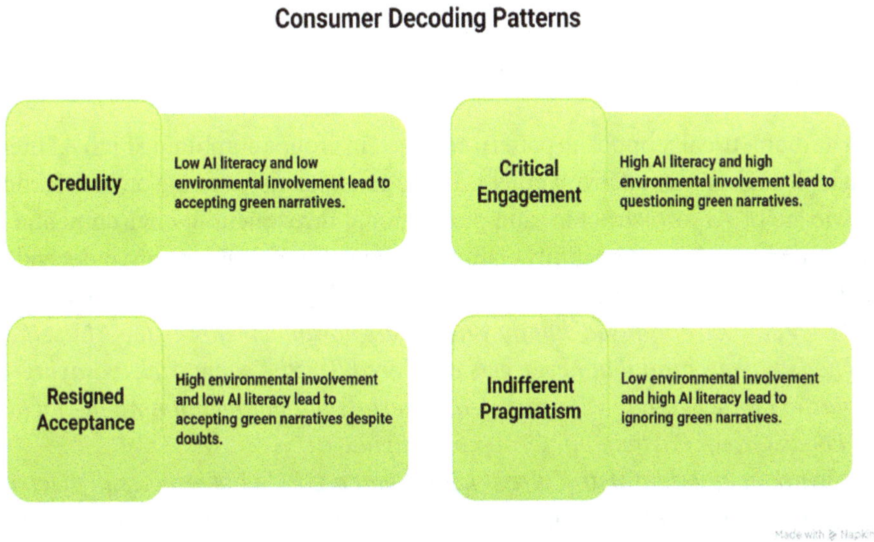

5.3.1 Pattern 1: Credulity (Low AI Literacy / Low Environmental Involvement)

Participants in groups characterised by low AI literacy and low environmental involvement were the most receptive to AI-inflected sustainability claims, exhibiting what this study terms credulity: uncritical acceptance of AI sustainability narratives, driven by the dual heuristics of technological optimism (the assumption that more advanced technology is inherently cleaner) and brand trust. These participants consistently interpreted AI references in sustainability communications as markers of genuine environmental commitment:

"If they're using AI, they've obviously thought about this more carefully. You don't use AI just to waste energy, the whole point is it's smarter." [Focus Group Participant, Group 1, Delhi]

The credulity pattern confirms the effectiveness of complexity opacity legitimation and proxy naturalisation as brand equity strategies among consumers with limited resources for evaluating AI-specific claims. However, the debriefing phase of focus groups in this category revealed the fragility of credulity-based brand associations: when presented with factual data on AI energy consumption, participants in credulity groups exhibited the sharpest brand trust declines, consistent with the asymmetric trust damage predicted by Yoon et al. (2006).

5.3.2 Pattern 2: Critical Engagement (High AI Literacy / High Environmental Involvement)

The diametrically opposite pattern emerged in groups combining high AI literacy with high environmental involvement. These participants, drawing on both technical knowledge of AI infrastructure and motivational investment in environmental outcomes, subjected AI sustainability claims to systematic critical scrutiny, independently identifying many of the discursive strategies catalogued in the CDA analysis:

"They say 'AI-optimised' but they never say how much energy the AI itself uses. That's the question that never gets answered. It's the same trick as saying your plastic bottle is made with recycled materials but not mentioning the factory." [Focus Group Participant, Group 5, Bengaluru]

"I work in cloud infrastructure. I know what a GPU cluster costs in electricity. When I see these claims I think — show me the Scope 2 numbers for your model training. They never do." [Focus Group Participant, Group 6, Bengaluru]

Critical engagement participants demonstrated active scepticism and expressed strong preferences for standardised, third-party verified AI environmental disclosures. This group's responses align with Newell et al.'s (1998) finding that domain knowledge is the primary moderator of greenwashing detection, extended to the AI-specific literacy dimension.

5.3.3 Pattern 3: Resigned Acceptance (High Environmental Involvement / Low AI Literacy)

A more nuanced pattern emerged among participants with high environmental involvement but low AI literacy, individuals who cared deeply about environmental outcomes but lacked the technical resources to evaluate AI-specific claims. This group exhibited what the analysis terms resigned acceptance: a sceptical but ultimately acquiescent stance characterised by expressed doubts about AI sustainability claims combined with an inability to substantiate those doubts through specific technical knowledge.

"I don't really trust these claims but I don't know enough about AI to say why not. It feels like every company is saying AI is green now and it can't all be true, but I can't point to what exactly is wrong." [Focus Group Participant, Group 3, Mumbai]

Resigned acceptance participants were the most responsive to simple, accessible information about AI energy consumption, with clear factual information about data centre water and electricity usage generating significant attitudinal shifts in the debriefing phase. This finding suggests that consumer education about AI's environmental footprint, rather than complex technical disclosure, may be the most effective intervention for this population segment.

5.3.4 Pattern 4: Indifferent Pragmatism (Low Environmental Involvement / High AI Literacy)

The fourth pattern, indifferent pragmatism, characterised participants with high AI literacy but low environmental involvement. These participants possessed the technical knowledge to identify potential gaps in AI sustainability claims but lacked motivational engagement with environmental outcomes, leading to a pragmatic indifference: they neither accepted nor actively resisted AI greenwashing, processing it as a generic marketing convention rather than an ethical concern:

> *"Of course they're spinning it. Everyone spins their sustainability stuff. I just want to know if the product works. The green story is for other people." [Focus Group Participant, Group 7, Delhi]*

These participants were the least affected by the AI energy consumption data presented in the debriefing phase, confirming that AI literacy without environmental salience does not translate into active resistance to greenwashing. This finding has implications for communication strategies aimed at mobilising consumer pressure for greater AI sustainability transparency.

Table 2. Consumer response patterns to AI-inflected green brand narratives

Pattern	AI Literacy	Environmental Involvement	Stance Toward AI Claims	Post-Debriefing Trust Impact
Credulity	Low	Low	Uncritical acceptance; technology optimism heuristic	Sharp decline (asymmetric damage)
Critical Engagement	High	High	Active scrutiny; independent strategy identification	Moderate decline (expectations partly met)
Resigned Acceptance	Low	High	Sceptical but unsubstantiated doubts	Significant shift with simple factual information
Indifferent Pragmatism	High	Low	Disengaged; greenwashing as marketing convention	Minimal impact

5.4 Theme 4: Green Brand Equity Dynamics Under AI Greenwashing Conditions

The integrated analysis across all three data sources enables theorisation of the green brand equity dynamics associated with AI greenwashing, extending Keller's (1993) CBBE model and Chen's (2010) Green Brand Equity framework into the AI sustainability domain. Four dynamics are identified:

First, AI greenwashing strategies temporarily inflate green brand associations among consumers exhibiting the Credulity and Resigned Acceptance patterns, artificially elevating perceived green brand image and green trust. This short-term equity enhancement is structurally fragile because it rests on brand associations that are not corroborated by verifiable environmental performance, rendering them acutely vulnerable to erosion when AI energy consumption data enters public discourse through investigative journalism, regulatory disclosure requirements, or consumer advocacy.

Second, among consumers exhibiting Critical Engagement, AI greenwashing produces negative brand equity effects even without disclosure events: active scepticism generates unfavourable brand associations, reduces green trust, and diminishes purchase intention. This finding extends Nyilasy et al.'s (2014) demonstration that perceived greenwashing reduces brand trust, showing that the mechanism is activated not only by external greenwashing revelations but by consumer-led scrutiny of brand communications, a dynamic of particular significance as AI literacy increases in the consumer population.

Third, the discrepancy between the green brand associations constructed through AI greenwashing and the verifiable environmental performance of AI-adopting brands creates what this study terms an association-performance gap, a structural vulnerability in the brand equity architecture of AI greenwashing firms. As regulatory requirements for AI environmental disclosure tighten and as third-party AI carbon accounting tools become more accessible, this gap is likely to widen, increasing the reputational risk of AI greenwashing strategies.

Fourth, interview data suggest that AI greenwashing erodes internal brand equity, the coherence of brand values as experienced by the employees responsible for sustainability communications. Participants expressing ethical dissonance described the psychological costs of producing communications they recognised as incomplete or misleading, with implications for employee engagement, talent retention, and internal brand culture. This internal dimension of brand equity damage is absent from existing green brand equity models and represents a theoretically significant extension of the CBBE framework.

6. DISCUSSION

The findings of this study invite rich theoretical dialogue with existing scholarship, while also pointing to significant extensions and revisions of established frameworks. This section situates the empirical results within the broader literature, evaluates the four conceptual propositions advanced in Section 4, and elaborates the theoretical, practical, and policy implications of the findings.

6.1 Theoretical Contributions

6.1.1 Extending Greenwashing Theory into the AI Domain

The five discursive strategies identified in this study: efficiency substitution framing, complexity opacity legitimation, temporal displacement, proxy naturalisation, and systemic responsibility deflection, both confirm and extend existing typologies of greenwashing communication. Efficiency substitution framing operationalises Lyon and Maxwell's (2011) 'selective disclosure' concept in an AI-specific communicative context, demonstrating how algorithmic efficiency metrics function as a technologically sophisticated variant of the classical greenwashing move of foregrounding favourable data while suppressing unfavourable information. Temporal displacement reproduces the 'aspirational future-orientation' mechanism documented by Szabo and Webster (2021), inflected through the specific temporal imaginaries of AI development ('as our models mature,' 'as computational efficiency improves').

Complexity opacity legitimation and systemic responsibility deflection represent qualitatively novel contributions to greenwashing theory. Unlike conventional greenwashing strategies that rely on consumer credulity regarding environmental science (the opacity of carbon chemistry, lifecycle analysis, or biodiversity metrics), AI greenwashing uniquely exploits consumers' limited AI literacy, their unfamiliarity with the energy economics of machine learning infrastructure, data centre operations, and model training processes. This creates what this study terms a double opacity: not only are environmental costs obscured by selective disclosure, but the very mechanisms of that disclosure are rendered opaque by technological complexity. This double opacity constitutes a qualitatively distinct mode of greenwashing that the existing typological literature has not anticipated and that warrants dedicated theoretical development.

The concept of systemic responsibility deflection also extends existing theory by highlighting the supply chain dimension of AI greenwashing, the distribution of environmental accountability across cloud providers, hardware manufacturers, and grid operators in ways that prevent any single actor from bearing full responsibility. This mechanism resonates with Buthe and Mattli's (2011) analysis of governance gaps in globalised supply chains but applies it to the specific context of AI infrastructure, where the dematerialised character of cloud computing facilitates particularly effective responsibility diffusion.

6.1.2 Extending the CBBE Model

The findings generate three theoretically significant extensions of Keller's (1993) CBBE model. First, they demonstrate that AI-specific brand associations constitute a distinctive category within the CBBE brand image architecture, one characterised by epistemic inaccessibility (consumers cannot directly verify AI capability claims), temporal orientation (AI associations are frequently future-directed rather than anchored in present performance), and cultural loading (AI carries powerful innovation and progress associations that transfer automatically to affiliated brand attributes including sustainability).

Second, the identification of internal brand equity, the coherence and psychological comfort of brand values as experienced by employees, as a dimension of equity susceptible to greenwashing damage extends the CBBE framework into the organisational domain. Keller's framework is explicitly consumer-focused; the present findings indicate that greenwashing strategies generate internal dissonance among employees with sustainability responsibilities, constituting an equity erosion mechanism that operates independently of consumer perceptions.

Third, the asymmetric trust damage dynamics documented in the focus group data — the finding that Credulity-pattern consumers exhibit sharper brand trust declines than Critical Engagement consumers following AI energy disclosure, refine and extend Yoon et al.'s (2006) asymmetry principle. The present findings suggest that the magnitude of trust asymmetry is itself moderated by the strength of prior AI sustainability associations, such that brands which have more aggressively deployed AI greenwashing strategies face disproportionately severe equity consequences when the association-performance gap is exposed.

6.1.3 Theoretical Propositions: Evaluation

Proposition 1 is strongly supported: the CDA analysis documents a coherent and prevalent repertoire of discursive strategies through which AI's energy footprint is marginalised in sustainable brand communications, confirming that AI greenwashing constitutes a distinctive discursive formation warranting dedicated theoretical treatment. The five strategies identified provide a more granular and empirically grounded account of AI greenwashing mechanics than the proposition anticipated.

Proposition 2 receives qualified support: the focus group data confirm that AI greenwashing temporarily inflates green brand associations among low-literacy and low-involvement consumers (Credulity and Resigned Acceptance patterns), and that these associations are structurally fragile and susceptible to asymmetric damage. However, the proposition's assumption that equity gains are uniform across consumer segments is not supported: Critical Engagement consumers exhibit negative brand equity responses to AI greenwashing even in the absence of disclosure events, suggesting that the equity effects of AI greenwashing are more heterogeneous than the proposition anticipated.

Proposition 3 is robustly supported: the three professional orientations identified, pragmatic optimism, strategic compliance, and ethical dissonance confirm that marketing professionals navigate the AI-sustainability tension through active legitimation strategies, and that these strategies are shaped by the institutional logics of commercial and legal functions rather than solely by individual ethical judgment. The ethical dissonance orientation, in particular, illuminates the internal institutional mechanisms through which AI greenwashing is reproduced even in the presence of individual resistance.

Proposition 4 is fully supported and extended: consumer responses are systematically moderated by AI literacy and environmental involvement, producing the four-pattern typology of Credulity, Critical Engagement, Resigned Acceptance, and Indifferent Pragmatism. The proposition's identification of AI literacy as a moderator is confirmed, and the interaction between AI literacy and environmental involvement produces more nuanced moderating dynamics than the proposition's bivariate formulation anticipated.

6.2 Practical Implications

The findings carry several urgent practical implications for marketing professionals engaged in AI-enabled sustainability communications. AI greenwashing poses a significant brand risk that outweighs short-term gains: as AI literacy and carbon accounting tools become mainstream, the association-performance gap will be exposed with increasing reputational severity. Organizations should instead build authentic AI sustainability narratives grounded in third-party verified energy disclosure, which generates durable brand equity and addresses the talent/engagement risk faced by sustainability professionals who experience ethical dissonance. Consumer segmentation is critical: Credulity-pattern consumers—vulnerable to trust collapse when greenwashing surfaces, they require durable, verifiable claims over impressionistic imagery, while Resigned Acceptance consumers are particularly responsive to simple, accessible factual information about AI energy consumption;

providing clear comparative energy data may build more authentic green trust than technical efficiency metrics.

Existing green marketing frameworks, including the EU Green Claims Directive, the UK Green Claims Code, and India's BRSR framework, are not adequately designed to address AI-specific environmental disclosure challenges. The absence of standardised measures for assessing AI lifecycle environmental impacts creates opportunities for selective disclosure and greenwashing. Therefore, regulatory and standards bodies should develop AI-specific environmental accounting standards and mandate disclosure requirements. In India, incorporating AI-related environmental guidance within the BRSR framework would help address a critical governance gap and improve transparency in rapidly expanding AI adoption contexts.

6.3 Limitations of the Study

Several limitations should be acknowledged when interpreting the findings. Although purposive sampling and theoretical saturation strengthen analytical richness, the results should not be interpreted as universally representative of all industries, markets, or consumer populations. Interview findings rely on self-reported professional perspectives and may reflect social desirability and selective disclosure biases. Rapid evolution in AI technologies, regulations, and sustainability practices may affect the long-term applicability of findings.

7. CONCLUSION

7.1 Summary of Principal Findings

This study has examined the phenomenon of AI greenwashing, the strategic deployment of AI-related rhetoric to construct or reinforce sustainable brand identities while obscuring the environmental costs of AI's energy footprint. Employing an integrated theoretical framework combining Keller's (1993) CBBE model, green marketing theory, and Fairclough's (2003) critical discourse framework, and drawing on three complementary qualitative methods the study has generated several substantive contributions to knowledge.

At the empirical level, the study has documented a coherent repertoire of five primary AI greenwashing discursive strategies: efficiency substitution framing, complexity opacity legitimation, temporal displacement, proxy naturalisation, and systemic responsibility deflection that collectively constitute AI greenwashing as a distinctive discursive formation. These strategies are highly prevalent in corporate AI sustainability communications, with the most common (efficiency substitution framing) present in 89% of the analysed corpus. Their co-occurrence and mutual reinforcement produce layered communicative effects of environmental legitimacy that are, under current regulatory and literacy conditions, largely effective in constructing favourable green brand associations among low-literacy and low-involvement consumer segments.

At the theoretical level, the study has made three principal contributions. First, it has introduced and theorised AI greenwashing as a qualitatively distinct variant of greenwashing communication, distinguished from conventional greenwashing by the double opacity mechanism, the simultaneous obscuring of environmental costs and the obscuring of the mechanisms of obscurement through technological complexity. Second, it has extended the CBBE model by identifying AI-specific brand associations as a distinctive category within the brand image architecture, by introducing internal brand equity as a dimension susceptible to greenwashing damage, and by demonstrating that the magnitude of asymmetric trust damage following greenwashing disclosure is itself moderated by the strength of prior AI sustainability associations. Third, it has generated a theoretically grounded consumer response typology, Credulity, Critical Engagement, Resigned Acceptance, and Indifferent Pragmatism that provides an empirically validated framework for understanding differential consumer responses to AI-inflected green brand narratives as a function of AI literacy and environmental involvement.

7.2 Future Research Directions

Several productive avenues for future research are suggested by the present findings. First, quantitative operationalisation of the AI greenwashing strategy taxonomy and the consumer response typology would enable large-scale experimental and survey research into the causal mechanisms and boundary conditions of the brand equity dynamics identified in this study. Second, longitudinal research tracking the brand equity consequences of AI greenwashing disclosure events when investigative journalism, regulatory action, or consumer advocacy exposes the gap between AI sustainability claims and energy consumption realities would provide direct empirical evidence of the trust asymmetry dynamics theorised in this study. Third, cross-cultural replication of the consumer focus group findings would establish the generalisability of the consumer response typology across different national markets, regulatory environments, and media consumption contexts. Fourth, research into the effectiveness of specific AI environmental disclosure interventions including standardised AI energy labels, mandatory AI carbon disclosures in sustainability reports, and consumer-facing AI energy comparison tools would provide evidence-based guidance for regulatory design and corporate communications practice. Finally, the internal brand equity dimension identified in this study the ethical

REFERENCES

Braun, V., & Clarke, V. (2022). Thematic analysis: A practical guide. *Sage (Atlanta, Ga.)*.

Brevini, B. (2021). *Is AI good for the planet?* Polity Press.

. Büthe, T., & Mattli, W. (2011). The new global rulers: The privatization of regulation in the world economy.

Carlson, L., Grove, S. J., & Kangun, N. (1993). A content analysis of environmental advertising claims: A matrix method approach. *Journal of Advertising*, *22*(3), 27–39. DOI: 10.1080/00913367.1993.10673409

Chen, Y.-S. (2010). The drivers of green brand equity: Green brand image, green satisfaction, and green trust. *Journal of Business Ethics*, *93*(2), 307–319. DOI: 10.1007/s10551-009-0223-9

Competition and Markets Authority. (2021). *Green claims code: Making environmental claims*. CMA.

Crawford, K. (2021). *Atlas of AI: Power, politics, and the planetary costs of artificial intelligence*. Yale University Press.

Creswell, J. W., & Plano Clark, V. L. (2018). *Designing and conducting mixed methods research* (3rd ed.). SAGE.

Dangelico, R. M., & Vocalelli, D. (2017). 'Green Marketing': An analysis of definitions, strategy steps, and tools through a systematic review of the literature. *Journal of Cleaner Production*, *165*, 1263–1279. DOI: 10.1016/j.jclepro.2017.07.184

Delmas, M. A., & Burbano, V. C. (2011). The drivers of greenwashing. *California Management Review*, *54*(1), 64–87. DOI: 10.1525/cmr.2011.54.1.64

Dulloo, R. (2026a). Harnessing artificial intelligence for global portfolio management: Enhancing agile practices in diverse enterprise environments. In *Harnessing artificial intelligence to ensure diverse global teams* (pp. 191–228). Academic Press., DOI: 10.1016/B978-0-443-34135-9.00015-5

Dulloo, R. (2026b). Sustainable adoption and continued use of voice-based AI: Examining the environmental impact through the lens of the Technology Acceptance Model. In *Congress on Control, Robotics, and Mechatronics* (Lecture Notes in Networks and Systems, Vol. 1850, pp. 45–57). Springer. DOI: 10.1007/978-981-96-9771-7_4

Dulloo, R., Appolloni, A., & Mathiyazhagan, K. (2025a). *AI-Powered Communication: Strategic Framework for Technological Innovation, Brand Perception, and Sustainable Marketing. Strategic Blueprints for AI-Driven Marketing in the Digital Era*. IGI Global Scientific Publishing., DOI: 10.4018/979-8-3373-3897-2.ch001

Dulloo, R., Biradar, K., & Srijaa, M. (2026d). Understanding Consumer Behavior in Sustainable Aviation: An Integrated Theoretical Model Examining Environmental Values, Technology Acceptance, and Travel Context. In *World Conference on Information Systems for Business Management* (Lecture Notes in Networks and Systems,vol 1744, pp. 375-385). Springer DOI: 10.1007/978-3-032-12996-3_36

Dulloo, R., Silic, M., & Appolloni, A. (2025b). *Key enablers of Marketing 6.0 in the metaverse: how technological, human-centric, sustainable, and seamless approaches drive performance outcomes. Leveraging AI-Powered Marketing in the Experience-Driven Economy*. IGI Global Scientific Publishing. DOI: 10.4018/979-8-3693-9561-5.ch016

Dulloo, R., Singh, A., & Kiran, V. (2026). AI-driven sustainable marketing in the metaverse and consumer purchase intentions: A TPB analysis. In *World Conference on Information Systems for Business Management* (Lecture Notes in Networks and Systems, Vol. 1742, pp. 376–386). Springer DOI: 10.1007/978-3-032-12987-1_35

Dunlap, R. E., Van Liere, K. D., Mertig, A. G., & Jones, R. E. (2000). Measuring endorsement of the New Ecological Paradigm: A revised NEP scale. *The Journal of Social Issues*, *56*(3), 425–442. DOI: 10.1111/0022-4537.00176

European Commission. (2023). *Proposal for a Directive on substantiation and communication of explicit environmental claims*. European Commission.

Fairclough, N. (1992). *Discourse and social change*. Polity Press.

Fairclough, N. (2003). *Analysing discourse: Textual analysis for social research*. Routledge. DOI: 10.4324/9780203697078

Fusch, P. I., & Ness, L. R. (2015). Are we there yet? Data saturation in qualitative research. *The Qualitative Report*, *20*(9), 1408–1416. DOI: 10.46743/2160-3715/2015.2281

Hartmann, P., & Apaolaza-Ibáñez, V. (2012). Consumer attitude and purchase intention toward green energy brands: The roles of psychological benefits and environmental concern. *Journal of Business Research*, *65*(9), 1254–1263. DOI: 10.1016/j.jbusres.2011.11.001

IEA. (2023). *Electricity 2024: Analysis and forecast to 2026*. International Energy Agency.

Isac, N., Javed, A., Radulescu, M., Cismasu, I. D. L., Yousaf, Z., & Serbu, R. S. (2025). Is greenwashing impacting on green brand trust and purchase intentions? Mediating role of environmental knowledge. *Environment, Development and Sustainability*, *27*(9), 21329–21346. DOI: 10.1007/s10668-023-04352-0

Jobin, A., Ienca, M., & Vayena, E. (2019). The global landscape of AI ethics guidelines. *Nature Machine Intelligence*, *1*(9), 389–399. DOI: 10.1038/s42256-019-0088-2

Jones, N. (2018). How to stop data centres from gobbling up the world's electricity. *Nature*, *561*(7722), 163–166. DOI: 10.1038/d41586-018-06610-y PMID: 30209383

Keller, K. L. (1993). Conceptualizing, measuring, and managing customer-based brand equity. *Journal of Marketing*, *57*(1), 1–22. DOI: 10.1177/002224299305700101

Keller, K. L. (2001). Building customer-based brand equity: A blueprint for creating strong brands. *Marketing Management*, *10*(2), 15–19.

Kim, J., Kim, J., & Yang, H. (2022). Loneliness and the use of social media to follow celebrities: A moderating role of social media literacy. *The Social Science Journal*, *59*(1), 116–130.

Konuk, F. A., Rahman, S. U., & Salo, J. (2015). Antecedents of green behavioral intentions: A cross-country study of Turkey, Finland and Pakistan. *International Journal of Consumer Studies*, *39*(6), 586–596. DOI: 10.1111/ijcs.12209

Kumar, P., & Ghodeswar, B. M. (2015). Factors affecting consumers' green product purchase decisions. *Marketing Intelligence & Planning*, *33*(3), 330–347. DOI: 10.1108/MIP-03-2014-0068

. Liao, L., & Li, Y. (2026). From compliance to capability: How environmental certification drives corporate ESG transformation. Energy & Environment, 0958305X261444556.

Lyon, T. P., & Maxwell, J. W. (2011). Greenwash: Corporate environmental disclosure under threat of audit. *Journal of Economics & Management Strategy*, *20*(1), 3–41. DOI: 10.1111/j.1530-9134.2010.00282.x

Newell, S. J., Goldsmith, R. E., & Banzhaf, E. J. (1998). The effect of misleading environmental claims on consumer perceptions of advertisement. *Journal of Marketing Theory and Practice*, *6*(2), 48–60. DOI: 10.1080/10696679.1998.11501795

Nyilasy, G., Gangadharbatla, H., & Paladino, A. (2014). Perceived greenwashing: The interactive effects of green advertising and corporate environmental performance on consumer reactions. *Journal of Business Ethics*, *125*(4), 693–707. DOI: 10.1007/s10551-013-1944-3

Parguel, B., Benoît-Moreau, F., & Larceneux, F. (2011). How sustainability ratings might deter 'greenwashing': A closer look at ethical corporate communication. *Journal of Business Ethics*, *102*(1), 15–28. DOI: 10.1007/s10551-011-0901-2

Patterson, D., Gonzalez, J., Le, Q., Liang, C., Munguia, L.-M., Rothchild, D., So, D., Texier, M., & Dean, J. (2021). Carbon and the broad question of AI's environmental impact. arXiv preprint arXiv:2104.10350.

Peattie, K., & Charter, M. (2003). Green marketing. In Baker, M. (Ed.), *The marketing book* (5th ed., pp. 726–755). Butterworth-Heinemann.

Pham, B. T., & Huynh, T. H. (2025). Greenwashing and brand avoidance: Examining green scepticism, green perceived risk, and green consumer confusion. *Journal of Services Marketing*, •••, 1–12.

Rust, R. T. (2020). The future of marketing. *International Journal of Research in Marketing*, *37*(1), 15–26. DOI: 10.1016/j.ijresmar.2019.08.002

Securities and Exchange Board of India. (2021). *Business Responsibility and Sustainability Report (BRSR).* SEBI.

Stadlthanner, K. A., Andreu, L., Font, X., Ribeiro, M. A., & Currás-Pérez, R. (2022). How environmental gain messages affect cause involvement, attitude and behavioural intentions: The moderating effects of CSR scepticism and biospheric values. *Corporate Communications*, *27*(4), 781–799. DOI: 10.1108/CCIJ-11-2021-0125

. Statista. (2024). Artificial intelligence in marketing — worldwide: Market data and analysis. Statista.

Strubell, E., Ganesh, A., & McCallum, A. (2019). Energy and policy considerations for deep learning in NLP. *Proceedings of the 57th Annual Meeting of the Association for Computational Linguistics*, 3645–3650. DOI: 10.18653/v1/P19-1355

Szabo, S., & Webster, J. (2021). Perceived greenwashing: The effects of green marketing on environmental and product perceptions. *Journal of Business Ethics*, *171*(4), 719–739. DOI: 10.1007/s10551-020-04461-0

Wang, B., Rau, P.-L. P., & Yuan, T. (2022). Measuring user competence in using artificial intelligence: Validity and reliability of artificial intelligence literacy scale. *Behaviour & Information Technology*, *42*(9), 1324–1337. DOI: 10.1080/0144929X.2022.2072768

Yoon, Y., Gürhan-Canli, Z., & Schwarz, N. (2006). The effect of corporate social responsibility (CSR) activities on companies with bad reputations. *Journal of Consumer Psychology*, *16*(4), 377–390. DOI: 10.1207/s15327663jcp1604_9

Zhong, B., Song, Y., Feng, G. C., Shi, J., Zhu, Y., Xie, L., Zhou, W. A., Yu, S., Lu, Y., Qin, Y., & Xiong, Z. (2025). AI imaginaries shape technological identity and digital futures. *Computers in Human Behavior*, *169*, 108682. DOI: 10.1016/j.chb.2025.108682

Zuboff, S. (2019). *The age of surveillance capitalism: The fight for a human future at the new frontier of power*. PublicAffairs.

APPENDIX

Table 3. Participant Profile Summary

Code	Sector	Role Level	Years Experience	Gender	Country
MP-01	Technology	Director	12	Male	India
MP-04	Technology	Director	15	Female	UK
MP-08	Energy	Director	9	Male	India
MP-11	Financial Services	Manager	7	Female	Singapore
MP-17	Technology	Head	18	Male	USA
MP-22	Consumer Goods	Head	11	Female	India
MP-26	Financial Services	Lead	8	Male	UK
MP-29	Retail	Director	14	Female	India
MP-31	Consumer Goods	Specialist	6	Female	India
MP-31+	Various (22 further participants)	Various	5–20	Mixed	Multi-country

Chapter 5
Critical Minerals for AI Hardware:
Material Dependencies, Environmental Dimensions, and Industrial Policy Responses in OECD Economies

Göran Roos
http://orcid.org/0000-0003-0943-3585
Adelaide University, UK

ABSTRACT

This chapter examines critical minerals required for AI hardware, including semiconductor fabrication and data centre infrastructure. While sustainability discussions have focused on operational energy consumption, minerals enabling machine learning generate substantial environmental impacts warranting parallel examination. The contribution connects three largely isolated literatures: environmental assessments of mineral extraction, critical minerals policy frameworks, and AI's environmental footprint. Covering developments from 2020 to 2025 across the European Union, United States, United Kingdom, Australia, Canada, Japan, and South Korea, the analysis addresses gigawatt-scale 'AI Factory' facilities and strategic weaponisation of mineral supply chains. Supply concentration emerges as defining, with China controlling 98 per cent of gallium production and 94 per cent of heavy rare earth processing. The chapter identifies tensions between supply security and environmental sustainability, examining burden-shifting risks and 'clean premium' strategies.

DOI: 10.4018/979-8-3373-8490-0.ch005

INTRODUCTION

Discussion of artificial intelligence and environmental sustainability has concentrated on the operational phase of AI systems—energy consumption during model training and inference, data centre cooling requirements, and associated carbon emissions (International Energy Agency, 2021; OECD, 2023). Less attention has been directed toward the material inputs required before any AI system becomes operational. The hardware enabling machine learning depends on minerals whose extraction and processing generate environmental impacts that merit examination alongside operational considerations (World Bank, 2020; Vera et al., 2023). This chapter addresses the material foundations underpinning computational infrastructure, examining the policy responses available to OECD governments as they attempt to manage tensions between AI infrastructure development and environmental commitments.

AI hardware production depends on a wide range of mineral resources, including silicon, copper, aluminium, gold, tin, cobalt, lithium, nickel, graphite, gallium, germanium, and various rare earth elements (International Energy Agency, 2021; European Commission, 2023). Many of these materials are also essential for other strategic sectors, notably renewable energy technologies, electric vehicles, and advanced manufacturing, creating overlapping demand pressures across policy domains (International Energy Agency, 2021; World Bank, 2020). AI hardware cannot be analysed in isolation from broader critical minerals dynamics, even where AI-specific demand represents only a subset of total consumption (OECD, 2023).

Recent foresight analysis undertaken for the European Commission demonstrates that the energy and digital transitions constitute a structural materials transition, with demand for several strategic and critical raw materials projected to increase multiple-fold by 2030 and 2050. A detailed assessment of fifteen strategic technologies shows that supply-chain vulnerability is systematically concentrated in upstream stages, particularly raw and processed materials, where the European Union's share of global production rarely exceeds single-digit percentages (Carrara et al., 2023). While downstream manufacturing capacity within the EU is comparatively stronger, this asymmetry implies that industrial resilience is ultimately constrained by upstream material bottlenecks rather than final assembly capabilities. Importantly, the analysis also indicates that recycling and secondary supply are unlikely to alleviate these constraints before 2030 due to insufficient end-of-life material volumes, underscoring the need for early and coordinated intervention across extraction, processing, substitution and demand-reduction strategies (Carrara et al., 2023).

This chapter changes the prevailing understanding of AI's environmental footprint along three lines. First, it shifts analytical attention upstream from operational energy consumption to material dependencies, showing that AI hardware imposes mineral-extraction obligations that precede and compound the data-centre electricity demand commonly cited in the literature. Second, it demonstrates that the mineral cluster supporting AI hardware overlaps with but is not identical to that supporting clean-energy or electric-vehicle deployment, so AI cannot be treated as a sub-case of the broader energy transition for criticality-assessment purposes. Third, it argues that the institutional and procurement dynamics through which AI hardware is acquired generate distinctive cross-jurisdictional supply-security and environmental-sustainability tensions that conventional critical-minerals frameworks, anchored in the 2010 rare-earth crisis, do not adequately capture.

The conceptual architecture of mineral criticality derives from a framework developed by the National Research Council (2008), which proposed a two-dimensional matrix plotting importance in use against supply restriction potential. Minerals appearing in the upper-right quadrant of this matrix—high importance combined with high supply risk—warrant classification as critical. Subsequent refinements have expanded the framework to incorporate environmental implications, creating three-dimensional criticality assessments (Graedel et al., 2012; Graedel et al., 2015; Graedel & Reck, 2016). Figure 1 visualises this three-dimensional criticality framework as adapted for AI hardware in the present chapter, with the three axes representing supply risk, economic importance and environmental impact.

Figure 1. Three-dimensional mineral criticality framework adapted for critical minerals for AI hardware

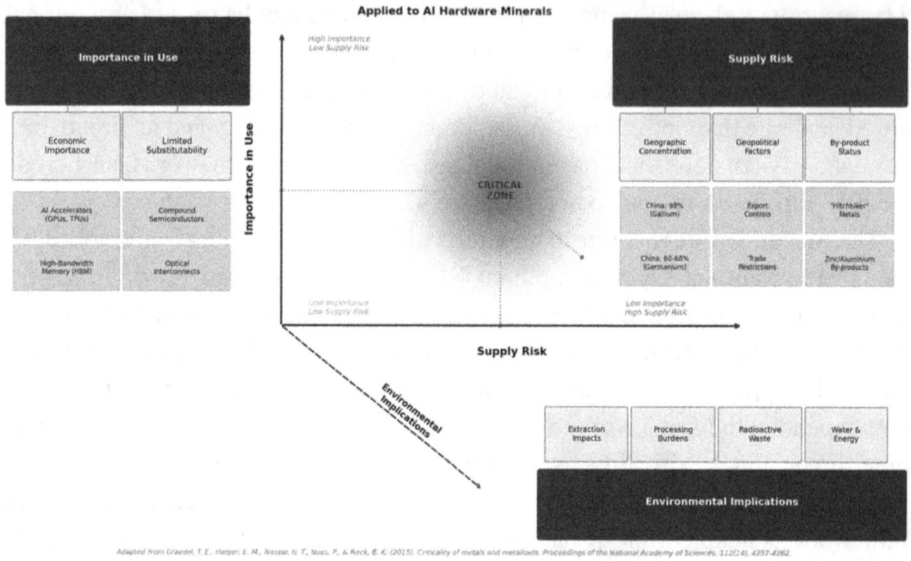

Note. Adapted from the National Research Council (2008) two-dimensional supply-risk-by-economic-impact framework, extended with a third dimension capturing the environmental implications of extraction and processing in line with European Commission Joint Research Centre methodology (Carrara et al., 2023; Georgitzikis et al., 2025). The framework is applied throughout this chapter to the AI-hardware-specific mineral cluster.

The European Commission has conducted periodic criticality assessments since 2011, with the most recent study identifying 34 critical raw materials based on economic importance and supply risk indicators (Grohol & Veeh, 2023). The United States maintains a list of 50 critical minerals pursuant to the Energy Act of 2020, updated through methodology developed by Nassar and Fortier (2021).

The Convergence of AI Expansion, Supply Chain Vulnerabilities, and Industrial Policy Activism

The rapid and accelerating deployment of artificial intelligence systems, particularly following the widespread adoption of large language models and generative AI applications in the early 2020s, has necessitated a restructuring of the physical infrastructure of global computation (FP Analytics, 2025). While the software layer of AI—comprising algorithms, weights, and training datasets—has received substantial public and academic scrutiny regarding its societal implications, the physical hardware layer that enables these computations remains comparatively opaque. This hardware infrastructure relies on a complex, materially intensive supply chain distinct from the established supply chains of traditional consumer electronics and general-purpose computing (FP Analytics, 2025; Baskaran & Schwartz, 2024).

The emergence of the 'AI Factory'—a dedicated, hyperscale facility optimised for generative artificial intelligence model training and inference—has disrupted the architectural and operational paradigms of digital infrastructure (Schneider Electric, 2025). Unlike traditional cloud data centres designed for mixed workloads and moderate power densities of typically 8–15 kW per rack, the AI Factory is defined by extreme density of 40–132 kW per rack, necessitating liquid cooling architectures, massive structural reinforcement, and unprecedented utility connections (Schneider Electric, 2025; Vertiv, n.d.). A 1 GW continuous IT load AI data centre is approximately five times larger than any single facility operating today, demanding 9.4–13.7 TWh annually depending on efficiency assumptions and consuming between 2.6 and 16.6 billion litres of water for cooling (World Economic Forum, 2025).

Table 1. Announced gigawatt-scale AI data centre projects (2025–2028)

Builder	Size (GW)	Location	Est. Operational	Est. Cost (US$)	Source
xAI	1.1	Memphis, TN / Southaven, MS	Q2 2027	$12 billion+ (seeking additional funding)	(SemiAnalysis, 2025)
Amazon Web Services	2.2	New Carlisle, IN	2025 (partially operational)	$11 billion (initial), $15 billion additional announced	(Sigalos, 2025)
Microsoft	Not specified (hundreds of thousands of GPUs)	Mount Pleasant, WI	Early 2026 (1st facility), 2027+ (2nd facility)	$7 billion+ ($3.3B + $4B)	(Microsoft, 2025)
Meta Platforms	1	El Paso, TX	2028	$1.5 billion	(Reuters, 2025)
OpenAI/Oracle/SoftBank (Stargate - Abilene, TX)	1.2	Abilene, TX	Mid-2026	$15 billion ($11.6B raised in 2nd phase)	(Crusoe, 2025)
OpenAI/Oracle (Stargate - Shackelford County, TX)	Part of 5.5 GW total	Shackelford County, TX	Not specified	Part of $300B Oracle-OpenAI deal	(OpenAI, 2025)

continued on following page

Table 1. Continued

Builder	Size (GW)	Location	Est. Operational	Est. Cost (US$)	Source
OpenAI/Oracle (Stargate - DoÃ±a Ana County, NM)	Part of 5.5 GW total	DoÃ±a Ana County, NM	Not specified	Part of $300B Oracle-OpenAI deal	(OpenAI, 2025)
OpenAI/Oracle (Stargate - Wisconsin)	~1.0	Wisconsin	2028	Part of $300B Oracle-OpenAI deal	(OpenAI, 2025; Vantage, 2025)
OpenAI/SoftBank (Stargate - Lordstown, OH)	Part of 1.5 GW total	Lordstown, OH	2026	Part of overall Stargate $500B	(OpenAI, 2025)
OpenAI/SoftBank (Stargate - Milam County, TX)	Part of 1.5 GW total	Milam County, TX	Not specified	Part of overall Stargate $500B	(OpenAI, 2025)
Tata Consultancy Services	1	India	2030-2032 (5-7 years from 2025)	$6.5 billion	(The Economic Times, 2025)
Google/NextEra Energy	Multiple GW-scale campuses	Multiple US locations	First 3 campuses in development	Not specified	(Carbon Credits, 2025)
Anthropic/Fluidstack	Multiple GW	Texas & New York	Throughout 2026	$50 billion	(Anthropic, 2025)
CoreWeave	1.3	Multiple US sites	Not specified	$9 billion (acquisition of Core Scientific)	(Acres (Land Values), 2025)
Crusoe Energy (Wyoming)	1.8	Wyoming	Not specified	Not specified	(SiliconANGLE, 2025)
Crusoe Energy/Blue Energy (Texas)	1.5	Victoria County, TX	Not specified	Not specified	(Crusoe, 2025)

Note: Table condensed for clarity. See original source documentation for complete project listings.

Environmental Dimensions of Hardware Production

The environmental implications of extracting and processing minerals for AI hardware are well documented. Mining and mineral processing are associated with land-use change, biodiversity loss, water consumption, water and soil contamination, waste generation, and greenhouse gas emissions, with impacts varying significantly by mineral type, extraction method, and regulatory context (International Energy Agency, 2021; World Bank, 2020; Vera et al., 2023). These impacts are predominantly concentrated upstream in extraction and refining regions, many of which are located outside OECD economies, while the downstream benefits of AI-enabled digital services accrue primarily in high-income countries (OECD, 2023).

Material intensity for AI-optimised facilities exceeds traditional hyperscale benchmarks by 20–40 per cent due to higher power densities and more robust electrical infrastructure (World Economic Forum, 2025). A 1 GW facility would require approximately 480,000–1,100,000 tonnes of concrete, 12,000–27,000 tonnes of copper, and 57,500–107,000 tonnes of steel for construction alone (World Economic Forum, 2025). The manufacturing of high-performance GPUs, High Bandwidth Memory, and solid-state storage is energy-intensive, with the embodied carbon of the IT equipment refresh cycle occurring every four years cumulatively exceeding the initial facility construction carbon by a factor of five over a 30-year lifespan (NVIDIA, 2025; Schneider Electric, 2025; Meta, 2024).

A representative 1 GW AI data centre campus would comprise approximately 20,000 high-density racks containing approximately 80,000 server nodes (NVIDIA, 2025). Each node carries an embodied carbon footprint estimated at 2,500 kg CO_2e, yielding a total IT equipment embodied carbon of approximately 200,000 tonnes (NVIDIA, 2025; Falk et al., 2025; Gupta et al., 2022). This embodied carbon is incurred every refresh cycle, typically every four years, meaning that over a 30-year facility lifespan, cumulative IT embodied carbon exceeds 2 million tonnes CO_2e—vastly exceeding the one-time embodied carbon of facility construction (NVIDIA, 2025; Schneider Electric, 2025; Meta, 2024).

Chapter Objectives and Contributions

This chapter is motivated by the observation that debates on the environmental footprint of AI remain incomplete if they exclude the material and mineral dimensions of AI hardware. While AI-related energy use is an important and visible concern, the environmental impacts embedded in hardware supply chains are less visible but structurally embedded in the expansion of AI capabilities (International Energy Agency, 2021; OECD, 2023). Addressing these impacts requires policy attention not only to energy systems but also to industrial policy, resource governance, and international coordination on critical minerals (European Commission, 2023).

The chapter pursues three interrelated objectives. First, it characterises the material dependencies of AI hardware by examining the minerals required for semiconductor fabrication and data centre infrastructure, with particular attention to supply concentration and criticality assessments (OECD, 2023; U.S. Geological Survey, 2024). Second, it surveys the industrial policy instruments currently deployed across OECD economies, including domestic extraction incentives, processing capacity development, recycling mandates, strategic stockpiling, and international partnership arrangements (European Commission, 2023; Department for Business and Trade,

2025). Third, it analyses the environmental dimensions of critical minerals extraction and processing, examining life cycle impacts, water and energy requirements, and the geographic distribution of environmental burdens between OECD consumption centres and non-OECD extraction sites (World Bank, 2020; Vera et al., 2023).

The contribution lies in connecting three bodies of work that have developed largely in isolation: environmental assessments of mineral extraction, critical minerals policy frameworks developed in response to supply chain concerns, and the emerging literature on AI's environmental footprint. The first addresses AI's environmental footprint, with studies quantifying training energy consumption, data centre emissions, and water use for cooling, establishing that AI systems carry environmental costs but focusing on the operational phase (International Energy Agency, 2021). The second comprises environmental assessments of mining and mineral processing, with life cycle assessment studies, environmental impact statements, and academic research documenting the consequences of extraction across mineral types and geographical contexts, though proceeding largely without reference to AI as an end use (Vera et al., 2023; World Bank, 2020). The third encompasses critical minerals policy analysis, with governments and international organisations producing strategies, regulations, and assessments addressing supply chain vulnerability, though environmental sustainability features in these documents but receives less attention than economic and strategic considerations (OECD, 2023; European Commission, 2023).

SCOPE AND ANALYTICAL BOUNDARIES

Definition of AI Hardware

AI hardware is understood in this chapter as the physical computing equipment that enables the training, deployment, and operation of AI models, including advanced semiconductors, specialised accelerators, servers, storage systems, and associated power and cooling components (International Energy Agency, 2021; European Commission, 2023; OECD, 2023). This definition encompasses the material infrastructure required for computational processes but excludes the software, algorithms, and AI applications that operate upon this infrastructure. The analytical focus is deliberately restricted to the upstream material inputs and manufacturing stages rather than the use-phase characteristics of AI systems.

The core components of AI hardware include graphics processing units (GPUs), tensor processing units (TPUs), and high-bandwidth memory (HBM) modules (FP Analytics, 2025). These components represent the specialised accelerators that distinguish AI workloads from general-purpose computing. A representative AI rack typically contains 4–8 GPU chassis, along with top-of-rack switches, power distribution units, and liquid cooling manifolds (NVIDIA, 2025). A single 8-GPU chassis weighs approximately 130 kg, and a fully integrated rack, including the rack frame, manifolds, and networking gear, can exceed 1,000–1,200 kg, totalling over 20,000–24,000 tonnes of high-value, diverse material for a 1 GW facility.

Analytical Exclusions

The analysis explicitly excludes operational energy consumption and use-phase emissions associated with AI systems, including data centre electricity demand, energy efficiency metrics, and carbon accounting for AI workloads. These issues are widely covered in the existing literature on digital sustainability and AI energy use (International Energy Agency, 2021; OECD, 2023). Operational electricity consumption for a 1 GW facility operating at a power usage effectiveness (PUE) of 1.20 would total approximately 10.5 terawatt-hours annually, equivalent to roughly 3.5 per cent of the United Kingdom's total electricity demand (Shehabi et al., 2024; National Energy System Operator, 2024). While these operational impacts are substantial, they fall outside the scope of this chapter, which concentrates on upstream material dependencies and the industrial policy responses they have occasioned.

Software, algorithms, and AI applications are addressed only insofar as they influence hardware demand and deployment intensity (OECD, 2023). The computational demands of deep learning—characterised by massive matrix multiplications and the need to move vast parameters between memory and processing units—shape the design requirements for AI hardware. Technical characteristics of specific AI models, training methodologies, and inference requirements are noted where relevant to understanding hardware demand but are not examined in their own right.

Geographic Scope

The geographic scope centres on OECD economies as sites of policy formation and consumption within advanced industrial economies (OECD, 2023). The discussion includes policy frameworks and industrial strategies from the European Union, the United States, the United Kingdom, Australia, Canada, Japan, and South Korea, each of which has published or updated critical minerals strategies with substantial public financing commitments (European Commission, 2023; Department for Business and Trade, 2025; Ministry of Economy, Trade and Industry, 2023; Australian Government, Department of Industry, Science and Resources,

2023; Government of Canada, Natural Resources Canada, 2022; Ministry of Trade, Industry and Energy, 2023).

China is considered as a contextual case due to its central role in global mineral processing and refining rather than as a subject of domestic policy analysis (European Commission, 2023; OECD, 2023). China's dominance extends across multiple stages of critical mineral value chains, with particularly pronounced concentration in gallium, germanium, and rare earth element processing. The scale and implications of this processing dominance, including the strategic weaponisation of export controls observed since 2023, are examined in detail in Section on China as Contextual Case

While extraction activities are frequently located outside OECD jurisdictions, these upstream realities are incorporated insofar as they shape policy responses, supply risks, and environmental outcomes relevant to OECD economies (World Bank, 2020; OECD, 2023). The Democratic Republic of the Congo accounts for approximately 70 per cent of global cobalt supply, Indonesia and the Philippines dominate nickel production, and Russia and South Africa are major palladium producers (FP Analytics, 2025; Davey, 2023). These geographic concentrations create supply chain dependencies that inform OECD policy responses even where the extraction sites themselves fall outside the primary analytical focus.

Temporal Scope

The temporal scope covers policy and technological developments from 2020 to late 2025. This period captures the inflection point marked by the enactment of the United States Energy Act of 2020, the subsequent Executive Orders on supply chains, the passage of the CHIPS and Science Act and the Inflation Reduction Act, and the early implementation phases of the European Union Critical Raw Materials Act (U.S. Congress, 2022a, 2022b; European Parliament and Council of the European Union, 2024). It also encompasses the launch of the Pax Silica initiative in late 2025, which represents the latest iteration of multilateral supply chain coordination (U.S. Department of State, 2025).

The period since 2022 has witnessed an acceleration of policy responses following Russia's invasion of Ukraine, which heightened concerns about supply chain resilience across multiple commodity categories. China's imposition of export licensing requirements for gallium and germanium in July 2023, followed by expanded restrictions in December 2024 covering additional minerals and eventually extending to components containing Chinese-origin materials, demonstrated the weaponisation of mineral supply for geopolitical purposes (International Energy Agency, 2023; International Energy Agency, 2025; Baskaran & Schwartz, 2024). These developments have intensified industrial policy efforts across OECD economies, with new legislation, expanded public financing, and international coordination mechanisms emerging in rapid succession.

Data Fragmentation and Evidentiary Constraints

Publicly available data on material flows through semiconductor and electronics supply chains are sparse and fragmented, meaning that estimates of AI-specific mineral consumption remain highly uncertain (OECD, 2023). Much literature treats critical minerals from the perspective of clean energy transitions; AI hardware accounts for only a portion of total demand and is rarely disaggregated (International Energy Agency, 2021). The lack of harmonised reporting on mineral flows, the opacity of multi-tier global supply chains, and the shared use of materials across multiple sectors constrain the precision with which AI-specific material impacts can be identified.

A closely related evidence base from the enterprise server domain illustrates why AI-hardware mineral accounting remains structurally challenging. In its analysis of material efficiency requirements for enterprise servers, the European Commission's Joint Research Centre (JRC) notes that reliable BoM data are difficult to assemble because manufacturers are often unwilling to disclose detailed material composition, and upstream suppliers may provide only partial information, resulting in fragmented and non-comparable datasets across studies (Talens Peiró & Ardente, 2015). The JRC report highlights that the lack of complete and verifiable material composition data constrains quantitative assessment, and that dispersed treatment of BoM, reuse, and recycling information across separate reports further weakens the evidence base needed for robust policy design (Talens Peiró & Ardente, 2015). For AI hardware, the implication is not merely that mineral flows are uncertain, but that uncertainty is partly an institutional and informational artefact of multi-tier component supply chains, requiring purposeful data architectures and disclosure mechanisms rather than relying on ad hoc proxies alone (Talens Peiró & Ardente, 2015).

Manufacturers such as NVIDIA and AMD do not publicly disclose the exact elemental composition of their leading-edge processors (FP Analytics, 2025). Consequently, quantitative data on mineral consumption per unit is derived from aggregated industry averages, technical proxies, and third-party teardown analyses rather than direct manufacturer disclosures. Explicit acknowledgement is made of the uncertainty inherent in scaling these unit-level estimates to global demand scenarios.

Environmental impact data for specific minerals are often based on life cycle assessments with limited geographical coverage and reliance on modelling assumptions (Luo et al., 2025). The inconsistency of boundaries, functional units, and data sources across studies complicates synthesis and comparison. Where quantitative figures are cited in this chapter, they are drawn from peer-reviewed or institutional sources and are used illustratively to characterise orders of magnitude or structural patterns, rather than to infer precise AI-driven demand trajectories. These limitations are treated not as peripheral caveats but as central analytical conditions shaping both research conclusions and policy design (International Energy Agency, 2021; European Commission, 2023; OECD, 2023).

BACKGROUND: CRITICAL MINERALS POLICY AND AI HARDWARE CONVERGENCE

The convergence of artificial intelligence hardware expansion with pre-existing concerns about critical minerals supply chains has prompted renewed industrial policy activism across OECD economies. This section traces the evolution of critical minerals policy frameworks from their origins in the 2010 rare earth crisis, examines the technological shifts that have created distinct material requirements for AI hardware, and analyses the geopolitical context of supply concentration that has intensified policy responses. Figure 2 traces the resulting rare earth elements price dynamics from 2008 to 2025, showing how supply concentration translates into recurrent price volatility.

Evolution of Critical Minerals Frameworks

The 2010 Rare Earth Crisis and Its Policy Legacy. Critical minerals policy emerged as a distinct field of governmental attention following supply disruptions in the early 2010s. China's restriction of rare earth element exports in 2010, prompted by a diplomatic dispute with Japan, demonstrated the vulnerability of technology supply chains to concentrated production (Sprecher et al., 2015; Mancheri et al., 2019). The incident catalysed systematic efforts to identify and monitor minerals presenting elevated supply risk. Prior to this disruption, rare earth elements had received limited policy attention despite their growing importance in high-technology applications, permanent magnets, and defence systems.

Figure 2. Rare earth elements price dynamics, 2008–2025

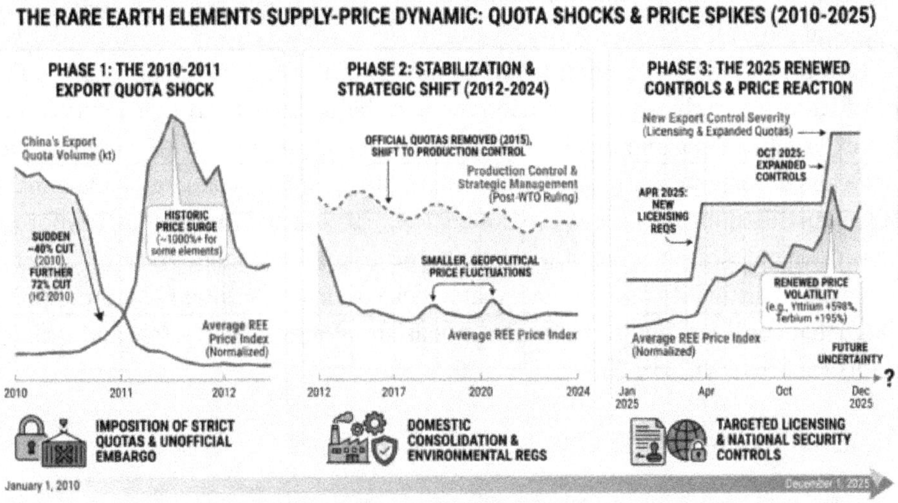

Note. Compiled from Airaudo et al. (2025), Baskaran (2024), Branigan (2010), Castilloux (2019), Davidson (2025), Office of the United States Trade Representative (2014) and Trading Economics (n.d.). The 2010–2011 price spike illustrates the rare-earth crisis whose policy legacy is examined in this section. Nominal US$ per kilogram; selected REE indicative of broader price movement.

The policy legacy of the 2010 crisis extended across multiple dimensions. The United States Geological Survey, European Commission, and national governments developed criticality assessment methodologies that evaluate minerals along dimensions including economic importance, supply concentration, substitutability, and recycling potential (National Research Council, 2008; Graedel et al., 2012; Schrijvers et al., 2020). Japan established strategic stockpiling programmes through the Japan Oil, Gas and Metals National Corporation (JOGMEC), maintaining 60–180 day reserves of critical materials (Ministry of Economy, Trade and Industry, 2020). The European Union initiated its first criticality assessment in 2011, establishing a periodic review process that would subsequently inform the Critical Raw Materials Act of 2024.

From Supply Risk to Three-Dimensional Assessments.

The evolution from supply risk-focused assessments toward three-dimensional frameworks reflects growing recognition that critical minerals policy cannot be divorced from environmental considerations (Graedel et al., 2012). The environmental impacts of extracting and processing critical minerals are well documented: mining and mineral processing are associated with land-use change, biodiversity loss, water consumption, water and soil contamination, waste generation, and greenhouse gas emissions, with impacts varying significantly by mineral type, extraction method, and regulatory context (International Energy Agency, 2021; World Bank, 2020; Vera et al., 2023).

This geographic and economic separation complicates accountability and policy coordination. Supply security objectives may be achieved through interventions that externalise environmental burdens to producing regions with weaker governance and enforcement capacity. Conversely, stringent environmental standards in OECD jurisdictions may constrain domestic extraction and processing, reinforcing import dependence on regions with less rigorous environmental oversight. Three-dimensional criticality assessments make these trade-offs explicit, enabling policy design that considers environmental implications alongside economic and strategic factors (Schrijvers et al., 2020).

The Shift to AI-Centric Compute Architectures

From Moore's Law to the 'Memory Wall'. The evolution of computing hardware has historically been driven by Moore's Law, which posits a doubling of transistor density approximately every two years. For decades, the central processing unit (CPU) was the locus of this innovation, with a focus on general-purpose logic operations. However, the specific computational demands of deep learning—characterised by massive matrix multiplications and the need to move vast parameters between memory and processing units—have exposed the limitations of traditional architectures (Waseem, 2025). The industry has encountered the 'memory wall', a phenomenon where the speed of the processor outpaces the rate at which data can be delivered from memory, leaving high-performance logic units idle while waiting for data.

Unlike general-purpose computing, which has historically prioritised logic density and transistor scaling, AI workloads are increasingly constrained by memory bandwidth. Training large language models requires moving trillions of parameters between memory and processing units, creating bottlenecks that cannot be resolved through faster processors alone. This technical constraint has driven the semiconductor industry towards heterogeneous computing architectures that pair specialised accelerators (GPUs, TPUs) with massive pools of high-speed memory.

High Bandwidth Memory and Advanced Packaging Technologies. High Bandwidth Memory addresses the memory wall by stacking dynamic random-access memory (DRAM) dies vertically, connecting them with thousands of microscopic channels called through-silicon vias (TSVs) (AnySilicon, n.d.). This architecture requires a silicon interposer—a bridge chip that connects the GPU and the memory stack within a single package. The interposer enables the dense interconnections necessary for high-bandwidth data transfer while managing the thermal and electrical challenges of three-dimensional integration. HBM achieves bandwidth exceeding one terabyte per second, compared to approximately 50 gigabytes per second for conventional DDR memory.

Advanced packaging technologies, particularly Chip-on-Wafer-on-Substrate (CoWoS), have emerged as enablers of AI hardware performance. CoWoS technology places multiple chiplets—including logic processors and HBM stacks—on a shared silicon interposer, which is then mounted on an organic substrate for connection to the broader system. This 2.5D packaging approach enables heterogeneous integration of components manufactured using different process nodes and materials. The technology has become a significant bottleneck for AI chip production, with limited global manufacturing capacity concentrated in a small number of fabrication facilities (FP Analytics, 2025).

Policy-facing strategic intelligence work within the European Commission's foresight ecosystem reinforces that the hardware enabling AI should be treated as a cross-cutting domain whose constraints extend beyond chip fabrication into adjacent areas such as advanced materials and energy uses and efficiency. In its synthesis of emerging technology signals, the Joint Research Centre observes that the 'Hardware & Semiconductors' category displays the strongest pattern of convergences with other domains, including advanced materials and energy efficiency, reflecting the extent to which performance gains in AI systems increasingly depend on packaging, thermal management, and power-delivery constraints rather than logic scaling alone (Bailey et al., 2024). This framing is relevant for critical minerals analysis because the minerals burden of AI hardware is shaped by these convergence dynamics: bottlenecks in energy efficiency and thermal management translate into material demand for conductors, thermal interface materials, and specialised components, whilst packaging and system integration can amplify demand for high-purity and high-reliability inputs (Bailey et al., 2024).

Material Implications of 2.5D/3D Silicon Packaging. The shift from two-dimensional printed circuit boards to 2.5D and 3D silicon packaging alters the material requirements of computing hardware. It increases the consumption of high-purity silicon for interposers, copper for through-silicon vias and micro-bumps, and specialised underfill materials (AnySilicon, n.d.). Through-silicon vias require copper filling of microscopic channels etched through silicon wafers, demanding high-purity copper and precise deposition techniques. Micro-bumps connecting dies to interposers employ tin-silver alloys, requiring materials with exacting purity specifications.

Furthermore, the thermal density of AI chips—often exceeding 700 watts per package—necessitates advanced thermal interface materials containing indium and high-performance tantalum capacitors for power regulation (FP Analytics, 2025). Indium's low melting point and excellent thermal conductivity make it essential for heat transfer between processor dies and cooling solutions. Tantalum capacitors provide the stable, high-capacitance power delivery required for consistent operation under variable computational loads. The physical form factor of an AI server is thus materially distinct from a standard web server, driving a divergent demand profile for critical minerals as outlined in Figure 3, Table 2, and Table 3.

Figure 3. Comparative architecture and material flow diagram: Web servers versus AI accelerator systems

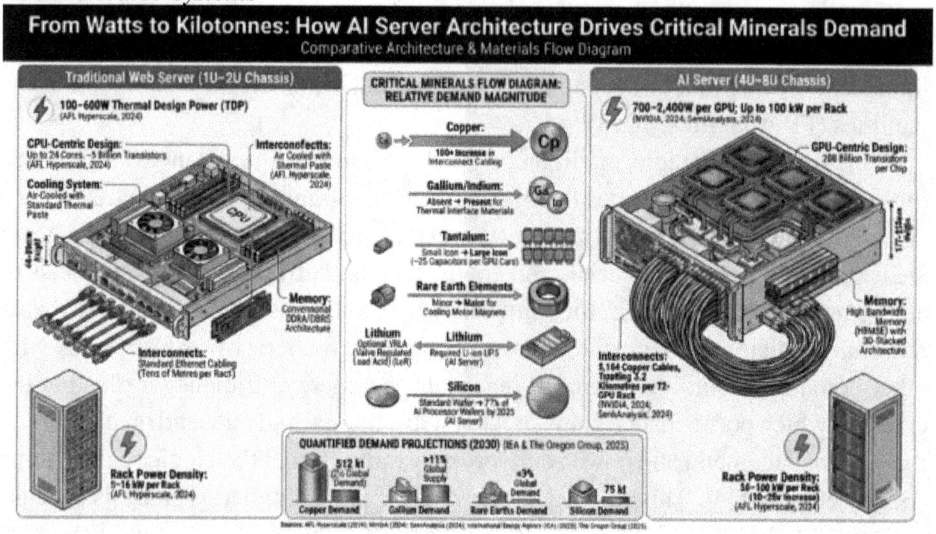

Note. Schematic comparison of the physical architecture and material flow in conventional web-server compute versus AI accelerator systems. Supporting quantitative data are presented in Table 2.

Table 2. Physical architecture comparison: Web servers versus AI Servers And Critical Minerals Implications

Physical Characteristic	Traditional Web Server	AI Server	Critical Minerals Implications
Form Factor	1U–2U rackmount chassis	4U–8U rackmount chassis (850mm × 438mm × 177mm typical)	Larger chassis requires additional aluminium, steel, and copper for structural and electrical systems
Processing Architecture	CPU-centric; up to 24 cores; ~3 billion transistors (AFL Hyperscale, 2024)	GPU-centric; thousands of cores; 208 billion transistors per chip (AFL Hyperscale, 2024)	70× increase in transistor density drives proportionally greater silicon, gallium (for compound semiconductors), and rare earth consumption in manufacturing equipment
Power Consumption	100–600W per server (AFL Hyperscale, 2024)	700–2,400W per GPU; 100 kW per rack (NVIDIA, 2024; Continuum Labs, n.d.)	Higher power density requires enhanced copper busbars, transformers with rare earth components, and lithium-ion UPS systems
Rack Power Density	5–15 kW per rack (AFL Hyperscale, 2024)	50–100 kW per rack; cutting-edge reaching 200 kW (AFL Hyperscale, 2024)	10–20× power density increase drives proportional increases in copper for power distribution and cooling infrastructure
Copper Interconnects	Standard Ethernet; tens of metres per rack	approximately 5,184 NVLink copper cables; 3.2 km per 72-GPU rack (SemiAnalysis, 2024; NVIDIA, 2024)	~100× increase in copper cabling per rack; Data centre copper demand projected at 512 kt by 2030 (IEA, 2025)
Memory Architecture	Conventional DDR4/ DDR5 memory	High Bandwidth Memory (HBM3E); 3D-stacked with TSVs; up to 192 GB per GPU (NVIDIA, 2024)	HBM requires advanced packaging (CoWoS) with gallium, indium, and significantly more silicon substrate material
Thermal Interface Materials	Silicon-based thermal paste (5–10 W/m·K conductivity)	Gallium-indium-tin liquid metal alloys (>25 W/m·K conductivity)	Creates new demand source for gallium (78.6%) and indium (21.4%) in thermal management applications
Cooling Systems	Air cooling with standard fans	Liquid cooling with high-performance BLDC motors containing NdFeB magnets (Continuum Labs, n.d.)	10–20× cooling requirements drive proportionally higher rare earth permanent magnet consumption (neodymium, dysprosium, praseodymium)
Capacitor Density	Moderate capacitor count for standard power management	~25 tantalum capacitors per GPU card; 10,000–20,000 MLCCs per board (Lessner, in Argus Media, 2025)	AI servers create significant new tantalum demand; supply chain concentration in Central Africa poses strategic risk

continued on following page

Table 2. Continued

Physical Characteristic	Traditional Web Server	AI Server	Critical Minerals Implications
Battery Backup (UPS)	Valve-regulated lead acid (VRLA) batteries adequate for steady loads	Lithium-ion required for rapid transients (1%–150% load swings); 2–3× energy density (Schneider Electric, 2025)	Shift to Li-ion creates material demand for lithium, cobalt, nickel, and associated battery minerals
Semiconductor Content	Mainstream silicon technology; minimal compound semiconductors	Advanced 4–5nm silicon; GaAs and GaN for power electronics and RF	Gallium demand from data centres projected to exceed 11% of global supply by 2030 (IEA, 2025)

Table 3. Projected critical minerals demand from data centres by 2030

Critical Mineral	2023 Baseline Demand (Approx. Data Center Share)	Projected 2030 Demand (AI Data Centers)	Share of Supply/ Demand Impact	Primary Application in AI Infrastructure	Primary Data Source(s)
Copper	~197,000 – 200,000 tonnes	512,000 – 1,000,000 tonnes	~2–3% of global demand; up to 1 Mt "AI delta"	Power distribution, busbars, cooling systems, grid connection	(International Energy Agency, 2025); (Trafigura, 2024); (MINING.COM, 2024)
Silicon (High Purity)	~30,000 tonnes (Est.)	~75,000 tonnes	~2% of global demand (specific grades)	GPU/CPU/NPU fabrication, HBM memory	(International Energy Agency, 2025); (Carbon Credits, 2025)
Rare Earth Elements (NdPr)	~5,000 tonnes	~12,000 tonnes	>3% of global demand	Permanent magnets in cooling fans, pumps, HDDs	(The Oregon Group, 2025); (Arafura Rare Earths Limited, n.d.)
Gallium	<50 tonnes	~120 tonnes	>11% of current global supply	Compound semiconductors (transceivers), thermal interface materials	(International Energy Agency, 2025); (Center for Strategic and International Studies, 2023)
Tantalum	~1,200 tonnes (Global electronics)	~1,700 tonnes (Allocated share)	Significant growth driver (>10% CAGR)	Power management capacitors on GPU cards	(Mordor Intelligence, 2025); (IMARC Group, 2024)

continued on following page

Table 3. Continued

Critical Mineral	2023 Baseline Demand (Approx. Data Center Share)	Projected 2030 Demand (AI Data Centers)	Share of Supply/ Demand Impact	Primary Application in AI Infrastructure	Primary Data Source(s)
Lithium (LCE)	~2,000 tonnes	15,000 – 25,000 tonnes	Growing share from UPS systems	Battery backup systems (BESS) for AI workloads	(Battery Council International, 2025); (International Council on Clean Transportation, 2024)
Cobalt	<500 tonnes	3,000 – 5,000 tonnes	High supply risk dependency	Cathodes in high-performance NMC Li-ion batteries for UPS	(International Council on Clean Transportation, 2024)
Indium	~200 tonnes	~450 tonnes	High supply risk (zinc by-product)	Thermal Interface Materials (TIMs) for high-heat GPUs	(Polaris Market Research, 2022); (Mordor Intelligence, 2025)
Nickel	~2,000 tonnes	10,000 – 15,000 tonnes	Medium supply impact	Battery cathodes; stainless steel cooling infrastructure	(International Council on Clean Transportation, 2024)

Geopolitical Context and Supply Concentration

China's Dominance in Processing and Refining. The supply chains for minerals enabling AI hardware architectures are characterised by extreme concentration. China's dominance in critical minerals extends across multiple stages of value chains. As of 2024, China accounted for 98 per cent of global raw gallium production and between 60 and 68 per cent of refined germanium output (Nassar et al., 2024; FP Analytics, 2025). For rare earth elements, China's share of global extraction has declined from earlier peaks but remains above 60 per cent, while its share of processing capacity, particularly for heavy rare earth elements, has increased to 94 per cent (International Energy Agency, 2025).

Trade-flow analysis further demonstrates that EU exposure to supply disruption is not limited to raw materials but extends across multiple stages of strategic value chains. Recent assessment of global trade patterns for strategic commodities and related components reveals a high degree of concentration in both exports and imports, frequently exceeding thresholds associated with systemic risk (Versino et al., 2025). In several cases, EU supply chains depend on a single external partner for more than half of total imports, not only for raw materials but also for processed inputs and specialised equipment. This pattern amplifies vulnerability to geopolitical shocks, trade restrictions and industrial policy interventions by third countries, reinforcing the need for risk-aware diversification, strategic stockpiling and coordinated international partnerships rather than reliance on market mechanisms alone (Carrara et al., 2023; Versino et al., 2025).

This concentration is not merely geological but industrial and economic. Gallium and germanium are not mined as primary products; they are recovered as by-products of aluminium (bauxite) and zinc refining, respectively (Berg et al., 2024). The economics of their extraction are inextricably tied to the capacity and profitability of the host metal sectors. As OECD economies have de-industrialised their primary aluminium and zinc smelting capacities over the past three decades due to energy costs and environmental regulations, they have inadvertently lost the capacity to produce these by-products.

The 'hitchhiker problem' describes this structural dependency: gallium, germanium, indium, and other technology-critical elements hitchhike on the production of commodity metals, meaning their supply is constrained by decisions made in unrelated markets (FP Analytics, 2025). Expanding gallium supply requires either increasing aluminium smelting capacity or improving recovery rates from existing operations—neither of which responds directly to gallium price signals.

Export Restrictions (2023–2025): Gallium, Germanium, and Rare Earths. The period since 2022 has witnessed an acceleration of supply chain tensions. China's imposition of export licensing requirements for gallium and germanium in July 2023 demonstrated the potential for mineral supply chains to be weaponised in broader technological disputes (International Energy Agency, 2023; Baskaran & Schwartz, 2024). These two elements are essential for compound semiconductors (gallium arsenide, gallium nitride) and high-speed optical transceivers used in data centres. The export controls required Chinese producers to obtain licences before shipping these materials, introducing uncertainty and administrative friction into supply chains that had previously operated with minimal regulatory oversight.

The restrictions expanded in December 2024, covering additional minerals and eventually extending to components containing Chinese-origin materials (International Energy Agency, 2025). This escalation demonstrated a willingness to use export controls as instruments of strategic competition, mirroring the technology export restrictions that the United States and allied nations had imposed on advanced semiconductors and manufacturing equipment destined for China. The reciprocal deployment of trade restrictions created a bifurcating supply chain architecture, with implications for material flows, pricing, and investment decisions across the critical minerals sector.

Market dynamics further complicate supply security. Prices for critical minerals exhibit high volatility due to tight supply–demand balances, speculative activity, and policy interventions. The International Energy Agency reports that after a period of sharp price increases driven by electric vehicle and battery demand, the market saw significant declines in 2023: lithium prices fell by 75 per cent and cobalt, nickel, and graphite prices by 30–45 per cent, even though underlying demand remained strong (International Energy Agency, 2024). Such volatility can deter investment in new projects because producers face uncertainty about future revenues. At the same time, supply bottlenecks and long development timelines make it difficult for production to respond quickly to price signals.

Strategic Weaponisation of Mineral Supply Chains. The 2023–2025 export restrictions imposed by China on gallium, germanium, and subsequently additional minerals represent a qualitative shift in critical minerals geopolitics. Unlike the 2010 rare earth crisis, which arose from a specific bilateral dispute with Japan, these restrictions are explicitly integrated with broader technology competition encompassing semiconductor manufacturing equipment, advanced chip exports, and artificial intelligence capabilities. The detailed chronology and specific measures are examined in Section "Export Controls and Strategic Weaponisation"; what matters for present purposes is the structural transformation these restrictions have occasioned in OECD policy approaches.

In response, OECD governments have abandoned the post-Cold War consensus of laissez-faire approaches to mineral supply chains in favour of active state intervention. The objective is no longer merely to obtain materials at the lowest market price but to secure 'sovereign' or 'allied' capacity that can withstand geopolitical shocks. This shift has profound implications for the mining and materials processing sectors, which now find themselves at the centre of high-level diplomatic and national security strategies.

The Pax Silica initiative launched in late 2025 exemplifies this reorientation (U.S. Department of State, 2025). Where earlier multilateral coordination efforts such as the Minerals Security Partnership focused on investment facilitation and supply diversification, Pax Silica frames critical minerals explicitly within a security architecture linking allied nations. The mining sector, historically peripheral to strategic policy discourse, has thus become a domain of active industrial policy intervention across OECD economies, with implications examined throughout Sections "Critical Minerals Policy" and "Industrial Policy Instruments".

While China dominates the public narrative on supply concentration, the geopolitics of AI-hardware critical minerals extend across several additional jurisdictions whose policy choices materially affect OECD supply security. Russia is a globally significant producer of palladium (approximately 38 per cent of mined supply in 2024) and an important source of high-grade nickel for battery and superalloy markets, with sanctions imposed after February 2022 introducing parallel-market dynamics that complicate Western OEM sourcing. Indonesia has become the world's largest nickel producer, accounting for above 50 per cent of mined nickel and a rapidly growing share of refined Class I nickel through Chinese-financed high-pressure-acid-leach and rotary-kiln-electric-furnace processing capacity in Sulawesi; Indonesia's intermittent restrictions on unprocessed ore export have driven this downstream build-out and reshaped global nickel cost curves (International Energy Agency, 2025; United States Geological Survey, 2025).

Sweden and Greenland represent the European frontier: the LKAB Per Geijer deposit in Kiruna, identified in 2023 as Europe's largest known rare-earth deposit, is being scoped for production in the late 2030s, and Greenland's Kvanefjeld and Kringlerne rare-earth resources have entered the strategic supply discussion under both the European Critical Raw Materials Act and recent United States diplomatic engagement. Canada has positioned itself as an upstream supplier of high-purity nickel, cobalt and rare earths to the United States through the 2020 Joint Action Plan on Critical Minerals and the 2024 Canada Critical Minerals Strategy, with capacity additions concentrated in Ontario, Saskatchewan and Quebec. Africa hosts a substantial share of the cobalt, copper and rare-earth resource base; the African Union's African Mining Vision (2009) and the subsequent African Green Minerals Strategy (2025) articulate a beneficiation-first development model that contrasts with the export-of-concentrates pattern still dominant in the Copperbelt and shapes the bargaining position of African states in critical-minerals partnerships with OECD economies (African Union, 2025).

MATERIAL DEPENDENCIES OF AI HARDWARE

AI hardware encompasses a complex assembly of materials spanning semiconductor devices, memory systems, advanced packaging, and supporting data centre infrastructure. This section examines the material dependencies at each layer of the AI hardware stack, from the ultra-high purity silicon and compound semiconductors that form the computational core, through the specialised materials enabling high-bandwidth memory and three-dimensional packaging, to the bulk metals and battery materials required for data centre construction and operation.

Technology-specific supply-chain analysis highlights that material criticality cannot be understood solely through aggregate lists of critical raw materials but must be assessed in relation to concrete technological applications and value-chain stages. Detailed mapping of strategic technologies such as wind turbines, batteries, photovoltaics and hydrogen systems shows that vulnerabilities frequently arise from highly specialised materials and components, including rare-earth permanent magnets, battery-grade graphite and platinum-group metals, for which substitution options are limited and processing capacity is geographically concentrated (Carrara et al., 2023). These material dependencies are often embedded in intermediate components rather than final products, rendering traditional trade diversification strategies insufficient on their own. Consequently, effective mitigation requires targeted intervention at specific points in the value chain, including processing capacity, component manufacturing and technology design choices (Carrara et al., 2023). Figure 4 represents this hierarchical material-dependency chain from semiconductor substrates through advanced packaging to data centre infrastructure, providing a single visual frame for the dependencies analysed in the following subsections.

Figure 4. Hierarchical material dependencies in AI hardware: From semiconductor substrates to data centre infrastructure

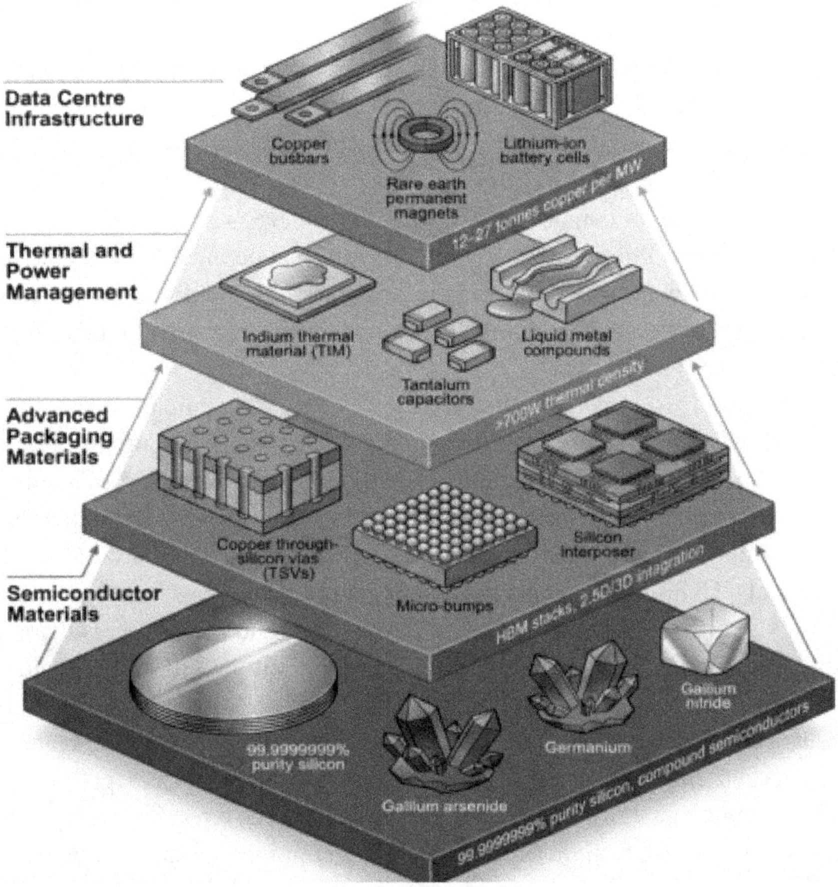

Note. The figure renders the material-dependency chain from the wafer substrate through advanced packaging and high-bandwidth memory to the surrounding data-centre infrastructure. The hierarchical structure underlies the chapter's argument that aggregate critical-raw-material lists must be disaggregated to technology-specific mineral clusters.

Semiconductor Manufacturing Materials

Silicon: Purity Requirements and Processing Capacity. Silicon forms the foundational substrate for virtually all semiconductor devices, including the graphics processing units, tensor processing units, and memory chips that constitute AI hardware (Baskaran & Schwartz, 2024). The silicon used in semiconductor manufacturing must achieve purity levels of 99.9999999 per cent (nine nines) or higher, far exceeding the requirements for metallurgical-grade silicon used in aluminium alloys or solar photovoltaic applications. This electronic-grade polysilicon is produced through energy-intensive purification processes, principally the Siemens process, which involves chemical vapour deposition of trichlorosilane onto heated silicon rods in a controlled atmosphere.

Global polysilicon production capacity has expanded substantially to serve solar photovoltaic demand. Semiconductor-grade polysilicon represents a distinct market segment with more stringent specifications and higher prices. China dominates polysilicon production, accounting for over 80 per cent of global capacity. AI hardware imposes additional silicon demands through the interposers required for advanced packaging. Silicon interposers—the bridge chips that connect graphics processing units to high-bandwidth memory stacks in 2.5D packaging configurations—consume high-grade polysilicon and require fabrication using semiconductor manufacturing techniques (AnySilicon, n.d.).

Gallium and Compound Semiconductors. Gallium is employed in compound semiconductors, particularly gallium arsenide (GaAs) and gallium nitride (GaN), which serve functions in AI hardware systems (FP Analytics, 2025). Gallium arsenide is used in high-frequency chips and optoelectronic devices, including the laser diodes and photodetectors that enable fibre-optic communication between data centre components. Gallium nitride is increasingly deployed in power electronics, where its ability to operate at higher voltages, frequencies, and temperatures than silicon enables more efficient power conversion and distribution within data centres (FP Analytics, 2025; Baskaran & Schwartz, 2024).

China controls approximately 98 per cent of global primary gallium production (FP Analytics, 2025; Nassar et al., 2024). This concentration reflects gallium's status as a by-product metal: it is not mined directly but is recovered from the processing of bauxite ore in aluminium production. The economics of gallium recovery are therefore coupled to the aluminium industry, with supply determined by smelter capacity and recovery infrastructure rather than gallium-specific demand signals.

Germanium: Fibre Optics, Photodetectors, and SiGe Transistors. Germanium serves multiple functions in AI hardware systems, including fibre-optic cables, infrared sensors, and silicon-germanium (SiGe) transistor technologies (FP Analytics, 2025). In fibre optics, germanium dioxide is used as a dopant in the silica glass cores of optical fibres, modifying the refractive index to enable light guidance. Germanium photodetectors are employed in optical receivers, where they convert light signals back to electrical impulses at the terminus of fibre-optic links.

China accounts for between 60 and 68 per cent of refined germanium production globally (FP Analytics, 2025; Nassar et al., 2024). Like gallium, germanium is predominantly recovered as a by-product—in this case, from zinc ore processing. The coupling of germanium supply to zinc production creates supply inelasticity: expanding germanium availability requires either increased zinc smelting or improved recovery rates from existing operations. The imposition of Chinese export licensing requirements for germanium in July 2023 highlighted the vulnerability created by this concentrated production (International Energy Agency, 2023; Baskaran & Schwartz, 2024). Figure 5 illustrates the hitchhiker problem by mapping each AI-hardware by-product metal—gallium, germanium, indium, tellurium and others—to its host industry and the consequent recovery pathway.

The 'Hitchhiker Problem': By-Product Metals and Host Metal Coupling. The 'hitchhiker problem' describes the structural dependency created when technology-critical elements are recovered as by-products of commodity metal production rather than mined as primary products (FP Analytics, 2025). Gallium hitchhikes on aluminium production, germanium on zinc, indium on zinc and lead, and tellurium on copper. The economics of by-product recovery depend on the profitability and scale of the host metal industry, meaning that supply of these elements is determined by decisions made in unrelated markets.

Figure 5. The hitchhiker problem: By-product metal recovery pathways and host industry dependencies

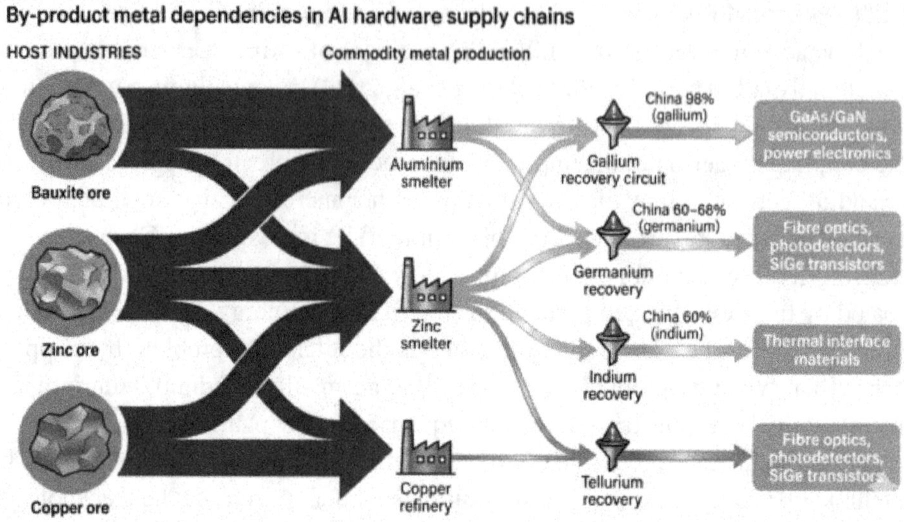

Note. Gallium, germanium, indium and similar by-product metals are recovered as adjuncts to host-metal production (zinc, alumina, copper, lead-zinc). Where host-metal demand is flat but technology-critical demand grows, the supply of by-product metals cannot expand proportionally, generating the structural coupling discussed in this section.

High Bandwidth Memory and Advanced Packaging

Copper and Through-Silicon Vias. Copper serves as the primary interconnect metal in advanced semiconductor packaging, particularly in the through-silicon vias (TSVs) that enable three-dimensional integration of memory and logic components (AnySilicon, n.d.). TSVs are vertical electrical connections that pass completely through a silicon wafer or die, enabling signals to travel between stacked chips or between a chip and its interposer. Each via is a microscopic channel, typically 5–50 micrometres in diameter, that must be filled with copper using electrochemical deposition techniques.

High-bandwidth memory stacks employ thousands of TSVs to achieve the data transfer rates required for AI workloads. A single HBM stack may contain eight or more DRAM dies connected by over 5,000 TSVs, each requiring precise copper filling to ensure reliable electrical connection. The copper intensity of AI hardware extends beyond TSVs to redistribution layers (RDLs) and the fine-pitch wiring within interposers.

Micro-Bumps and Silicon Interposers. Micro-bumps provide the mechanical and electrical connections between dies and interposers in advanced packaging configurations. These solder joints, typically 20–50 micrometres in diameter, employ tin-silver alloys that must be deposited with precise control over composition, height, and placement. The transition to lead-free solders, mandated by environmental regulations in most jurisdictions, has standardised tin-silver-copper (SAC) alloys for most packaging applications.

Silicon interposers are passive substrates that enable heterogeneous integration of multiple chiplets—including logic processors and HBM stacks—within a single package (AnySilicon, n.d.). The area of silicon interposers in AI accelerator packages substantially exceeds the area of the logic die they support. Manufacturing such large silicon pieces with acceptable yield requires stringent control of defect density and process uniformity. Figure 6 provides a cross-sectional view of a representative 2.5D AI accelerator package, locating each critical material—copper interconnects, silicon interposer, high-bandwidth-memory dies, micro-bumps and thermal-interface layers—within the integrated structure.

Figure 6. Cross-sectional view of a 2.5D AI accelerator package: Material locations in high-bandwidth memory integration

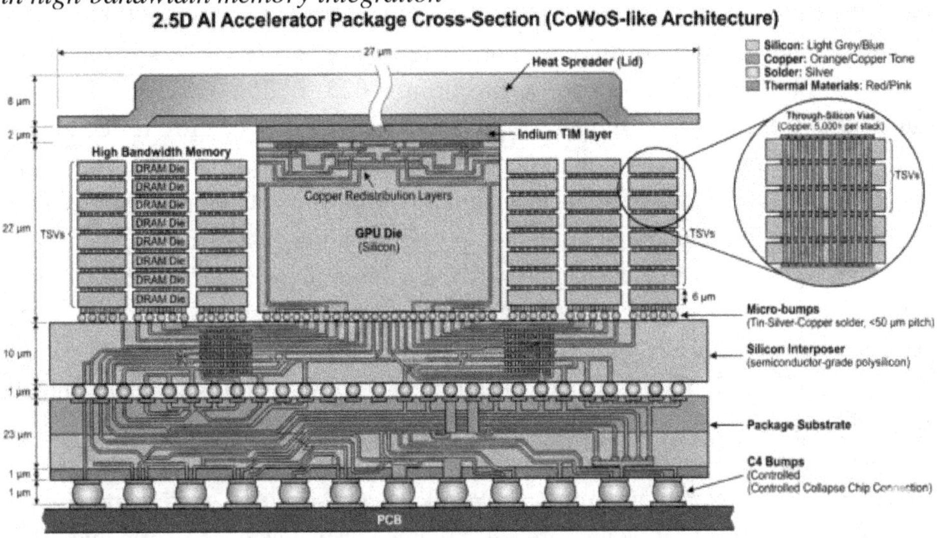

Note. Schematic cross-section showing the location of high-bandwidth memory stacks, through-silicon vias, micro-bumps with indium-containing thermal interface materials, the silicon interposer, organic substrate and the surrounding heat-spreader assembly. Material concentrations of indium, copper and specialty solders are highlighted.

Thermal Interface Materials and Capacitors. The thermal density of AI accelerator packages—often exceeding 700 watts per package—necessitates advanced thermal interface materials to transfer heat from the die to cooling solutions (FP Analytics, 2025). Indium-based thermal interface materials are widely employed due to indium's low melting point (157°C), excellent thermal conductivity, and ability to conform to surface irregularities. Liquid metal thermal compounds, typically gallium-indium alloys, offer even higher thermal conductivity but require careful handling due to their reactivity with aluminium.

Indium is another by-product metal, recovered primarily from zinc ore processing (FP Analytics, 2025). China accounts for approximately 60 per cent of global indium production. Tantalum capacitors provide the stable, high-capacitance power delivery required for consistent operation of AI accelerators under variable computational loads. The tantalum supply chain presents distinct concerns: tantalum is primarily extracted from columbite-tantalite (coltan) ores, with significant production from artisanal mining in the Democratic Republic of the Congo, raising due diligence and responsible sourcing considerations.

Data Centre Infrastructure Materials

Copper: Electrical Distribution and Heat Dissipation. Copper is the dominant material for electrical distribution within data centres, serving in busbars, cables, transformers, and switchgear from the point of grid connection through to the server power supplies. A 1 GW AI data centre requires approximately 12,000–27,000 tonnes of copper for construction, representing an intensity of 12–27 tonnes per megawatt of IT load (World Economic Forum, 2025; BHP, 2025). This intensity substantially exceeds requirements for general-purpose data centres, reflecting the enhanced electrical distribution systems necessary for high-density AI workloads.

Rare Earth Elements: Permanent Magnets in Motors and Cooling Systems. Rare earth elements, particularly neodymium, praseodymium, dysprosium, and terbium, are employed in the permanent magnets that enable high-efficiency electric motors and cooling fans throughout data centre infrastructure (FP Analytics, 2025). Neodymium-iron-boron (NdFeB) magnets provide the highest energy density of any commercially available permanent magnet technology, enabling compact motors with high power-to-weight ratios.

The rare earth supply chain presents concentration concerns. China's share of global rare earth extraction remains above 60 per cent, while its share of processing capacity, particularly for heavy rare earth elements such as dysprosium and terbium, has increased to 94 per cent (International Energy Agency, 2025). Heavy rare earths are essential for high-temperature magnet applications, where dysprosium additions prevent demagnetisation at elevated operating temperatures. Figure 7 maps the resulting geographic concentration of processing capacity for AI-hardware critical materials, making visible the share of global refining capability located in each principal jurisdiction.

Figure 7. Geographic concentration of AI hardware critical material processing capacity

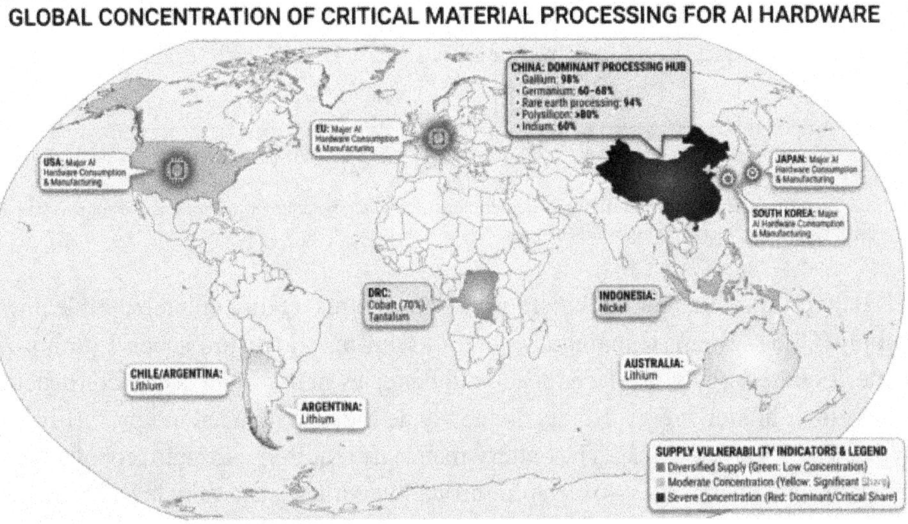

Note. Country-level shares of refining and processing capacity for the principal AI-hardware critical minerals. Figures correspond to the 2024 base year from the United States Geological Survey Mineral Commodity Summaries (2025) and to European Commission Joint Research Centre data (Georgitzikis et al., 2025). Detailed numerical values are presented in Table 4.

Table 4. Global supply concentration for AI hardware critical minerals (2024)

Mineral	China's Share of Extraction	China's Share of Processing	Alternative Sources
Gallium	98%	98%	Japan, Russia, Ukraine
Germanium	60%	60–68%	Belgium, Canada, USA
Rare earths (light)	60%	87%	Australia, USA, Myanmar
Rare earths (heavy)	90%	94%	Myanmar, limited alternatives
Silicon (polysilicon)	79%	85%	Germany, USA, South Korea
Graphite (natural)	77%	95%	Mozambique, Brazil, Madagascar
Cobalt	2%	74%	DRC (extraction), Finland
Lithium	19%	65%	Australia, Chile, Argentina

continued on following page

Table 4. Continued

Mineral	China's Share of Extraction	China's Share of Processing	Alternative Sources
Copper	8%	42%	Chile, Peru, DRC, Australia
Tantalum	<5%	~40%	DRC, Rwanda, Australia, Brazil

Source: International Energy Agency (2025); Nassar et al. (2024); FP Analytics (2025); U.S. Geological Survey (2024).

Note: Processing shares reflect refining and intermediate product manufacturing. Alternative sources listed represent current or emerging production centres.

Battery Materials for Backup Power. Data centres require uninterruptible power supply (UPS) systems to maintain operations during grid disturbances. Lithium-ion batteries have largely displaced lead-acid batteries in new data centre construction due to their higher energy density, longer cycle life, and reduced footprint (FP Analytics, 2025; Davey, 2023). The battery materials required—lithium, cobalt, nickel, graphite, and manganese—overlap significantly with electric vehicle battery supply chains. The Democratic Republic of the Congo accounts for approximately 70 per cent of global cobalt supply, while lithium production is concentrated in Australia, Chile, and Argentina (FP Analytics, 2025; Davey, 2023).

Table 5. Critical minerals for AI hardware by application category

Application	Critical Minerals	Function
Logic semiconductors	Si, Ga, Ge, In, As, P	Wafer substrates, compound semiconductors, dopants
Memory (HBM/ DRAM)	Si, Cu, W, Co, Ta	Capacitors, through-silicon vias, interconnects
Advanced packaging	Cu, Au, Ag, Sn, In	Hybrid bonding, solder bumps, thermal interface
Passive components	Ta, Nb, Ni, Pd, Ag	Capacitors, resistors, inductors
EUV lithography	Mo, Si, Ti, Sn, Zr	Multilayer mirrors, photomask substrates, tin plasma source
Printed circuit boards	Cu, Au, Ni, Pd, Ag, Sn	Conductive traces, surface finishes, solder
Thermal management	Cu, Al, In, Ga, Ag	Heat spreaders, thermal interface materials, liquid cooling
Permanent magnets	Nd, Pr, Dy, Tb, B, Fe	Motors in cooling systems, generators
Optical interconnects	Si, Ge, InP, GaAs, Er	Transceivers, photodetectors, fibre amplification
Power electronics	Si, SiC, GaN, Cu, Ag	Voltage regulation, power conversion, distribution

Source: Compiled from Berg et al. (2024); Baskaran & Schwartz (2024); NVIDIA (2024); AnySilicon (n.d.).

Note: Element symbols: Si=Silicon, Ga=Gallium, Ge=Germanium, In=Indium, As=Arsenic, P=Phosphorus, Cu=Copper, W=Tungsten, Co=Cobalt, Ta=Tantalum, Au=Gold, Ag=Silver, Sn=Tin, Nb=Niobium, Ni=Nickel, Pd=Palladium, Mo=Molybdenum, Ti=Titanium, Zr=Zirconium, Al=Aluminium,

Nd=Neodymium, Pr=Praseodymium, Dy=Dysprosium, Tb=Terbium, B=Boron, Fe=Iron, InP=Indium Phosphide, GaAs=Gallium Arsenide, Er=Erbium, SiC=Silicon Carbide, GaN=Gallium Nitride.

Shared Dependencies and Overlapping Demand

Intersection with Clean Energy and Electric Vehicle Supply Chains. The materials used in AI hardware are drawn from global supply chains that are deeply integrated with other industrial sectors. Copper, aluminium, and steel are widely used across construction, transport, and energy systems, while lithium, nickel, cobalt, and graphite are essential for battery technologies in electric vehicles and energy storage (International Energy Agency, 2021; World Bank, 2020). Rare earth elements play a role in permanent magnets used in motors and electronic components, as well as in clean energy technologies including wind turbine generators and electric vehicle traction motors (European Commission, 2023).

This overlap means that AI-related material demand interacts with, rather than displaces, other structural drivers of mineral consumption (OECD, 2023). The International Energy Agency projects that minerals demand for clean energy technologies could increase four-fold by 2040 under a net-zero emissions scenario, with particularly acute growth for lithium, cobalt, nickel, and rare earth elements (International Energy Agency, 2021). AI infrastructure expansion adds to these projections rather than being captured within them.

The temporal coincidence of AI infrastructure buildout with clean energy transition amplifies supply chain stress. Both sectors require expanded copper production, rare earth processing capacity, and lithium-ion battery manufacturing. Competition for the same materials, processing facilities, and skilled workforce may create bottlenecks that constrain progress in both domains. Figure 8 presents the overlapping AI-infrastructure, clean-energy and electric-vehicle demand profiles for the principal AI-hardware critical minerals, highlighting where the three sectors compete for the same upstream supply.

Figure 8. Overlapping critical mineral demand: AI infrastructure, clean energy and electric vehicle sectors

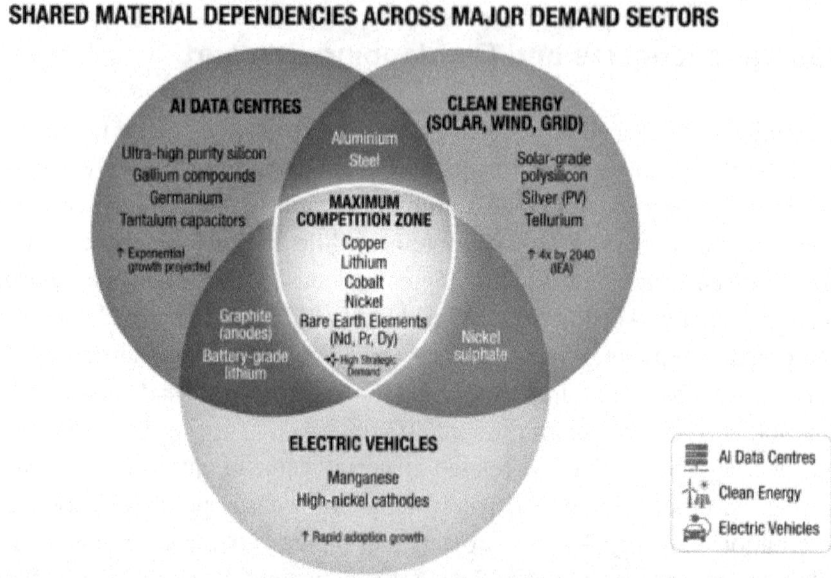

Note. Sector-by-mineral demand-overlap matrix. Copper, nickel, rare earths and lithium emerge as the highest-overlap minerals across the three sectors and are therefore the principal candidates for cross-sectoral supply-stress dynamics examined in this section.

Allocation Challenges Across Electronics, Automotive, and Energy Sectors. Available data on mineral use are typically aggregated across electronics, energy, automotive, and industrial sectors, making disaggregation to AI-specific demand analytically uncertain (International Energy Agency, 2021; OECD, 2023). Production statistics and trade data are reported at aggregated product or sectoral levels, without distinguishing between AI hardware, general computing, consumer electronics, and other end uses. This data fragmentation constrains the precision with which AI-specific material impacts can be identified.

The allocation challenge is compounded by the shared nature of manufacturing facilities and supply chains. A semiconductor fabrication plant may produce both AI accelerators and smartphone processors from the same wafer starts, using the same materials and equipment. A copper refinery supplies metal to wire manufacturers serving data centres, automotive harness producers, and construction cable makers without differentiation.

This chapter does not attempt to produce original estimates of mineral quantities attributable solely to AI hardware given these methodological constraints. Where quantitative figures are cited, they are drawn from peer-reviewed or institutional sources and are used illustratively to characterise orders of magnitude or structural patterns, rather than to infer precise AI-driven demand trajectories. The shared dependencies and overlapping demand documented in this section demonstrate why AI hardware cannot be analysed in isolation from broader critical minerals dynamics (OECD, 2023).

Comparative Material Intensity of AI-Optimised Facilities. The material intensity of AI-optimised data centre facilities substantially exceeds that of traditional hyperscale infrastructure, reflecting the distinct physical requirements of high-density AI workloads. Table 6 summarises key differences across construction materials, equipment specifications, and operational requirements.

Table 6. Comparative material intensity: AI-optimised versus traditional hyperscale data centres

Material/Metric	AI-Optimised Facility	Traditional Hyperscale	Source
Rack power density	40–132 kW/rack	8–15 kW/rack	Schneider Electric, 2025; Vertiv, n.d.
Copper intensity (t/MW IT)	12–27 t/MW	8–12 t/MW	World Economic Forum, 2025; BHP, 2025
Concrete (1 GW facility)	480,000–1,100,000 t	300,000–500,000 t	World Economic Forum, 2025
Steel (1 GW facility)	57,500–107,000 t	40,000–60,000 t	World Economic Forum, 2025
Material intensity vs baseline	+20–40%	Baseline	World Economic Forum, 2025
Rack weight (integrated)	1,000–1,500 kg	300–500 kg	NVIDIA, 2025; Dell, 2024
Server embodied carbon	~2,500 kg CO_2e/node	~500–800 kg CO_2e/node	NVIDIA, 2025; Falk et al., 2025
Cooling architecture	Liquid cooling required	Air cooling typical	Schneider Electric, 2025
Backup generators (1 GW)	400–640 units @ 2.5 MW	Proportionally fewer	World Economic Forum, 2025
Annual water (cooling)	2.6–16.6 billion L	Variable by design	World Economic Forum, 2025

Note: Traditional hyperscale figures represent industry benchmarks for general cloud computing facilities. AI-optimised figures reflect requirements for facilities designed primarily for AI training and inference workloads. Figure 9 compares material intensity per gigawatt of installed capacity for AI-optimised facilities against traditional hyperscale benchmarks, making the divergence in copper, rare earths and indium requirements directly visible.

Figure 9. Comparative material intensity: AI-optimised versus traditional hyperscale data centre facilities (per 1 GW of installed capacity)

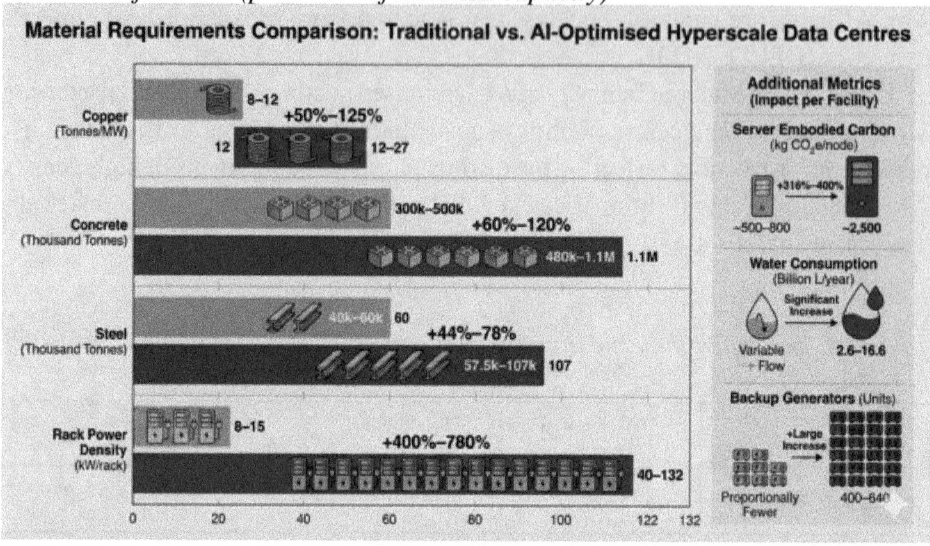

Note. Traditional hyperscale figures represent published industry benchmarks for general cloud-computing facilities; AI-optimised figures are synthesised from facility-disclosure data for representative gigawatt-scale AI campuses commissioned or under construction in 2024 and 2025. Material categories shown are copper, aluminium, steel, rare earths, indium and gallium per gigawatt of installed capacity.

The cross-sectoral framing should not be reduced to competition alone. Two important complementarities and temporal dynamics qualify the simple-competition reading. First, AI-enabled control systems for electricity grids, building energy management and battery management can deliver efficiency gains that moderate cumulative mineral demand from the broader energy transition (Kaack et al., 2022) - for example, by extending battery cycle life, optimising grid-storage dispatch and reducing curtailment of variable renewables. The net effect of these complementarities depends on the rebound elasticities of demand and is currently underdetermined empirically. Second, the temporal profile of demand differs across sectors: electric-vehicle battery demand is projected to peak in absolute terms in the early to mid-2030s under most scenarios, while AI accelerator demand for copper, rare earths and indium is expected to continue rising into the late 2030s, generating sectoral demand timing mismatches that may partially relieve simultaneous-peak supply stress. These dynamics complicate but do not invalidate the overlap-demand findings reported above.

Substitution Pathways and Alternative Materials

The material dependencies documented in the preceding sections are not immutable; substitution pathways exist at various stages of technological maturity that could reduce reliance on the most concentrated supply chains. The European Commission's Joint Research Centre has systematically assessed substitution and reduction solutions across clean energy technologies, providing a framework applicable to AI hardware components (Georgitzikis et al., 2025). Solutions range from commercial-stage alternatives with technology readiness levels exceeding Technology Readiness Level (TRL) 9 to early-stage research below TRL 8, with significant implications for policy timelines and investment priorities.

For rare earth permanent magnets used in data centre cooling systems, several substitution pathways merit consideration. Ferrite-based alternatives using strontium or barium compounds are commercially available and eliminate rare earth dependencies entirely, though with power density reductions of 50 to 70 per cent that may constrain applications requiring compact, high-efficiency motors (Georgitzikis et al., 2025). High-temperature superconductor systems operating at 30 to 50 Kelvin can generate magnetic fields 100 times stronger than copper windings for equivalent dimensions, potentially enabling smaller and lighter cooling infrastructure, though these remain at TRL 6-7 with significant engineering challenges in cryogenic thermal management (Georgitzikis et al., 2025). More immediately realisable is grain boundary diffusion processing, a commercial-stage technique that concentrates dysprosium at magnet grain boundaries rather than distributing it uniformly, reducing heavy rare earth content by up to 50 per cent while maintaining coercivity and thermal stability (Georgitzikis et al., 2025).

Copper substitution in electrical distribution presents different trade-off profiles. Aluminium conductors offer lower cost and weight but require larger cross-sections for equivalent current-carrying capacity due to higher resistivity, necessitating design modifications in busbar and cabling systems. For backup power systems, alkaline electrolysers and fuel cells eliminate platinum group metal dependencies present in proton exchange membrane technologies, offering comparable efficiency with higher durability—up to 90,000 operating hours compared to 60,000 for PEM systems—though with larger physical footprints and slower dynamic response times that may affect suitability for uninterruptible power supply applications (Georgitzikis et al., 2025).

Battery chemistry alternatives for data centre energy storage demonstrate the trade-offs inherent in substitution decisions. Lithium iron phosphate cathodes eliminate cobalt, manganese, and nickel dependencies with superior thermal stability and cycle life exceeding 2,000 cycles, though at the cost of 20 to 30 per cent lower gravimetric energy density compared to nickel-manganese-cobalt chemistries (Georgitzikis et al., 2025). Sodium-ion batteries, now entering commercial production, eliminate lithium entirely using abundant sodium, achieving energy densities of 160 Wh/kg with cycle lives of 3,000 to 4,000 cycles and enhanced safety from the ability to discharge to zero volts without degradation (Georgitzikis et al., 2025). For stationary storage applications where energy density constraints are less binding, these alternatives offer viable pathways to reduced critical mineral exposure.

The assessment of substitution pathways requires systematic evaluation across multiple dimensions: performance relative to incumbent technologies, operational lifetime and degradation characteristics, environmental and social impact profiles, cost implications across capital and operational phases, and net effects on critical raw material usage including potential increases in alternative materials (Georgitzikis et al., 2025). A substitution that eliminates one critical material while introducing dependencies on others with equally concentrated supply chains may not improve overall supply security. This multi-dimensional framework should inform policy priorities, directing support toward substitution pathways that offer genuine improvements in supply resilience rather than merely shifting vulnerabilities.

Emerging Frontier: Deep-Sea Mining of Polymetallic Nodules.

An emerging frontier in critical-mineral supply for AI hardware is deep-sea mining of polymetallic nodules in the abyssal zones of the Pacific and Indian Oceans. The Clarion-Clipperton Zone of the central Pacific holds nodules concentrated in nickel, cobalt, copper and manganese - all materials embedded in the AI hardware mineral cluster examined in this chapter. The International Seabed Authority, which has regulatory authority over mineral resources in international waters under the United Nations Convention on the Law of the Sea, has continued multilateral negotiation of an exploitation code through 2024 and 2025 without finalising the regulatory framework, leaving open a legal route by which a member state's contractor could submit a commercial application in advance of agreed rules (International Seabed Authority, 2025; Pew Charitable Trusts, 2025).

Two specific developments highlight the AI-hardware relevance of deep-sea mining. The Metals Company - through its subsidiary Nauru Ocean Resources Inc. (NORI) - submitted in 2025 the first major commercial-exploitation application for nodule recovery in the Clarion-Clipperton Zone, triggering renewed attention to whether nodule mining could supply battery-grade nickel and cobalt at the scale required by both electric-vehicle and AI-infrastructure demand projections through 2030. Independent peer-reviewed assessments raise substantial concerns about benthic biodiversity loss, sediment plume dispersion and the multi-decade timescale of ecosystem recovery, with several scientific bodies and a growing coalition of states calling for a precautionary pause until governance arrangements are finalised.

For the criticality framework adopted in this chapter, deep-sea mining sits at the highest-uncertainty extreme of supply-source diversification: it could potentially relieve land-based concentration pressures on cobalt and nickel, but at environmental cost-of-extraction profiles that are imperfectly characterised and subject to active regulatory contestation.

ENVIRONMENTAL DIMENSIONS OF CRITICAL MINERALS EXTRACTION

The environmental impacts associated with the extraction and processing of minerals used in AI hardware have been extensively documented in the environmental science and resource economics literature (World Bank, 2020; International Energy Agency, 2021). These impacts arise primarily at the upstream stages of the value chain and vary by mineral type, extraction method, processing technology, and regulatory context (OECD, 2023). This section synthesises evidence on the principal environmental dimensions of critical minerals extraction, examining land degradation, water impacts, greenhouse gas emissions, hazardous waste streams, and human rights concerns.

Land Degradation and Biodiversity Loss

Open-Pit Mining and Habitat Fragmentation. Land use and biodiversity impacts are among the most visible environmental consequences of mineral extraction. Both open-pit and underground mining operations require substantial land disturbance, including vegetation removal, soil excavation, and infrastructure development (World Bank, 2020). Mining activities represent a significant driver of land-use change and habitat fragmentation in several regions supplying critical minerals, with implications for biodiversity conservation and ecosystem services (OECD, 2023). Minerals relevant to AI hardware, such as copper, nickel, cobalt, and rare earth elements, are often extracted from large-scale operations with extensive surface footprints, particularly in tropical and sub-tropical regions (International Energy Agency, 2021).

The spatial extent of disturbance extends beyond the mine pit itself to encompass access roads, processing facilities, waste rock dumps, tailings storage areas, and worker accommodation (Luo et al., 2025). Habitat fragmentation isolates wildlife populations, disrupting migration corridors and reducing genetic diversity. The removal of topsoil and vegetation eliminates seed banks and soil biota, complicating post-mining restoration.

Tropical Forest Impacts. Gallium and germanium extraction, as by-products of bauxite and zinc mining, contribute to large-scale habitat disruption because bauxite mining frequently occurs in tropical forests (Luo et al., 2025). Bauxite deposits underlie extensive areas of tropical rainforest in Guinea, Brazil, Indonesia, and Australia, where extraction involves complete removal of forest cover and upper soil horizons. In the Democratic Republic of the Congo, cobalt mining has cleared significant areas of the Congo Basin forest, a store of approximately 60 billion tonnes of carbon, undermining climate mitigation efforts (Davey, 2023).

Rare earth mining has similarly been associated with tropical deforestation, particularly in Myanmar, where illicit extraction operations have expanded into forested border regions (Hess, 2024). These unregulated operations typically employ rudimentary techniques with minimal environmental controls, maximising habitat destruction relative to ore extracted.

Tailings Management and Long-Term Remediation. Tailings storage facilities can pose risks of leakage or failure, with potentially severe consequences for downstream ecosystems and communities (World Bank, 2020). Legacy contamination from historical mining illustrates the long-lived nature of these risks, particularly for minerals requiring complex chemical processing, such as rare earth elements (OECD, 2023). Abandoned mine sites continue to generate acid mine drainage and release heavy metals for decades or centuries after operations cease.

Abbadi and Mucsi (2024) documented that rare earth element beneficiation plants achieve yields of only 36 to 78 per cent, leaving significant rare earth content in tailings alongside other potentially valuable materials. The valorisation of tailings represents a potential secondary supply source while also addressing environmental remediation requirements.

Water Use, Contamination, and Hydrological Stress

Water Intensity of Mining and Mineral Processing. Water use and water quality impacts are central concerns in mineral extraction. Mining and mineral processing consume substantial volumes of water for ore beneficiation, dust suppression, and chemical separation processes (World Bank, 2020). The water intensity of mining varies by mineral type, ore grade, processing technology, and local climate. Water contamination occurs when tailings seep or overflow, releasing heavy metals and acidic drainage into rivers and groundwater.

Lithium Brine Extraction and Andean Water Conflicts. Lithium extraction from brine deposits, concentrated in the high-altitude salt flats of South America, presents distinct water concerns. Brine extraction involves pumping lithium-rich groundwater to the surface and concentrating through evaporation or chemical treatment (Vera et al., 2023). The 'lithium triangle' spanning Chile, Argentina, and Bolivia contains the world's largest brine-based lithium reserves beneath high-altitude salt flats in one of Earth's driest environments.

Jerez et al. (2021) documented groundwater level declines exceeding 10 metres over 15 years in the Salar de Atacama, with salt flat surfaces subsiding at one to two centimetres annually from brine extraction. Indigenous communities have characterised lithium extraction as 'water mining' that threatens traditional livelihoods and cultural practices tied to fragile desert ecosystems.

Rare Earth Processing Wastewater. Rare earth element processing is particularly water-intensive and chemically complex, involving multiple stages of crushing, leaching, and solvent extraction that produce large volumes of tailings and chemically contaminated residues (International Energy Agency, 2021; OECD, 2023). Processing wastewater typically contains acids, heavy metals, and in some cases radioactive materials. The concentration of processing capacity in China has externalised these impacts from OECD consumption centres, creating what Jerez et al. (2021) term a 'colonial shadow' in the geography of environmental burdens.

Table 7. Environmental impact profiles by mineral extraction method

Extraction Method	Representative Minerals	Land Impact	Water Impact	Carbon Intensity
Open-pit mining	Cu, Ni, REEs, Co	Very High	High	Moderate–High
Underground mining	Cu, Ta, W, Sn	Moderate	Moderate	Moderate
Brine extraction	Li, B	Low–Moderate	Very High	Low
Hard rock mining (spodumene)	Li	High	Moderate	Moderate–High
Ion-adsorption clay	Heavy REEs	High	Very High	Moderate
By-product recovery	Ga, Ge, In, Co	Indirect	Indirect	Low (allocated)
Artisanal mining	Co, Ta, Sn, Au	Variable	Variable	Low
Tailings reprocessing	REEs, Cu, Co	Low (remediation)	Moderate	Moderate

Source: Compiled from International Energy Agency (2021); World Bank (2020); Jerez et al. (2021); Vera et al. (2023); Abbadi & Mucsi (2024).

Note: Impact ratings are relative assessments. 'Indirect' indicates impacts allocated to primary product. Artisanal mining impacts vary significantly by regulatory context and practices.

Greenhouse Gas Emissions and Air Pollution

Energy Intensity of Smelting and Refining. Greenhouse gas emissions and energy intensity are significant across mineral supply chains. Mining equipment, ore processing, and transportation consume energy, with associated emissions depending on the fuel mix of the local or national electricity grid and the diesel consumption of mobile equipment (International Energy Agency, 2021). Materials commonly used in AI hardware, including aluminium, copper, and steel, are associated with substantial embodied emissions (World Bank, 2020; OECD, 2023).

Aluminium production is among the most energy-intensive industrial processes, with smelting consuming approximately 14 megawatt-hours per tonne of primary metal. Where smelters are powered by coal-fired electricity, as in much of China and Australia, the carbon footprint of aluminium exceeds 15 tonnes of CO_2 per tonne of metal. Aluminium produced using hydroelectric power achieves substantially lower emissions intensities, illustrating the importance of grid decarbonisation for reducing the embodied carbon of metals used in AI hardware.

Fluorinated Gases in Semiconductor Manufacturing. Semiconductor fabrication uses fluorinated gases that have global warming potentials thousands of times greater than carbon dioxide (Hess, 2024). These gases, including perfluorocarbons (PFCs), sulphur hexafluoride (SF_6), and nitrogen trifluoride (NF_3), are used in plasma etching and chamber cleaning processes essential to chip manufacturing. While the absolute quantities released are small relative to CO_2 emissions from energy production, the extreme potency and atmospheric persistence of these compounds make them significant contributors to the climate footprint of semiconductor production.

Carbon Footprint of Advanced Packaging. Gallium is recovered primarily as a by-product from Bayer-process liquors during alumina refining, with subsequent purification by zone refining or fractional crystallisation and final electrolytic refining and chlorination. The high-temperature processes for compound semiconductors increase energy demand relative to silicon wafers (Hess, 2024). The advanced packaging technologies essential for AI hardware—including CoWoS interposers and high-bandwidth memory stacks—add processing steps and material consumption beyond those required for standard chip packaging. Extreme ultraviolet (EUV) lithography, the enabling technology for leading-edge chip production, is exceptionally energy-intensive, with a single EUV scanner consuming approximately one megawatt of electrical power during operation.

Beyond the extraction-stage emissions analysed above, the downstream semiconductor fabrication step exposes workers and surrounding communities to a distinctive set of process chemicals and ultrafine particulates. Historical contamination cases at facilities in California and Taiwan - including IBM and Hewlett-Packard sites investigated during the 1980s and 1990s for solvent contamination of groundwater - inform contemporary concerns about per- and polyfluoroalkyl substances (PFAS), fluorinated etchants and engineered nanoparticles in advanced fabs, and inform the trajectory of national and supranational regulation of semiconductor-fabrication emissions through to 2026 (Eduljee et al., 2024). The geographic concentration of advanced fabs in a small number of clusters (Taiwan, Korea, the United States, Japan and emerging European facilities) raises broader environmental-justice questions about the distribution of fabrication-stage exposure that the chapter notes here without resolving.

Radioactive and Hazardous Waste Streams

Thorium and Uranium Co-Occurrence in Rare Earth Deposits. Rare earth element deposits frequently contain thorium and uranium as co-occurring radioactive elements, creating regulatory and waste management challenges that distinguish rare earth processing from other mineral industries (International Energy Agency, 2021; OECD, 2023). The concentration of thorium in particular increases through the beneficiation process, resulting in waste streams with elevated radioactivity that require special handling, storage, and disposal procedures.

The Lynas Malaysia Case.

The Lynas rare earth processing facility in Malaysia provides a prominent case study of the regulatory and social challenges associated with radioactive waste from rare earth processing. Lynas, an Australian mining company, established a separation plant in Kuantan, Malaysia, to process rare earth concentrate from its Mount Weld mine in Western Australia. The facility attracted sustained opposition from local communities and environmental groups concerned about radioactive waste storage and potential contamination. Malaysian regulatory authorities imposed conditions requiring Lynas to remove radioactive residues from the country or identify permanent disposal solutions. Subsequent developments confirm both the resilience and the regulatory salience of the facility: Lynas commissioned its dysprosium-oxide circuit in May 2025, making it the world's only commercial heavy rare earth producer outside China, and the Malaysian Atomic Energy Licensing Board issued a further ten-year operating licence in March 2026, conditional on continued progress towards relocating cracking-and-leaching upstream to Western Australia.

PFAS in Semiconductor Fabrication.

Per- and polyfluoroalkyl substances (PFAS) are used in semiconductor manufacturing for photoresist processing, equipment cleaning, and cooling system applications. These 'forever chemicals' are characterised by exceptional persistence in the environment and have been associated with adverse health effects including cancer, immune system disruption, and developmental impacts. Semiconductor fabrication facilities have been identified as sources of PFAS contamination in groundwater and surface water near manufacturing sites. Regulatory attention to PFAS has intensified across OECD jurisdictions, with the European Union proposing restrictions under REACH that would affect semiconductor manufacturing processes.

Human Rights and Social Justice Dimensions

Extraction of critical minerals often entails labour exploitation, conflict financing, and human rights abuses. In the Democratic Republic of the Congo, cobalt mining is associated with child labour, hazardous working conditions, and modern slavery (Davey, 2023). Artisanal and small-scale mining accounts for a significant proportion of Congolese cobalt production, with thousands of diggers extracting ore using rudimentary tools under dangerous conditions. Workers, including children, descend into unstable hand-dug tunnels without protective equipment, facing risks of collapse, respiratory disease from dust inhalation, and toxic exposure to cobalt-containing materials.

Cobalt mining pollutes streams and lakes with toxic waste and dust, killing fish and contaminating water sources consumed by communities; high levels of cobalt exposure are linked to birth defects and other health problems (Davey, 2023). Banza Lubaba Nkulu et al. (2018) documented severe water pollution near cobalt mining operations, with at least 22 scientific studies confirming contamination impacts. Muimba-Kankolongo et al. (2022) extended this analysis to examine impacts on food crops and ambient air quality.

Rare earth mining in Myanmar has expanded significantly in recent years, with production concentrated in Kachin State along the Chinese border. Illicit mining operations have proliferated in this conflict-affected region, where ethnic armed organisations and military-aligned entities compete for control of lucrative deposits. Rare earth extraction in Myanmar has been linked to environmental devastation, including deforestation, water contamination, and the destruction of agricultural land, with impacts falling disproportionately on ethnic minority communities already displaced by decades of armed conflict.

Human rights issues intersect with gender, health, and governance dimensions that complicate supply chain accountability. Women and children often bear disproportionate burdens from mining impacts, including displacement from agricultural land, exposure to contaminated water and soil, and disruption of household economies. The OECD Due Diligence Guidance for Responsible Supply Chains of Minerals from Conflict-Affected and High-Risk Areas provides recommendations for companies to identify and mitigate human rights risks (Organisation for Economic Co-operation and Development, 2011). However, many supply chains remain opaque, and accountability gaps persist.

While discussion of labour conditions in critical minerals supply chains appropriately emphasises extraction regions where human rights concerns are most acute, the expansion of processing and recycling capacity within OECD economies creates labour market implications that policy frameworks should anticipate. Evidence from technology adoption across industrial sectors indicates heterogeneous effects on employment quality and distribution, with positive aggregate productivity relationships coexisting with widening wage gaps and disproportionate impacts on specific demographic groups (European Commission, Joint Research Centre, 2025). Critical minerals processing facilities may generate similar patterns, particularly where automation and advanced process control reduce demand for production workers while increasing requirements for technical specialists.

Skills requirements for critical minerals processing span chemical engineering, materials science, metallurgy, and environmental management—specialisations with limited existing workforce capacity in many OECD jurisdictions that have outsourced processing activities over recent decades. Research suggests that green skills are often associated with lower-productivity occupations, creating tension between environmental policy objectives and labour market quality considerations (European Commission, Joint Research Centre, 2025). Recycling operations, while environmentally beneficial, may generate employment concentrated in lower-wage sorting and dismantling activities unless deliberately structured to incorporate higher-value material recovery and processing functions. Regional employment effects depend on facility location decisions that policy can influence through infrastructure investment, workforce development programmes, and permitting priorities.

Complementary education and training investments are essential to realise workforce development benefits from critical minerals capacity building. Targeted staff programmes facilitate technology adoption by ensuring that regional workforces possess requisite skills, while their absence may constrain facility operations or necessitate workforce importation that limits local employment gains (European Commission, Joint Research Centre, 2025). Job quality safeguards—including wage standards, occupational health protections, and worker voice mechanisms—should accompany industrial policy support to ensure that supply security objectives do not come at the cost of labour market degradation. The intersection of critical minerals policy with broader just transition frameworks merits attention, particularly in regions where processing facilities may locate in communities affected by decarbonisation-driven industrial restructuring.

Responsible-sourcing certification schemes have proliferated as a market-based response to the human-rights and environmental risks documented above, but their effectiveness in the cobalt and rare-earth supply chains relevant to AI hardware is contested. The Initiative for Responsible Mining Assurance (IRMA) Standard is the most developed multi-stakeholder mining-standard, with 50 standard indicators across environmental, social and governance dimensions, achievement levels (transparent / 50 / 75 / 100), and independent third-party verification by accredited audit firms;

uptake among the largest AI-hardware-relevant mines remains limited (Initiative for Responsible Mining Assurance, 2025). The persistent gap between certification status and on-the-ground outcomes is most visible in the cobalt supply chain of the Democratic Republic of the Congo, where artisanal and small-scale mining sites - documented to involve child labour, occupational health hazards from cobalt-laden dust and elevated metal body-burden in surrounding populations (Banza Lubaba Nkulu et al., 2018; Muimba-Kankolongo et al., 2022) - operate alongside, and at times within, the supply catchments of certified large-scale operations. Chain-of-custody verification across the multi-tier supply chains characteristic of AI hardware is complicated by mass-balance accounting at refining and smelting nodes, where streams from differently sourced concentrates are combined. These limitations do not negate the value of certification but indicate that procurement systems relying solely on certification status without complementary supplier audits and operational due-diligence are unlikely to deliver the human-rights and environmental performance their declared standards imply.

The application of life cycle assessment to critical minerals extraction has expanded substantially, though methodological inconsistencies limit cross-study comparisons. Sonderegger et al. (2020) reviewed 27 different life cycle impact assessment methods for mineral resources, identifying four method families based on depletion, future efforts, thermodynamic, and supply risk approaches. The diversity of methodological choices contributes to significant variation in results across studies examining nominally similar production systems.

A 2025 analysis in ACS Engineering Au reviewed 74 peer-reviewed critical mineral mining and processing studies, identifying persistent gaps in coverage of water impacts across critical minerals extraction (Luo et al., 2025). The study found that water consumption and water quality impacts were inconsistently quantified, with different studies employing incompatible metrics that preclude aggregation across the supply chain. Pell et al. (2019) documented harmonisation efforts and identified potential improvements through integration with process simulation.

Overall, the environmental impacts associated with the extraction and processing of minerals used in AI hardware are substantial, heterogeneous, and largely concentrated upstream. While AI hardware does not introduce new categories of environmental impact, its contribution to cumulative mineral demand reinforces existing pressures on ecosystems, water resources, and climate mitigation efforts (International Energy Agency, 2021; OECD, 2023). Measures such as electrification of mining equipment, water recycling, tailings reprocessing, and enhanced due diligence frameworks offer pathways for reducing the environmental footprint of critical minerals supply chains, though implementation requires sustained policy attention and investment across producing and consuming regions.

CRITICAL MINERALS POLICY

OECD economies have increasingly framed critical minerals policy as an integral component of industrial strategy, linking resource access to competitiveness in advanced technologies, including AI hardware and semiconductors (European Commission, 2023; Department for Business and Trade, 2025; Ministry of Economy, Trade and Industry, 2023). These strategies typically combine measures to diversify import sources, stimulate domestic extraction and processing, promote recycling and circular economy approaches, and strengthen international partnerships with resource-producing countries (OECD, 2023). In the EU policy context, strategic autonomy is increasingly understood not as self-sufficiency but as the capacity to act and make policy choices under conditions of interdependence. Recent conceptual work distinguishes between innovation autonomy, relating to knowledge generation and technological capability, and economic autonomy, relating to production capacity, supply-chain control and exposure to external risk (Edler et al., 2023). From this perspective, open strategic autonomy involves selectively reducing critical dependencies while maintaining openness to trade and collaboration where these enhance resilience and technological performance. This distinction is particularly relevant for strategic technologies whose value creation occurs within complex industrial ecosystems, where upstream material dependencies may undermine downstream strengths if left unaddressed (Edler et al., 2023). This section examines the policy frameworks adopted by major OECD economies.

European Union: The Critical Raw Materials Act

Quantitative Benchmarks. The European Union enacted the Critical Raw Materials Act in 2024, establishing binding domestic extraction, processing, and recycling benchmarks that represent the most comprehensive regulatory framework for critical minerals among OECD economies (European Parliament and Council of the European Union, 2024). The Act sets quantitative targets to be achieved by 2030: at least 10 per cent of the EU's annual consumption of strategic raw materials should be extracted within the Union, at least 40 per cent should be processed domestically, and at least 25 per cent should be recovered from recycling.

Although circular economy principles are increasingly embedded in EU industrial and sustainability policy, recent empirical evidence suggests that their capacity to offset import dependence is often overstated. Analysis of circular economy research and innovation activity across strategic technologies reveals a pronounced concentration on lithium-ion battery systems, with significantly lower levels of research intensity directed towards equally critical components such as permanent magnets, photovoltaic cells and hydrogen electrolysers (Baldassarre & Carrara, 2025; Baldassarre et al., 2023). Moreover, demand-side pressures arising from accelerated deployment of clean and digital technologies imply that even substantial improvements in recycling rates would only marginally reduce primary material demand in the medium term (Baldassarre & Carrara, 2025). These findings indicate that circular economy strategies must extend beyond end-of-life recycling to include material efficiency, product lifetime extension and demand moderation if they are to contribute meaningfully to strategic resilience

Import Diversification Requirements. The Critical Raw Materials Act establishes import diversification requirements specifying that not more than 65 per cent of the Union's annual consumption of any strategic raw material should originate from a single third country (European Parliament and Council of the European Union, 2024). This threshold targets extreme concentration of supply chains, particularly for materials where Chinese production or processing exceeds current thresholds. Implementation necessitates development of alternative supply sources through international partnerships, investment in emerging producer countries, and potential substitution where single-source dependencies cannot be resolved.

Strategic Project Designations and Permitting Reform. The Act establishes a framework for designating Strategic Projects that benefit from accelerated permitting procedures and priority access to financing mechanisms. The Act establishes maximum timeframes for permit decisions: 27 months for extraction projects and 15 months for processing and recycling projects, with provisions for extension in exceptional circumstances. These timelines are substantially shorter than historical permitting durations for mining projects in many European jurisdictions.

Tensions Between Supply Security and Nature Protection. The Critical Raw Materials Act's relationship with existing EU environmental legislation has generated debate. The European Union maintains extensive nature protection frameworks through the Birds and Habitats Directives, which designate protected areas (Natura 2000 sites) and restrict activities that could harm protected species or habitats. Mining and processing activities may conflict with these protections, particularly where mineral deposits coincide with areas of high biodiversity value.

The Act includes provisions allowing Strategic Projects to be designated as being of 'overriding public interest' for the purposes of environmental legislation, potentially enabling projects to proceed despite impacts on protected areas where mitigation and compensation measures are implemented. Environmental organisations have criticised this provision as creating loopholes that could undermine nature protection progress. The tension illustrates the fundamental trade-offs that critical minerals policy must manage.

United States: Security-Driven Industrial Policy

The United States has deployed a combination of tax incentives, production credits, and trade restrictions through the Inflation Reduction Act and CHIPS and Science Act (U.S. Congress, 2022a, 2022b). This legislative architecture represents the most substantial federal industrial policy intervention in decades, mobilising hundreds of billions of dollars in public and additional private investment toward domestic manufacturing capacity in semiconductors, clean energy, and critical minerals.

The CHIPS and Science Act, enacted in August 2022, provides approximately $52 billion in subsidies for domestic semiconductor manufacturing, including $39 billion in direct manufacturing incentives and $13 billion for research and workforce development. The Inflation Reduction Act provides substantial incentives for clean energy and electric vehicle manufacturing, with provisions that directly affect critical minerals supply chains through domestic content and critical mineral sourcing requirements.

Section 45X of the Internal Revenue Code establishes Advanced Manufacturing Production Tax Credits for domestic production of specified clean energy components and critical minerals. The credits provide per-unit incentives for domestic extraction and processing of critical minerals, with rates calibrated to offset cost disadvantages relative to foreign production. For critical minerals, the credit equals 10 per cent of production costs for materials extracted and processed in the United States.

The Inflation Reduction Act establishes Foreign Entity of Concern (FEOC) provisions that exclude vehicles containing battery components or critical minerals from certain entities from eligibility for consumer tax credits. The FEOC designation encompasses entities owned, controlled, or subject to the jurisdiction of specified foreign governments, with China, Russia, North Korea, and Iran identified as countries of concern. Implementation has proven complex, requiring detailed supply chain mapping to identify the origin of materials at each stage of processing.

The Pax Silica initiative, launched in late 2025, represents the latest iteration of multilateral supply chain coordination among allied nations (U.S. Department of State, 2025). The initiative builds on earlier frameworks including the Minerals Security Partnership, establishing a more formalised structure for coordinating investment, harmonising standards, and sharing information across critical minerals supply chains. The 'friend-shoring' concept underlying these initiatives seeks to restructure supply chains within networks of trusted partners rather than pursuing full domestic self-sufficiency.

Australia: Resource-Led Strategy

Australia has published and updated its national critical minerals strategy with substantial public financing commitments, positioning the country as a major supplier of responsibly produced critical minerals to global markets (Australian Government, Department of Industry, Science and Resources, 2023). The Critical Minerals Strategy 2023–2030 identifies priority minerals based on Australian resource endowments and global demand projections, with emphasis on lithium, rare earth elements, cobalt, nickel, and other battery and electronics materials. The strategy positions Australia as a 'trusted supplier' to allies seeking diversification from Chinese-dominated supply chains.

Australia's resource endowment provides a strong foundation for this positioning. The country holds significant reserves of lithium, rare earth elements, cobalt, nickel, copper, and other critical minerals, with established mining operations and exploration potential. The strategy seeks to extend Australian participation in value chains beyond extraction to encompass processing and component manufacturing, capturing greater value while contributing to allied supply security. However, Australia's high energy costs and limited processing infrastructure present barriers to downstream value addition.

The Australian Government has established the Critical Minerals Facility, a $2 billion financing mechanism to support project development across the critical minerals value chain. The Facility provides loans, equity investments, and guarantees to projects that contribute to supply security objectives while meeting environmental and social governance standards. Priority is given to projects that advance domestic processing capacity, address supply chain gaps identified through market analysis, and demonstrate commercial viability with appropriate government support.

Additional financing mechanisms include the Northern Australia Infrastructure Facility, which supports enabling infrastructure for resource projects in remote regions, and Export Finance Australia, which provides concessional finance for critical minerals projects with export orientation. The aggregated public financing commitment exceeds $4 billion across these mechanisms.

Australia's critical minerals strategy emphasises environmental, social, and governance (ESG) leadership as a competitive differentiator. The strategy positions Australian production as meeting higher environmental and social standards than competing sources, particularly those in jurisdictions with weaker regulatory frameworks. This positioning seeks to attract buyers willing to pay higher prices for certified responsible production, potentially offsetting Australia's higher cost structure.

First Nations engagement represents a significant dimension of Australian critical minerals governance. Many prospective mining areas are located on or near Indigenous lands, and project development requires negotiation of land access agreements, heritage protection, and benefit-sharing arrangements. The strategy commits to meaningful consultation and partnership with Indigenous communities, recognising both the legal requirements under the Native Title Act and the commercial necessity of obtaining social licence for project development.

Social licence constraints can represent a binding condition for critical minerals supply expansion, particularly where projects intersect with Indigenous land rights and heritage protections. In the Australian context, recent synthesis indicates that a substantial share of prospective critical mineral projects are located within formally recognised Indigenous lands, rising further when native title claims are included, implying that project timelines and feasibility are closely tied to consent processes and benefit-sharing arrangements rather than technical or financial parameters alone (Roos, 2025b). The same evidence base notes that this is not unique to Australia: a significant proportion of global energy transition mineral projects are located on or near Indigenous peoples' lands, reinforcing the need for policy frameworks that treat consultation, consent, and social partnership models as core elements of supply security strategy rather than as peripheral compliance functions (Roos, 2025b).

Australia has established bilateral critical minerals partnerships with the United States, European Union, Japan, South Korea, and India, creating frameworks for investment coordination, technology sharing, and market access. The Australia-US Climate, Critical Minerals, and Clean Energy Transformation Compact establishes preferential treatment for Australian critical minerals under US legislation, including eligibility for Inflation Reduction Act benefits. Similar agreements with other partners create diversified market access and reduce dependence on any single trading relationship.

The practical design of 'trusted supplier' partnerships increasingly relies on a limited set of commercial structures intended to reduce price and financing risk for new extraction and processing projects. Evidence from the Australia–EU partnership context indicates that long-term offtake agreements are treated as foundational instruments for underwriting project bankability, sometimes complemented by hybrid arrangements that combine equity participation with downstream purchase commitments to align incentives across the value chain (Roos, 2025b). The same evidence base suggests a growing emphasis on government-backed financial architecture—loans, guarantees, and strategic project designations—designed to lower the cost of capital and compress development timelines for midstream capacity, particularly where private investors remain cautious due to commodity price volatility and policy uncertainty (Roos, 2025b). In this framing, partnership effectiveness depends less on headline memoranda and more on whether these de-risking mechanisms translate into credible investment pipelines for processing and refining capacity that can compete under higher-cost, higher-standard production conditions (Roos, 2025b).

Canada: Upstream Supplier Positioning

Canada has published a national Critical Minerals Strategy with substantial public financing commitments, identifying 31 critical minerals based on supply risk assessment and strategic importance (Government of Canada, Natural Resources Canada, 2022). The strategy positions Canada as a secure, reliable, and sustainable source of critical minerals for North American and allied supply chains. Priority minerals include lithium, graphite, nickel, cobalt, rare earth elements, and copper, reflecting both Canadian resource endowments and demand projections from clean energy and electronics applications.

Canada's proximity to the United States and integration through the US-Mexico-Canada Agreement (USMCA) provides advantageous market access for North American supply chain development. Critical minerals extracted and processed in Canada qualify as 'domestic' under US content requirements, enabling seamless integration into US manufacturing supply chains. This regulatory alignment creates competitive advantages relative to more distant suppliers and has attracted investment from companies seeking to serve the North American market while avoiding FEOC restrictions.

Canada provides tax incentives for mineral exploration through the Mineral Exploration Tax Credit, which allows flow-through share arrangements that transfer deduction benefits to investors. The credit has been enhanced for critical minerals exploration, providing additional incentive for early-stage project development. The flow-through share mechanism enables junior exploration companies to raise capital by passing tax deductions to investors, effectively subsidising exploration risk through the tax system.

The Strategic Innovation Fund has allocated substantial resources to critical minerals processing projects, supporting feasibility studies, pilot plants, and commercial-scale facilities. Fund contributions have supported rare earth separation facilities, lithium hydroxide plants, and battery precursor manufacturing. The programme emphasises projects that advance Canadian participation in downstream value chains rather than simple extraction.

The volume of public research expenditure on critical minerals and materials substitution provides an incomplete measure of policy effectiveness; research quality and innovation productivity determine whether investments translate into deployable solutions that reduce supply vulnerabilities. Evidence from the European Conference on Corporate R&D and Innovation indicates declining ratios of patents to R&D expenditure and productivity gains to R&D investment in Europe relative to the United States, suggesting that research funding increases have not proportionally expanded the frontier of commercially viable technologies (European Commission, Joint Research Centre, 2025). This productivity gap has direct implications for critical minerals policy, where breakthrough substitution technologies rather than incremental improvements offer the greatest potential for supply chain transformation.

Research on innovation dynamics suggests that firm scale affects research quality through mechanisms relevant to critical minerals development. Large R&D investors demonstrate capacity to undertake riskier, higher-quality projects with longer development horizons and greater uncertainty—precisely the characteristics of transformational materials substitution research (European Commission, Joint Research Centre, 2025). Small firms and startups contribute complementary strengths in agility and novel approaches but may lack resources for the sustained investment required to advance materials technologies from laboratory demonstration to commercial scale. Policy frameworks should avoid structures that inadvertently disadvantage either category; support mechanisms that crowd out private investment from large firms while failing to provide patient capital for small firm scale-up may reduce overall research productivity.

Universities and public research centres play essential roles in long-term fundamental research that private firms underinvest in due to appropriability challenges and extended payback periods (European Commission, Joint Research Centre, 2025). Materials science advances underpinning next-generation permanent magnets, battery chemistries, and semiconductor compounds typically emerge from academic research programmes operating on decadal timescales incompatible with commercial R&D planning horizons. Strengthening collaboration between public research institutions and firms—including spin-off formation, personnel mobility, and joint research programmes—can accelerate translation of fundamental discoveries into commercial applications. Critical minerals strategies should explicitly address the research ecosystem architecture, ensuring that public funding complements rather than substitutes for private investment while maintaining the long-term fundamental research capacity that commercial pressures erode.

Indigenous partnership represents a distinctive feature of Canadian critical minerals governance. The duty to consult and accommodate Indigenous peoples on resource development projects affecting their rights is constitutionally protected, and modern treaties and land claim agreements in many regions establish Indigenous consent requirements or co-management arrangements. The Critical Minerals Strategy commits to partnership approaches that respect Indigenous rights while enabling project development.

Successful partnership models have included Impact and Benefit Agreements that provide employment, contracting, and revenue-sharing arrangements; equity participation enabling Indigenous ownership stakes in projects; and co-management of environmental monitoring and cultural heritage protection. These arrangements add complexity and time to project development but create more durable social licence and distribute benefits more equitably.

Canada has invested in supply chain traceability systems to verify the provenance and production conditions of critical minerals, enabling differentiation based on ESG performance. Blockchain-based tracking systems, third-party certification, and chain of custody documentation provide assurance to buyers seeking to demonstrate responsible sourcing. These systems respond to regulatory requirements including FEOC provisions, conflict mineral due diligence obligations, and emerging sustainability disclosure mandates.

The 'clean premium' strategy positions Canadian production as commanding price premiums based on environmental and social performance. This positioning acknowledges that Canadian production costs are higher than many competing sources and seeks to recover these costs through differentiated market positioning rather than cost competition. Evidence suggests that some buyers, particularly those with strong sustainability commitments or regulatory obligations, are willing to pay premiums for certified supply, though the magnitude and durability of such premiums remains uncertain.

Japan and South Korea: Import-Dependent Strategies

Japan has developed critical minerals policy through the Japan Oil, Gas and Metals National Corporation (JOGMEC), which maintains strategic stockpiling programmes for critical materials (Ministry of Economy, Trade and Industry, 2020). Japan's approach reflects its position as a resource-poor, import-dependent economy that cannot achieve meaningful domestic production for most critical minerals. The strategy emphasises supply diversification, strategic reserves, recycling, and international partnerships to manage supply risk without domestic extraction capacity.

JOGMEC maintains stockpiles of rare earth elements, rare metals, and other critical materials, with reserve levels typically calibrated to provide 60–180 days of domestic consumption. The stockpiling programme provides buffer capacity to absorb short-term supply disruptions while diversification and partnership strategies address longer-term supply security. Stockpile composition is reviewed periodically based on supply risk assessments and industrial demand patterns.

Japan enacted the Economic Security Promotion Act in 2022, establishing a comprehensive framework for managing supply chain risks across strategic sectors including semiconductors and critical minerals. The Act designates 'specified critical materials' subject to enhanced monitoring, requires supply chain surveys for designated industries, and provides support for supply chain resilience measures. The legislation reflects Japan's experience with supply disruptions, including the 2010 rare earth crisis, and positions supply chain security as an economic security priority.

The Act establishes mechanisms for government-industry coordination on supply chain monitoring, early warning systems for potential disruptions, and contingency planning for crisis response. Companies in designated sectors are required to report on supply chain dependencies and vulnerability assessments, enabling government analysis of aggregate exposure. The framework provides authority for government support to reduce identified vulnerabilities, including financing for diversification investments, stockpile access during emergencies, and coordination with allied governments on shared supply security concerns.

South Korea has published and updated its national critical minerals strategy with substantial public financing commitments, reflecting the country's position as a major semiconductor and battery manufacturer with high exposure to critical minerals supply risks (Ministry of Trade, Industry and Energy, 2023). The strategy identifies strategic minerals, sets diversification targets, and establishes financing mechanisms to support supply chain restructuring. South Korea's approach emphasises partnership with resource-rich allies, investment in overseas mining and processing, and development of recycling capacity.

The strategy establishes specific targets for reducing dependence on concentrated supply sources, with particular attention to materials where Chinese processing dominates. Investment support is available for Korean companies participating in overseas mining and processing projects, creating offshore supply sources under Korean corporate control. The Korea Mine Rehabilitation and Mineral Resources Corporation (KOMIR) and Korea Resources Corporation support overseas investment and technology development for critical minerals applications.

China as Contextual Case

China is considered as a contextual case due to its central role in global mineral processing and refining rather than as a subject of domestic policy analysis (European Commission, 2023; OECD, 2023). China's imposition of export licensing requirements for gallium and germanium in July 2023, followed by expanded restrictions in December 2024 covering additional minerals and eventually extending to components containing Chinese-origin materials, demonstrated the weaponisation of mineral supply for geopolitical purposes (International Energy Agency, 2023; International Energy Agency, 2025; Baskaran & Schwartz, 2024).

The export restrictions represent strategic weaponisation of mineral supply chains—the deliberate use of supply concentration as a tool of statecraft. This weaponisation differs from the 2010 rare earth crisis, which arose from a specific bilateral dispute, in its explicit integration with broader technology competition. The restrictions on gallium and germanium were imposed in the context of ongoing disputes over semiconductor manufacturing equipment, advanced chip exports, and artificial intelligence capabilities. Mineral supply has thus become entangled with technological and geopolitical competition between major powers.

China's dominance in critical minerals extends across multiple stages of value chains. As of 2024, China accounted for 98 per cent of global raw gallium production and between 60 and 68 per cent of refined germanium output (Nassar et al., 2024; FP Analytics, 2025). For rare earth elements, China's share of global extraction remains above 60 per cent, while its share of processing capacity, particularly for heavy rare earth elements, has increased to 94 per cent (International Energy Agency, 2025). This concentration reflects decades of industrial policy, infrastructure investment, and tolerance for environmental externalities that OECD producers have been unable or unwilling to replicate.

The policy implications of Chinese processing dominance shape OECD responses across all dimensions of critical minerals policy. Diversification targets must be calibrated against the reality of limited alternative capacity, with achievable timelines extending over decades rather than years for materials requiring substantial processing infrastructure development. Investment incentives must account for the cost disadvantage relative to Chinese production, which benefits from scale economies, lower environmental compliance costs, and established infrastructure. International partnerships must coordinate limited allied capacity to avoid duplication while ensuring supply resilience.

Supply chain restructuring to reduce dependence on concentrated foreign sources may inadvertently create new concentration risks within OECD economies if competition policy fails to account for innovation dynamics in critical minerals processing. Evidence on technological mergers and acquisitions indicates that post-acquisition innovation typically declines in acquired firms without corresponding increases in acquirer innovation output, suggesting that consolidation reduces rather than enhances sectoral research productivity (European Commission, Joint Research Centre, 2025). Where acquisitions involve firms developing substitution technologies, recycling processes, or alternative material pathways, the innovation losses may directly undermine supply security objectives that motivate industrial policy intervention.

The mechanism through which technological acquisitions affect innovation merits attention for critical minerals policy. Acquiring firms gain market power through intellectual property consolidation rather than efficiency improvements, potentially blocking alternative development pathways that would increase supply diversity (European Commission, Joint Research Centre, 2025). A rare earth processing firm acquiring a startup developing rare-earth-free permanent magnet technology may rationally suppress rather than commercialise the acquired innovation if it threatens existing business models. Competition authorities assessing mergers in critical minerals sectors should incorporate innovation considerations alongside conventional market share analyses, recognising that supply security depends not only on current production capacity but on the continued development of alternatives that reduce long-term vulnerability.

The tension between scale economies in processing—which favour consolidation—and innovation dynamics—which favour competitive markets with multiple independent development pathways—requires explicit policy attention. Processing facilities for rare earth separation, platinum group metal refining, and battery material production exhibit significant economies of scale that create natural tendencies toward concentration. Achieving efficient scale while maintaining competitive innovation dynamics may require structural interventions such as mandatory licensing of publicly funded intellectual property, restrictions on vertical integration between mining and processing, or requirements that subsidised facilities maintain open access for toll processing. The objective is not to prevent efficient consolidation but to ensure that consolidation does not foreclose the innovation pathways essential for long-term supply security.

Multilateral Coordination

The Minerals Security Partnership (MSP) was established in 2022 as a US-led initiative to coordinate critical minerals investment and policy among like-minded partners. Initial membership included Australia, Canada, Finland, France, Germany, Japan, the Republic of Korea, Sweden, the United Kingdom, the United States, and the European Commission. The Partnership aims to catalyse public and private investment in responsible critical minerals supply chains, with focus on diversifying sources, developing processing capacity, and establishing high environmental and social standards.

The MSP operates through working groups addressing specific minerals and cross-cutting issues, with coordination of investment priorities and project support across partner governments. The Partnership has identified priority projects for coordinated support, seeking to avoid duplication of investment while ensuring geographic diversification. The initiative represents an institutionalisation of friend-shoring approaches, creating frameworks for allies to share information, coordinate policy, and align investment incentives in pursuit of shared supply security objectives.

The OECD Due Diligence Guidance for Responsible Supply Chains of Minerals from Conflict-Affected and High-Risk Areas provides recommendations for companies to identify and mitigate human rights and environmental risks in mineral supply chains (Organisation for Economic Co-operation and Development, 2011). The Guidance establishes a five-step due diligence framework encompassing supply chain mapping, risk identification and assessment, risk mitigation, independent audit, and public reporting. Originally developed for conflict minerals in the Great

Lakes Region of Africa, the framework has been extended to cover broader supply chain risks including child labour, environmental damage, and governance failures.

The Guidance has been incorporated into regulatory requirements across multiple jurisdictions, including the EU's Conflict Minerals Regulation, US Dodd-Frank Act provisions on conflict minerals, and emerging due diligence legislation in several European countries. Companies sourcing critical minerals are expected to implement due diligence systems consistent with the OECD framework, creating baseline expectations for supply chain transparency and risk management. However, implementation varies widely across companies and sectors, with coverage gaps particularly for materials and regions not specifically targeted by regulatory requirements.

The International Energy Agency has developed substantial analytical capacity on critical minerals for clean energy technologies, producing regular assessments of supply, demand, and policy developments. The IEA's annual Critical Minerals Market Review provides comprehensive analysis of market conditions, supply chain developments, and policy responses across major economies. The agency's scenario modelling projects future mineral demand under various energy transition pathways, providing evidence base for policy development and investment decisions.

While the IEA's mandate centres on energy security and clean energy transition, its critical minerals work increasingly encompasses the digital economy given the overlap in material dependencies. The agency's analysis informs policy development across OECD member states, providing common analytical foundations that support coordination of national strategies. The IEA has called for expanded international cooperation on critical minerals, including information sharing, joint research, and coordination of stockpiling arrangements (International Energy Agency, 2021; International Energy Agency, 2025).

Three international law and governance frameworks intersect with AI-hardware mineral sourcing beyond the OECD due-diligence guidance examined above. The United Nations Guiding Principles on Business and Human Rights (United Nations Human Rights Council, 2011) establish a corporate responsibility to identify, prevent and account for adverse human-rights impacts across business activities and value chains; their state-duty and remedy pillars apply directly to mineral-sourcing decisions by AI-hardware OEMs operating in jurisdictions with extractive-sector human-rights risk, an issue rendered particularly acute by the geographic concentration of energy-transition mineral deposits on land occupied by or adjacent to land-connected peoples (Owen et al., 2023). The Extractive Industries Transparency Initiative provides a payment-disclosure and beneficial-ownership standard adopted by 57 implementing countries by 2025, including major critical-minerals producers; its limitations include patchy implementation, voluntary status and partial coverage of artisanal and small-scale mining, but it operationalises the transparency layer on which downstream due-diligence depends (Extractive Industries Transparency Initiative, 2025). The Kunming-Montreal Global Biodiversity Framework, adopted under the Convention on Biological Diversity in December 2022, includes Target 14 (mainstreaming biodiversity in business decisions) and Target 7 (reducing pollution risks) of direct relevance to mining-induced land-use change and tailings management; its operationalisation through national biodiversity strategies through 2030 generates parallel reporting obligations that intersect with AI-hardware supplier due-diligence and recycled-content auditing.

INDUSTRIAL POLICY INSTRUMENTS

The industrial policy frameworks examined in the preceding section deploy a range of instruments to achieve supply security objectives. This section evaluates these instruments through the lens of environmental sustainability, examining the extent to which policy measures address environmental risks alongside strategic supply concerns (OECD, 2023). The analysis considers recycling and circular economy approaches, domestic processing incentives, regulatory reform, strategic stockpiling, trade policy, and international partnerships, assessing each instrument category for its environmental implications and potential unintended consequences. The section concludes with an assessment of policy coherence and the risk of burden-shifting across jurisdictions with divergent governance capacity.

Recycling and Circular Economy Instruments

Recycling and circular economy approaches feature prominently in OECD critical minerals strategies, reflecting recognition that secondary supply can reduce dependence on primary extraction and its associated environmental impacts (OECD, 2023; European Commission, 2023). The European Union's Critical Raw Materials Act establishes a 25 per cent recycling target for strategic raw materials by 2030, positioning circular economy as a pillar of supply security alongside domestic extraction and processing (European Parliament and Council of the European Union, 2024). Similar commitments appear in national strategies across OECD economies, though quantitative targets and implementation mechanisms vary substantially.

Policy support for recycling takes multiple forms: extended producer responsibility schemes that assign collection and recycling costs to manufacturers; investment incentives for recycling infrastructure; research funding for recycling technology development; and regulatory requirements for recycled content in new products. The Inflation Reduction Act includes provisions supporting battery recycling, while the European Battery Regulation establishes recycled content requirements for battery materials. These frameworks create both regulatory push and economic pull for recycling capacity development.

Current recycling rates for critical minerals used in AI hardware reveal substantial gaps between policy ambitions and industrial reality, with technical barriers that complicate aspirational targets. Copper recycling represents a relative success, with established collection and processing infrastructure achieving recovery rates that partially offset primary production requirements; however, the magnitude of demand growth projected for data centre infrastructure means that even high recycling rates cannot substitute for expanded primary supply in the medium term (Georgitzikis et al., 2025; World Bank, 2020). For specialty metals such as gallium, germanium, and indium, and for rare earth elements, the situation is considerably more challenging, with recycling rates typically below 1 per cent globally, reflecting both technical difficulties and unfavourable economics at current prices and scales. Rare earth element beneficiation plants achieve yields of only 36 to 78 per cent during primary processing, leaving significant rare earth content in tailings alongside potentially hazardous co-occurring elements (Abbadi & Mucsi, 2024). This upstream inefficiency compounds the downstream recycling challenge, as materials that never enter the use phase cannot subsequently be recovered. The installed base of AI hardware that could serve as feedstock for urban mining remains limited, as the sector's rapid growth means most equipment is still in active use.

Technical barriers to recycling critical minerals fall into several categories that policy interventions must address differentially (Georgitzikis et al., 2025). Low concentrations of target materials in electronic waste streams—often measured in parts per million for specialty metals—render conventional mechanical separation economically unviable, requiring energy-intensive hydrometallurgical or pyrometallurgical processes with their own environmental footprints and hazardous waste generation (Luo et al., 2025). Product heterogeneity complicates automated sorting, as AI accelerators from different manufacturers and generations contain varying material compositions without standardised labelling or design-for-recycling features. Dissipative uses present fundamental recovery barriers; materials dispersed into products or environments from which collection is technically infeasible—such as thermal interface compounds, thin-film coatings, or materials lost through wear—cannot be recycled regardless of processing technology advancement.

The European Union's Critical Raw Materials Act establishes a 25 per cent recycling target for strategic raw materials by 2030, a benchmark that must be contextualised against these technical realities. For materials with established recycling infrastructure and concentrated end-of-life streams—copper from cables, aluminium from casings—this target appears achievable with sufficient collection system development. For rare earth elements with current recycling rates below 1 per cent globally, achieving 25 per cent recovery within six years would require transformational advances in collection logistics, separation technology, and processing capacity that current investment trajectories do not support (International Energy Agency, 2025). The transition from aspirational targets to operational recycling systems represents a multi-decade challenge that cannot substantially contribute to supply security in the near term. Policy frameworks should accordingly distinguish between materials where recycling can make meaningful contributions to supply security and those where substitution, demand reduction, or supply diversification represent more realistic pathways to reduced vulnerability Figure 10 sets out the circular economy pathways available for the critical minerals embedded in AI hardware, from product-as-a-service and refurbishment through component recovery and material recycling, with indicative end-of-life flow shares.

Figure 10. Circular economy pathways for critical minerals in AI hardware supply chains

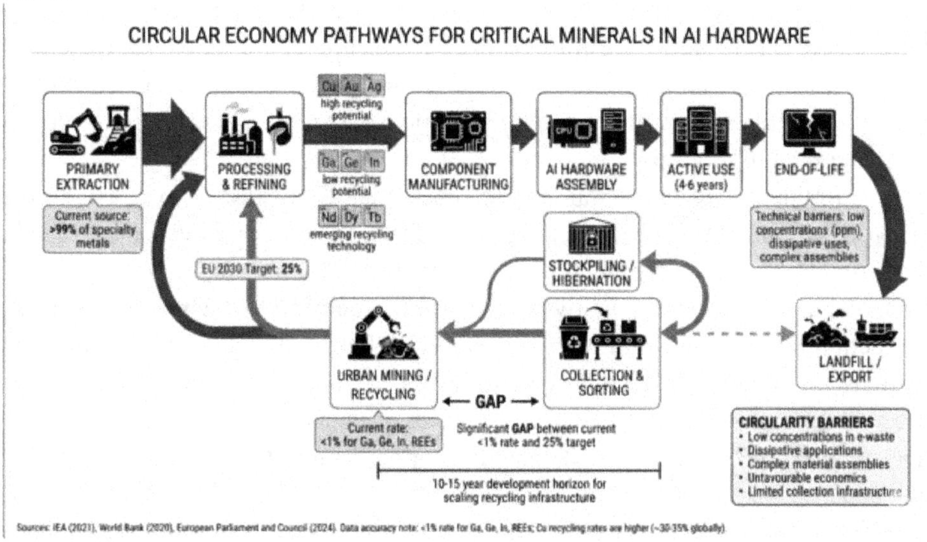

Note. The diagram illustrates material flows from primary extraction through manufacturing to end-of-life, comparing current recycling rates (typically below 1 per cent for specialty metals) with the European Union's 25 per cent recycling target for 2030. Arrow thickness indicates relative material-flow volumes. Barriers to circularity include low concentrations in electronic waste, dissipative uses and unfavourable economics at current prices.

Technical challenges constrain recycling potential for many critical minerals. Low concentrations of target materials in electronic waste streams—often measured in parts per million for specialty metals—require processing of large volumes of material to recover commercially meaningful quantities. The heterogeneity of electronic waste, with thousands of different product types containing varying material compositions, complicates automated sorting and requires manual disassembly for effective recovery. These factors drive processing costs that often exceed the value of recovered materials at current prices.

Dissipative uses present barriers to recycling for certain materials. When materials are dispersed into products or environments from which recovery is technically or economically infeasible—such as thin films, coatings, or materials that wear away during use—the circular economy loop cannot be closed regardless of policy support. Some critical minerals used in AI hardware are employed in applications where recovery is theoretically possible but practically challenging, such as the indium in thermal interface materials or the silver in solder joints. Understanding the distinction between technically recyclable and practically recoverable materials is essential for realistic assessment of circular economy potential.

'Urban mining'—the recovery of materials from discarded products and waste streams—represents a potential secondary supply source that could reduce pressure on primary extraction. Abbadi and Mucsi (2024) documented that rare earth element beneficiation plants achieve yields of only 36 to 78 per cent, leaving significant rare earth content in tailings alongside other potentially valuable materials. The valorisation of tailings and electronic waste represents both an environmental remediation opportunity and a potential supply source, though economics remain challenging in most cases.

The limitations of urban mining are substantial. The stock of materials in the urban environment is finite and cannot grow faster than new products are added to the installed base; for a rapidly growing sector like AI hardware, secondary supply from end-of-life products will always lag primary demand by the product lifetime. Collection rates for electronic waste remain well below 100 per cent even in jurisdictions with mandatory take-back schemes, as products are stockpiled, informally processed, or exported to developing countries. The quality of recovered materials may not meet specifications for demanding applications without additional refining that adds cost and environmental impact.

Table 8. Current recycling rates and circular economy potential for AI hardware materials

Material	Current Recycling Rate	Technical Potential	Key Barriers
Copper	30–35%	High (>50%)	Collection infrastructure, alloy separation
Gold	20–30%	High (>40%)	Informal sector competition, low concentrations
Silver	15–20%	Moderate (30–40%)	Dispersed applications, collection costs
Rare earth elements	<1%	Low–Moderate (10–20%)	Technical complexity, unfavourable economics

continued on following page

Table 8. Continued

Material	Current Recycling Rate	Technical Potential	Key Barriers
Gallium	<1%	Low (<10%)	Dissipative uses, very low concentrations
Germanium	<1%	Low–Moderate (10–15%)	Small volumes, dispersed in electronics
Indium	<1%	Low (<10%)	Thin-film applications, dissipative uses
Tantalum	10–20%	Moderate (25–35%)	Capacitor recovery, product complexity
Cobalt	~32%	High (>50%)	Battery chemistry changes, collection rates
Lithium	<5%	Moderate (20–30%)	Process economics, pyrometallurgical losses

Source: Compiled from International Energy Agency (2021, 2025); World Bank (2020); Abbadi & Mucsi (2024); European Commission (2023).

Note: Current recycling rates are global estimates and vary by region. Technical potential assumes development of commercially viable recovery processes. EU CRMA 2030 target is 25% recycling across strategic raw materials.

Domestic Processing Incentives

Domestic processing incentives aim to build OECD capacity for mineral refining and component manufacturing, reducing dependence on concentrated foreign supply sources. However, reshoring processing capacity also means reshoring the environmental burdens associated with these activities. The environmental trade-offs inherent in 'onshoring' mineral processing include the management of radioactive waste streams associated with rare earth element separation and the high energy intensity of purification processes (Berg et al., 2024). Processing that currently occurs in jurisdictions with weaker environmental standards would be subject to more stringent controls in OECD locations, potentially improving global environmental outcomes but also revealing costs that have historically been externalised.

Water treatment and tailings management represent significant cost components for operations meeting OECD environmental standards, contributing to the competitive disadvantage of production outside China (Jerez et al., 2021). Rare earth element processing is particularly water-intensive and chemically complex, involving multiple stages of crushing, leaching, and solvent extraction that produce large volumes of tailings and chemically contaminated residues (International Energy Agency, 2021; OECD, 2023). The concentration of processing capacity in China has externalised these impacts from OECD consumption centres, creating what Jerez et al. (2021) term a 'colonial shadow' in the geography of environmental burdens.

The cost structure of mineral processing in OECD jurisdictions substantially exceeds that of Chinese operations, reflecting higher labour costs, more stringent environmental compliance requirements, and the absence of scale economies that benefit established producers. Section 45X production tax credits of 10 per cent of production costs provide meaningful support but may be insufficient to close cost gaps that often exceed this level. The sustained subsidisation requirements implied by current cost differentials raise questions about fiscal sustainability and the distortionary effects of long-term production support.

The production tax credit structure represents a departure from previous industrial policy approaches, which typically relied on grants, loans, or demand-side procurement preferences. The tax credit mechanism provides ongoing operational support rather than one-time capital subsidies, reducing project risk by ensuring revenue support throughout the production period. However, critics note that the credits create ongoing fiscal obligations without guaranteed supply outcomes, and the level of support required to achieve competitiveness with established producers may exceed what is politically sustainable over the multi-decade timelines required for supply chain restructuring.

Processing in OECD jurisdictions would be subject to environmental regulatory frameworks that impose controls absent or weakly enforced in many current producing regions. Air quality standards, water discharge limits, hazardous waste management requirements, and worker safety regulations add costs but also improve environmental and social outcomes. The net environmental effect depends on whether reshoring substitutes for foreign production or adds to global capacity, and whether displaced foreign production would continue to serve other markets.

The Lynas Malaysia case illustrates the regulatory complexity of rare earth processing. Malaysian regulatory authorities imposed conditions requiring Lynas to remove radioactive residues from the country or identify permanent disposal solutions, effectively demanding that the 'cracking and leaching' stages—which generate the radioactive waste—be relocated back to Australia or elsewhere. While Lynas was eventually granted a reprieve to continue importing concentrate while it constructed cracking and leaching facilities in Australia, the episode demonstrates how community opposition and regulatory requirements in host countries can constrain the geographic distribution of processing capacity.

Critical minerals policy discourse emphasises large-scale mining and processing facilities, yet innovative firms developing recycling technologies, traceability systems, and substitution materials represent essential components of supply chain transformation. Startups and spinoffs from university research programmes often pioneer novel approaches that established firms lack incentives to develop, including technologies that threaten incumbent business models. Evidence from European innovation ecosystems indicates that high-potential startups relocate in pursuit of resources—talent, finance, and supportive regulatory environments—suggesting that jurisdictions seeking to build critical minerals innovation capacity must attend to ecosystem characteristics rather than merely offering financial incentives (European Commission, Joint Research Centre, 2025).

The financing landscape for critical minerals startups presents particular challenges. Europe relies more heavily than the United States on public venture capital for early-stage funding, with private investors entering at later stages when technology and market risks have substantially diminished (European Commission, Joint Research Centre, 2025). This pattern may reflect risk aversion in European private capital markets or gaps in the institutional infrastructure connecting early research to commercial deployment. Public venture interventions should mobilise private capital through mixed funding models rather than substituting for private investment, ensuring that public resources address genuine market failures rather than crowding out private risk-taking. The extended development timelines characteristic of materials technologies—often exceeding a decade from laboratory demonstration to commercial production—require patient capital structures that conventional venture models, oriented toward five-to-seven-year exit horizons, struggle to provide.

Scale-up represents a critical transition point where promising technologies frequently fail to achieve commercial viability. The 'valley of death' between successful pilot demonstration and commercial-scale production is particularly deep for materials technologies requiring capital-intensive processing infrastructure. European firms face additional barriers from fragmented capital markets that impose a 'glass ceiling' limiting growth potential relative to firms with access to deeper US capital pools (European Commission, Joint Research Centre, 2025). Place-based policies should reflect local ecosystem characteristics—the presence of relevant research institutions, availability of skilled personnel, access to necessary infrastructure—rather than applying uniform approaches across diverse regional contexts. Critical minerals strategies that focus exclusively on attracting large-scale foreign investment may neglect the startup ecosystem development essential for long-term innovation capacity and supply chain resilience.

Regulatory and Permitting Reform

Permitting reform represents one of the most contentious elements of critical minerals policy. The Critical Raw Materials Act establishes maximum timeframes for permit decisions: 27 months for extraction projects and 15 months for processing and recycling projects, with provisions for extension in exceptional circumstances (European Parliament and Council of the European Union, 2024). These timelines are substantially shorter than historical permitting durations for mining projects in many jurisdictions, where environmental impact assessment, public consultation, and appeals processes have typically extended approval periods to five years or longer.

Critics argue that accelerated permitting may compromise environmental assessment quality, reducing the thoroughness of impact analysis and limiting opportunities for public participation and legal challenge. Environmental organisations have characterised some permitting reforms as prioritising economic interests over environmental protection, potentially enabling projects to proceed despite significant ecological risks. Proponents contend that existing timelines are unnecessarily prolonged due to regulatory inefficiency rather than substantive assessment requirements, and that streamlined processes can maintain assessment quality while reducing delays.

Permitting authority for mining and processing projects is typically distributed across multiple levels of government, creating coordination challenges that accelerated timelines must address. In federal systems such as Australia, Canada, and the United States, primary permitting authority often rests with sub-national governments, while federal legislation may impose additional requirements for projects with national significance or environmental impacts affecting federal interests. The Critical Raw Materials Act's timeline requirements apply at the EU level but depend on implementation by member state authorities with varying administrative capacities and procedural traditions.

Indigenous rights and consultation requirements add additional layers of governance in jurisdictions with constitutional or treaty protections for Indigenous peoples. In Australia and Canada, project development requires negotiation of land access agreements, heritage protection measures, and benefit-sharing arrangements that cannot be compressed into accelerated timelines without compromising the quality of engagement. The duty to consult and accommodate Indigenous peoples reflects constitutional protections that cannot simply be overridden by economic urgency, creating tension between permitting reform objectives and rights-based governance frameworks.

Substantial gaps exist between policy ambitions expressed in targets and timelines and the implementation capacity available to achieve them. The European Union's 2030 targets for extraction, processing, and recycling require the development of projects that currently do not exist, through permitting processes that have not been reformed, using financing mechanisms that are still being established. Even with accelerated permitting, the timeline from project conception through exploration, feasibility, financing, construction, and ramp-up typically exceeds a decade for major mining projects.

Policy analysis is constrained by the lag between legislative enactment and the availability of empirical outcome data; evaluation must focus on structural incentives, announced allocations, and early regulatory actions rather than long-term impact assessments. Industrial policy measures are evolving rapidly, and national strategies are often aspirational rather than implemented programmes. Current assessments are therefore provisional and the effectiveness of policy instruments will only become apparent over multi-year implementation periods.

Policy frameworks for critical minerals increasingly reference alternative materials and substitution technologies without systematic assessment of their readiness for deployment at scale. A technology readiness level framework, adapted from aerospace and defence procurement practices, provides essential discipline for distinguishing between solutions available for immediate implementation, those requiring final commercialisation efforts, and those still in research phases with uncertain development trajectories (Georgitzikis et al., 2025). Misalignment between policy timelines and technology maturity creates implementation gaps that undermine credibility and may misdirect investment.

Commercial-stage solutions at TRL 9 and above can be deployed within existing industrial ecosystems with standard procurement and installation processes. For critical minerals relevant to AI hardware, this category includes grain boundary diffusion processing for rare earth magnet optimisation, which reduces heavy rare earth content by up to 50 per cent using adaptations to existing magnet manufacturing lines (Georgitzikis et al., 2025). Ferrite permanent magnets and alkaline electrolyser systems similarly represent mature technologies with established supply chains, though their performance characteristics may constrain applicability to specific use cases. Lithium iron phosphate battery chemistries have achieved commercial scale with multiple manufacturers, demonstrating that supply chain transitions are achievable when technology maturity aligns with market incentives.

Early commercialisation technologies at TRL 8-9 require final engineering, manufacturing scale-up, and supply chain development before widespread deployment. Sodium-ion batteries exemplify this category, with initial commercial production announced by multiple manufacturers but volumes insufficient to materially affect lithium demand in the near term (Georgitzikis et al., 2025). High-temperature superconductor systems for motors and generators have demonstrated performance in prototype installations but face cost and reliability challenges that limit adoption. Perovskite solar cells have achieved laboratory efficiencies exceeding 26 per cent but encounter stability and lifetime concerns requiring resolution before utility-scale deployment. Policy support for early commercialisation technologies appropriately emphasises manufacturing demonstration, supply chain coordination, and demand aggregation rather than basic research.

Research-stage technologies below TRL 8 remain subject to fundamental uncertainties regarding performance, cost, and manufacturability that complicate inclusion in near-term policy targets. Iron nitride permanent magnets, frequently cited as potential rare earth alternatives, have not progressed beyond laboratory demonstration despite decades of research interest (Georgitzikis et al., 2025). Quantum dot solar cells and metal-air batteries similarly represent scientifically promising concepts whose commercialisation timelines remain indeterminate. Policy frameworks that assume availability of research-stage technologies by 2030 risk creating dependencies on solutions that may not materialise, potentially delaying investment in proven alternatives. Realistic technology assessment should inform differentiated policy approaches: commercial technologies warrant deployment incentives and procurement preferences; early commercialisation technologies merit manufacturing support and patient capital; research technologies require sustained fundamental research funding without inclusion in binding supply security targets.

Strategic Stockpiling

Strategic stockpiling provides buffer capacity to absorb short-term supply disruptions while longer-term diversification and capacity-building strategies mature. Stockpiles may be structured as defence reserves, maintained to ensure continued supply of military-essential materials during conflict, or as economic reserves, designed to moderate price volatility and prevent supply disruptions from affecting broader industrial production. The distinction has implications for stockpile composition, release criteria, and governance arrangements.

Defence stockpiles are typically managed by military or national security agencies with release authority tied to declared emergencies or conflict scenarios. Economic stockpiles may be managed by dedicated agencies or central banks with release triggered by market conditions rather than security crises. The appropriate scale, composition, and governance of stockpiles depends on the objectives they are designed to serve, with defence applications requiring smaller quantities of specific military-critical materials while economic stabilisation may require larger holdings of industrially significant commodities.

Japan's approach to strategic stockpiling, administered through the Japan Oil, Gas and Metals National Corporation (JOGMEC) as detailed in Section "JOGMEC and Strategic Stockpiling", exemplifies a hybrid model that combines elements of both defence and economic stockpiling rationales (Ministry of Economy, Trade and Industry, 2020). This hybrid character merits examination as a potential template for other import-dependent OECD economies considering stockpile programme design.

The distinguishing feature of Japan's model lies in its release criteria, which encompass both security emergencies and severe market disruptions affecting industrial operations. Where purely defence-oriented stockpiles typically reserve materials exclusively for military contingencies, and purely economic stockpiles focus on price stabilisation through counter-cyclical releases, Japan's framework permits intervention across a broader spectrum of supply disruptions. This flexibility reflects Japan's experience with the 2010 rare earth crisis, which demonstrated that supply restrictions can inflict severe economic damage without rising to the level of a military emergency (Sprecher et al., 2015; Mancheri et al., 2019).

The hybrid model carries implications for stockpile governance and composition. Materials must be selected not only for military criticality but also for industrial significance, potentially expanding the range of commodities held in reserve. JOG-MEC maintains stockpiles of rare earth elements, rare metals, and other critical materials, with reserve levels typically calibrated to provide 60–180 days of domestic consumption (Ministry of Economy, Trade and Industry, 2020). The stockpile composition is reviewed periodically based on supply risk assessments and industrial demand patterns, with adjustments to reflect changing technological requirements and supply chain conditions.

For OECD economies contemplating stockpile programmes, Japan's hybrid model offers an alternative to the binary choice between defence-only and economic stabilisation approaches. The model is particularly relevant for economies with limited domestic extraction capacity and high dependence on concentrated supply sources, where the practical imperative of maintaining industrial continuity may outweigh analytical distinctions between economic disruption and security threat.

The United States has invoked the Defense Production Act (DPA) to support critical minerals production, authorising Title III investments in domestic mining and processing capacity. The DPA mechanism also provides authority for stockpile acquisition, enabling federal purchase of materials for strategic reserves. Recent policy developments have emphasised expansion of the National Defense Stockpile to include materials essential for both defence applications and civilian industrial production, reflecting the broader conception of economic security that now encompasses supply chain resilience.

The U.S. Energy Act of 2020 and subsequent Executive Orders defined critical minerals as those essential to economic or national security and vulnerable to disruption. This definition has been operationalised through stockpile planning that extends beyond traditional defence materials to encompass semiconductor inputs, battery materials, and other minerals essential for advanced technology production. The expanded scope increases both the ambition and complexity of stockpile management, requiring assessment of requirements across diverse industrial applications.

Table 9. Comparative analysis of OECD industrial policy instruments

Instrument	EU	United States	Australia	Canada	Japan/Korea
Extraction targets	10% by 2030	No formal target	Resource-led expansion	Exploration incentives	N/A (import focus)
Processing targets	40% by 2030	45X tax credits	Value-add focus	SIF funding	Tech development
Recycling targets	25% by 2030	IRA incentives	Emerging programs	Emerging programs	High priority
Diversification threshold	65% max single source	FEOC exclusions	Partnership-based	USMCA integration	Explicit targets
Permitting reform	27/15 month limits	Limited federal role	State-led processes	Provincial jurisdiction	N/A
Strategic stockpiling	Under consideration	DPA authority	Limited	Limited	JOGMEC 60-180d
Public financing	Strategic Projects	$52B+ (CHIPS/IRA)	$4B+ facilities	SIF + tax credits	Overseas investment
Indigenous engagement	Member state-led	Tribal consultation	First Nations focus	Co-management models	N/A
ESG positioning	Regulatory standards	FEOC compliance	Clean premium	Traceability focus	Supplier standards

Sources: European Parliament and Council (2024); U.S. Congress (2022a, 2022b); Australian Government (2023); Government of Canada (2022); Ministry of Economy, Trade and Industry (2020); Ministry of Trade, Industry and Energy (2023).

Beyond public financing, private capital flows materially affect the development trajectory of critical-mineral supply for AI hardware. Private equity and infrastructure funds have committed an estimated US$15 billion to critical-mineral projects globally between 2023 and 2025, with a concentration of investments in lithium, copper and rare-earth processing capacity outside China. Development finance institutions including the United States Development Finance Corporation, the European Investment Bank's Strategic European Investment Facility and Export Development Canada have provided concessional debt and equity guarantees that lower the cost of capital for greenfield production in OECD-aligned jurisdictions. Environmental, social and governance investment criteria, increasingly codified through the European Union Sustainable Finance Disclosure Regulation, the Corporate Sustainability Reporting Directive and the parallel International Sustainability Standards Board frameworks, have begun to differentiate mining-project access to capital based on disclosed environmental and human-rights performance, though their empirical effect on operational practice remains contested in the peer-reviewed literature.

Trade Policy and Export Restrictions

The Inflation Reduction Act's clean vehicle tax credits incorporate demand-side preferences that effectively mandate supply chain restructuring. Vehicles are eligible for credits only if specified percentages of battery components and critical minerals are sourced from North America or countries with free trade agreements with the United States. These requirements create market incentives for supply chain restructuring without imposing direct import restrictions, using tax expenditure policy to achieve supply diversification objectives.

The Foreign Entity of Concern provisions extend this approach by excluding materials from specified countries entirely, regardless of where final processing occurs. Implementation of FEOC provisions has proven complex, requiring detailed supply chain mapping to identify the origin of materials at each stage of processing. The Treasury Department has issued guidance specifying that materials must be traceable to their extraction and processing locations, with exclusion triggered by any FEOC involvement at any stage. These provisions have accelerated investment in supply chain visibility systems and created demand for certified 'FEOC-free' material streams.

Demand-side policies designed to stimulate markets for low-carbon industrial products can also generate indirect but material impacts on the availability and pricing of critical minerals used in AI hardware. Evidence from OECD economies shows that instruments such as green public procurement, long-term offtake agreements, advance market commitments, and carbon border adjustment mechanisms create guaranteed demand for low-emission metals, particularly steel and aluminium, by underwriting price premiums and reducing market uncertainty for producers (Roos, 2025c). These instruments are effective precisely because they address multiple market failures simultaneously, including price externalities, asset specificity, and demand uncertainty, thereby accelerating investment in new production capacity (Roos, 2025c). However, the resulting scale-up of low-carbon metal production increases upstream demand for shared critical inputs such as copper, nickel, aluminium, graphite, and specialised alloys used in electrification, hydrogen infrastructure, and advanced manufacturing equipment (Roos, 2025c; International Energy Agency, 2021). For AI hardware supply chains, this implies that demand-side decarbonisation policies operating outside the digital sector can nonetheless intensify competition for critical minerals, reinforcing the need to treat AI hardware not as an isolated demand domain but as part of a wider system of policy-induced material demand interacting across sectors.

China's imposition of export licensing requirements for gallium and germanium in July 2023, followed by expanded restrictions in December 2024 covering additional minerals and eventually extending to components containing Chinese-origin materials, demonstrated the weaponisation of mineral supply for geopolitical purposes (International Energy Agency, 2023; International Energy Agency, 2025; Baskaran & Schwartz, 2024). Supply-side restrictions create price differentials between markets with and without access to restricted materials, with implications for manufacturing competitiveness that extend beyond the directly affected sectors.

The 2025 trajectory of these controls is consequential for any forward-looking assessment of AI-hardware mineral security and is captured in three discrete moves. In April 2025, China added seven heavy and medium rare earth elements—samarium, gadolinium, terbium, dysprosium, lutetium, scandium and yttrium—to its export-licensing regime, targeting magnet-grade and sputtering-target materials of direct relevance to AI accelerator manufacture (Davidson, 2025). On 9 October 2025, the Ministry of Commerce expanded the regime further to include holmium, erbium, thulium, europium and ytterbium, introduced a 50 per cent extraterritorial rule extending controls to non-Chinese products containing as little as 0.1 per cent Chinese-origin input, and added permanent-magnet manufacturing equipment to the list of controlled items (International Energy Agency, 2025). Following the Xi–Trump meeting in Busan on 30 October 2025, China announced on 9–10 November 2025 the suspension of both the December 2024 gallium-germanium-antimony ban and the October 2025 rare earth expansion through 27 November 2026, framed as part of a wider tariff and export-controls de-escalation (Davidson, 2025). The suspension is conditional and time-limited, and the underlying legal architecture remains in place—an important caveat for OEMs designing multi-year supply contracts and for governments calibrating strategic stockpile and friend-shoring policy. The cumulative effect is to make policy-shock probability, not steady-state market access, the relevant planning variable for AI-hardware mineral supply chains over the 2025–2030 horizon.

An additional trade-risk channel arises from unevenness in sustainability policy regimes across supplier regions, which can affect both market access and cost structures for mineral and metal inputs into AI hardware supply chains. Evidence from the ASEAN context indicates that divergent adoption of carbon pricing, reporting systems, and sustainability standards can create a patchwork of compliance costs, potentially discouraging investment where standards are unclear and increasing the risk of trade diversion toward suppliers able to demonstrate conformity with import-market requirements (Roos, 2025a). Where importing jurisdictions implement border measures and due diligence expectations, suppliers without equivalent domestic frameworks may face effective cost penalties or heightened documentation burdens, whilst suppliers with more coherent sustainability regimes may gain relative advantage (Roos, 2025a). For AI hardware minerals, this suggests that supply security strategies premised on diversification need to consider not only geological endowments and processing capacity, but also the likelihood that prospective supplier regions can meet evolving sustainability-related trade requirements without introducing prohibitive compliance friction (Roos, 2025a).

Export restrictions by dominant producers can have multiple effects: increasing prices for importers, reducing prices for domestic consumers of the restricted materials, and creating incentives for alternative supply development. The 2010 rare earth crisis demonstrated that supply restrictions can catalyse investment in alternative sources, though the timeline for bringing new supply online extends over years rather than months. The longer-term effects of current restrictions on gallium and germanium will depend on whether alternative supply sources can be developed at scale and at what cost premium relative to pre-restriction prices.

The combination of demand-side preferences in importing countries and supply-side restrictions from dominant producers is creating bifurcated market structures for some critical minerals. Markets are segmenting into 'allied' supply chains serving OECD economies with FEOC-compliant materials and 'non-aligned' supply chains serving other markets through Chinese or Chinese-origin production. This bifurcation has implications for pricing, investment allocation, and the geographic distribution of processing capacity development.

Friend-shoring initiatives seek to channel investment toward supply sources within trusted partner networks, accepting potentially higher costs in exchange for supply security. The objective is no longer merely to obtain materials at the lowest market price but to secure 'sovereign' or 'allied' capacity that can withstand geopolitical shocks (Vivoda, 2023; Vivoda & Matthews, 2024). This shift has profound implications for the mining and materials processing sectors, which now find themselves at the centre of high-level diplomatic and national security strategies. The efficiency losses from abandoning lowest-cost global sourcing represent a policy choice to prioritise resilience over efficiency.

Table 10. Comparative analysis of OECD critical minerals policy frameworks (2020–2025)

Jurisdiction	Primary Legislation	Extraction Target	Processing Target	Key Financing	Distinctive Feature
European Union	CRMA (2024)	10% by 2030	40% by 2030	EIB, InvestEU	Diversification cap (65%)
United States	IRA, CHIPS Act (2022)	No explicit target	No explicit target	§45X credits, DPA Title III	FEOC exclusions
Australia	CM Strategy (2023)	Export leadership	Downstream expansion	>$4B across facilities	First Nations engagement
Canada	CM Strategy (2022)	31 priority minerals	Value chain integration	METC, SIF	Indigenous co-management

continued on following page

Table 10. Continued

Jurisdiction	Primary Legislation	Extraction Target	Processing Target	Key Financing	Distinctive Feature
Japan	Economic Security Act (2022)	N/A (import-dependent)	Recycling emphasis	JOGMEC	60–180 day stockpiles
South Korea	CM Strategy (2023)	Overseas investment	Diversification targets	KOMIR, KRC	Semiconductor linkage
United Kingdom	CM Strategy (2025)	Under development	Under development	UKEF, BBB	Vision 2035 framework

Source: European Parliament and Council (2024); U.S. Congress (2022a, 2022b); Australian Government (2023); Government of Canada (2022); Ministry of Economy, Trade and Industry (2020, 2023); Ministry of Trade, Industry and Energy (2023); Department for Business and Trade (2025).

Note: Abbreviations: CRMA=Critical Raw Materials Act; IRA=Inflation Reduction Act; FEOC=Foreign Entity of Concern; METC=Mineral Exploration Tax Credit; SIF=Strategic Innovation Fund; JOGMEC=Japan Oil, Gas and Metals National Corporation; KOMIR=Korea Mine Rehabilitation and Mineral Resources Corporation; KRC=Korea Resources Corporation; UKEF=UK Export Finance; BBB=British Business Bank.

The export-restriction narrative is not exclusively a Chinese story. Several OECD jurisdictions have also imposed export controls on critical minerals: the United States added antimony and tungsten to its strategic-export-restriction list under successive Department of Commerce actions through 2024 and 2025, Australia maintains export-permit regimes for uranium and rare-earth concentrates, and Canada has restricted the export of certain rare-earth and uranium intermediates under the Investment Canada Act amendments of 2024. Recognising this reciprocal dynamic is important for two reasons. First, the cumulative effect of restrictions imposed by both producer and consumer jurisdictions raises the risk of fragmentation in critical-minerals trade rather than a unilateral capture of bargaining advantage by any single state. Second, it reframes the policy-design question from "how to respond to Chinese export controls" to "how to coordinate a rules-based trading order for critical minerals" - a question that the multilateral institutions examined earlier in this chapter are currently ill-equipped to answer.

International Partnerships and Friend-Shoring

Australia has established bilateral critical minerals partnerships with the United States, European Union, Japan, South Korea, and India, creating frameworks for investment coordination, technology sharing, and market access. The Australia-US Climate, Critical Minerals, and Clean Energy Transformation Compact establishes preferential treatment for Australian critical minerals under US legislation, including eligibility for Inflation Reduction Act benefits. Similar agreements with other partners create diversified market access and reduce dependence on any single trading relationship.

Canada's proximity to the United States and integration through the US-Mexico-Canada Agreement (USMCA) provides advantageous market access for North American supply chain development. Critical minerals extracted and processed in Canada qualify as 'domestic' under US content requirements, enabling seamless integration into US manufacturing supply chains (Government of Canada, Natural Resources Canada, 2022). This regulatory alignment creates competitive advantages relative to more distant suppliers and has attracted investment from companies seeking to serve the North American market while avoiding FEOC restrictions.

The Minerals Security Partnership represents an institutionalisation of friend-shoring approaches, creating frameworks for allies to share information, coordinate policy, and align investment incentives in pursuit of shared supply security objectives. The Partnership has identified priority projects for coordinated support, seeking to avoid duplication of investment while ensuring geographic diversification. Initial assessments suggest that coordinated investment could substantially increase allied production capacity for several critical minerals over the coming decade, though not to levels that would achieve full independence from Chinese supply.

The production potential of MSP-aligned supply chains depends on the pace of project development, the scale of investment mobilised, and the competitiveness of costs relative to incumbent producers. Even with substantial policy support, development timelines for major mining and processing projects extend over years, and cost competitiveness may require sustained subsidisation. The Partnership's value lies in coordination that avoids duplicative investment and ensures that limited capacity is allocated efficiently across partner needs, rather than in achieving autarky that remains beyond reach.

Friend-shoring strategies face significant limitations. The cost structures of mineral production in OECD jurisdictions and allied partners substantially exceed those of Chinese operations, reflecting higher labour costs, more stringent environmental compliance requirements, and the absence of scale economies. China's dominance in critical minerals reflects decades of industrial policy, infrastructure investment, and tolerance for environmental externalities that OECD producers have been unable or unwilling to replicate (Berg et al., 2024). Closing these cost gaps requires either sustained subsidisation or acceptance of higher material costs that flow through to end products.

The geographic scope of friend-shoring partnerships has drawn criticism for potentially excluding developing countries from supply chain opportunities. The emphasis on 'trusted partners' and governance standards that reflect OECD norms may limit participation by countries with weaker institutional frameworks, even where they possess significant resource endowments. This exclusion could perpetuate patterns of resource extraction without domestic value addition in developing countries, while also limiting the supply diversification achievable through friend-shoring approaches.

The 'clean premium' strategy positions production from OECD jurisdictions and aligned partners as commanding price premiums based on environmental and social performance. This positioning acknowledges that production costs are higher than competing sources and seeks to recover these costs through differentiated market positioning rather than cost competition. Success depends on buyer willingness to pay premiums for certified responsible production, which varies across market segments and economic conditions.

Major technology companies and data centre operators—hyperscalers—have emerged as significant actors in critical minerals supply chains, with purchasing commitments and investment decisions that can influence market development. Several hyperscalers have announced commitments to responsible sourcing and have invested in supply chain transparency systems that enable verification of environmental and social performance. These private sector interventions complement public policy by creating market demand for differentiated supply that policy instruments alone may not generate. The combination of public policy and private purchasing power may prove more effective than either approach alone in driving supply chain restructuring.

Environmental Outcomes Assessment

Evaluating critical minerals policy instruments requires a systematic framework that makes trade-offs explicit rather than treating them as unfortunate complications to be acknowledged but not resolved. The European Commission's Joint Research Centre has developed a multi-dimensional assessment methodology that evaluates substitution and reduction solutions against five criteria: performance relative to benchmark technologies, operational lifetime and degradation characteristics, environmental and social impact profiles across the full life cycle, cost implications spanning capital expenditure and ongoing operations, and net effects on critical raw material usage including potential dependencies introduced by alternative materials (Georgitzikis et al., 2025). This framework, originally developed for clean energy technologies, offers direct applicability to AI hardware policy analysis. Each policy instrument—whether recycling mandates, domestic processing incentives, or strategic stockpiling—produces differentiated effects across these dimensions that aggregate into overall value propositions for different stakeholder perspectives. Supply security gains from reshoring processing capacity may come at environmental cost if OECD facilities operate with higher energy intensity than displaced production; conversely, environmental improvements from substituting toxic materials may introduce supply concentration in alternative material chains. Explicit multi-dimensional assessment disciplines policy discourse, requiring advocates of particular instruments to acknowledge trade-offs rather than presenting partial analyses that emphasise benefits while obscuring costs. The sections that follow apply this framework to evaluate tensions between supply security and environmental sustainability, policy coherence across domestic and international dimensions, and risks of burden-shifting to jurisdictions with weaker governance capacity.

Critical minerals policy must manages tensions between supply security and environmental sustainability objectives. Supply security imperatives may be achieved through interventions that externalise environmental burdens to producing regions with weaker governance and enforcement capacity. Conversely, stringent environmental standards in OECD jurisdictions may constrain domestic extraction and processing, reinforcing import dependence on regions with less rigorous environmental oversight (OECD, 2023). Three-dimensional criticality assessments make these trade-offs explicit, enabling policy design that considers environmental implications alongside economic and strategic factors, though the operationalisation of environmental criteria remains methodologically contested (Graedel et al., 2012).

The Act includes provisions allowing Strategic Projects to be designated as being of 'overriding public interest' for the purposes of environmental legislation, potentially enabling projects to proceed despite impacts on protected areas where mitigation and compensation measures are implemented. Environmental organisations have criticised this provision as creating loopholes that could undermine decades of nature protection progress. The tension between supply security and environmental protection illustrates the trade-offs that critical minerals policy must manage, and the resolution of specific cases will shape the practical meaning of policy frameworks.

Table 11. Industrial policy instruments: Environmental and supply security trade-offs

Policy Instrument	Supply Security Effect	Environmental Effect	Timeline	Fiscal Cost
Domestic extraction incentives	Moderate positive	Negative (local)	10–15 years	High
Processing capacity investment	High positive	Mixed (reshoring)	5–10 years	Very High
Strategic stockpiling	High positive (short-term)	Neutral	1–3 years	Moderate
Recycling mandates/targets	Low–Moderate positive	Positive	10–20 years	Moderate
Permitting reform	Moderate positive	Risk of negative	2–5 years	Low

continued on following page

Table 11. Continued

Policy Instrument	Supply Security Effect	Environmental Effect	Timeline	Fiscal Cost
Friend-shoring partnerships	Moderate positive	Mixed (burden-shifting)	5–15 years	Variable
Demand-side preferences (FEOC)	High positive	Indirect positive	3–7 years	Low (tax expenditure)
R&D support (substitution)	Variable (long-term)	Potentially positive	10–20+ years	Moderate
Due diligence requirements	Neutral	Positive (governance)	2–5 years	Low
'Green premium' certification	Moderate positive	Positive	3–10 years	Low

Source: Author's synthesis based on policy analysis in Sections "Critical Mineral Policy" and "Industrial Policy Instruments".

Note: Assessments are qualitative and indicative. 'Mixed' indicates that outcomes depend on implementation details and geographic distribution of activities. Timelines represent periods to achieve meaningful impact on supply security or environmental outcomes.

Policy coherence requires alignment of domestic and international dimensions of critical minerals governance. Domestic policies that promote extraction and processing create environmental impacts within OECD jurisdictions, where robust regulatory frameworks and civil society oversight may ensure relatively high standards. International partnerships and import diversification strategies engage supply chains in jurisdictions with varying regulatory capacity and enforcement, potentially exposing OECD consumers to embedded environmental impacts that would not be tolerated domestically.

The OECD Due Diligence Guidance for Responsible Supply Chains provides a framework for extending governance expectations to international supply chains, but implementation varies widely across companies and sectors, with coverage gaps particularly for materials and regions not specifically targeted by regulatory requirements (Organisation for Economic Co-operation and Development, 2011). Supply chain transparency remains limited, and the ability of importing countries to influence practices in producing regions depends on market power and the availability of alternative sources. Policy coherence requires both strengthening domestic governance and extending influence through international engagement mechanisms.

A significant risk of current policy approaches is burden-shifting: achieving OECD supply security through supply chains that concentrate environmental impacts in jurisdictions with weaker governance capacity. The geographic and economic separation between sites of environmental impact and sites of consumption complicates accountability and policy coordination across the AI value chain (OECD, 2023). Many mining operations supplying critical minerals are located outside OECD economies, in jurisdictions with differing regulatory capacities, environmental standards, and enforcement mechanisms, resulting in heterogeneous environmental outcomes (World Bank, 2020).

Diversification away from Chinese supply may shift demand to producing regions with less developed environmental governance, potentially worsening aggregate environmental outcomes even as OECD supply security improves. The expansion of cobalt mining in the Democratic Republic of Congo, rare earth extraction in Myanmar, and nickel mining in Indonesia illustrates how demand growth can outpace governance capacity development in producing regions. Addressing burden-shifting requires policy attention that extends beyond supply security to encompass environmental outcomes across the full supply chain, regardless of geographic location.

Overall, the environmental impacts associated with the extraction and processing of minerals used in AI hardware are substantial, heterogeneous, and largely concentrated upstream (International Energy Agency, 2021; OECD, 2023). While AI hardware does not introduce new categories of environmental impact, its contribution to cumulative mineral demand reinforces existing pressures on ecosystems, water resources, and climate mitigation efforts. Despite these challenges, the literature also highlights opportunities for impact mitigation through improved technologies and governance. Measures such as electrification of mining equipment, water recycling, tailings reprocessing, and enhanced due diligence frameworks offer pathways for reducing the environmental footprint of critical minerals supply chains, though implementation requires sustained policy attention and investment across producing and consuming regions.

Policy Evaluation Frameworks

The proliferation of critical minerals strategies across OECD economies has generated diverse policy experiments whose effectiveness remains largely unevaluated. Evidence from industrial policy research indicates that policy impacts are highly heterogeneous and context-dependent, varying with industrial structure, institutional capacity, and market conditions in ways that generic policy prescriptions fail to capture (European Commission, Joint Research Centre, 2025). Monitoring and evaluation frameworks are essential for adaptive policymaking that learns from implementation experience rather than persisting with ineffective interventions or abandoning promising approaches prematurely.

Effective policy evaluation requires metrics aligned with ultimate objectives rather than intermediate outputs. Processing capacity additions, while easily measured, provide incomplete indicators of supply security if capacity remains uncompetitive or dependent on continued subsidisation. Cost competitiveness metrics should track progress toward self-sustaining production that does not require perpetual public support. Environmental outcome measures—greenhouse gas intensity of production, water consumption, waste generation—enable assessment of whether supply security gains come at acceptable environmental cost. Employment quality indicators—wages relative to regional benchmarks, skill composition, workplace safety records—capture labour market effects that aggregate employment figures obscure.

Data-driven analytical tools can support evaluation while informing ongoing policy development. The Economic Complexity framework provides methods for assessing productive capabilities and identifying diversification opportunities aligned with existing industrial strengths (European Commission, Joint Research Centre, 2025). Strategic Dependency Indices quantify vulnerability to supply disruptions across material categories, enabling prioritisation of policy attention. Import concentration measures track progress toward diversification objectives. These tools require sustained investment in data collection and analytical capacity that current policy frameworks often neglect in favour of immediate programme implementation. The importance of measuring outcomes to ensure adaptive policymaking should be embedded in programme design from inception, with evaluation requirements and sunset clauses that force periodic reassessment rather than indefinite continuation of potentially ineffective interventions.

Fourth, friend-shoring partnerships with binding mineral-supply commitments rather than aspirational declarations, supported by long-term offtake guarantees, concessional financing through development finance institutions and bilateral environmental-standard convergence. Fifth, due-diligence and transparency: mandate supply-chain mapping through to refining-stage origin for AI-hardware procurement above defined value thresholds, supported by independent audit and integrated with corporate-sustainability-reporting obligations. Sixth, multilateral coordination: pursue plurilateral agreements on critical-minerals trade rules including rules-of-origin definitions, export-restriction notification, environmental standards equivalence and crisis-response mechanisms; the existing patchwork of the Minerals Security Partnership, the OECD due-diligence guidance and bilateral arrangements is insufficient to govern the supply order on which AI infrastructure now depends.

The comparative analysis developed across §§5-7 of this chapter supports six policy recommendations directed at the OECD jurisdictions whose mineral-supply choices most directly affect AI-hardware availability. These recommendations are presented as proximate priorities for the 2025-2030 period rather than long-run prescriptions; their substantive content reflects the trade-offs identified in the mineral-AI governance trilemma developed above. First, permitting reform: shorten the time from mineral-discovery to operational production through pre-construction environmental-impact-assessment streamlining and explicit critical-mineral designation in domestic permitting frameworks, while preserving substantive environmental and free-prior-and-informed-consent protections. Second, recycling and circular-economy mandates: combine end-of-life electronic-waste collection targets with technology-neutral recycled-content requirements that escalate to twenty-five per cent by 2030, supported by extended-producer-responsibility regimes covering data-centre infrastructure as well as consumer electronics. Third, strategic stockpiling: establish jurisdictional stockpiles of the highest-criticality minerals (gallium, germanium, indium, dysprosium, terbium and neodymium) at levels equivalent to twelve to twenty-four months of domestic AI-hardware demand, with explicit drawdown protocols tied to disruption-event triggers.

Key Policy Recommendations

Seven recommendations follow from the comparative analysis of policy instruments developed in this chapter. They are framed as design principles rather than prescriptions, since the appropriate instrument mix in any jurisdiction is conditional on existing industrial capability, governance quality and trade-policy posture.

First, quantitative targets should be calibrated against processing-capacity reality rather than consumption-share aspirations alone. The European Union's 10 per cent extraction, 40 per cent processing and 25 per cent recycling targets for strategic raw materials by 2030 (European Parliament and Council of the European Union, 2024) are useful as direction-setting commitments but, on present capacity trajectories, exceed what installed and pipeline projects can deliver within the Act's timeframe (Baldassarre & Carrara, 2025). Targets should therefore be paired with rolling capacity-gap reports that identify the specific material, processing-stage and geographic combinations where ambition exceeds feasibility, so that policy attention and capital can be directed to binding constraints rather than to aggregate compliance. Where binding constraints cannot be resolved domestically within the policy horizon, friend-shoring partnerships under the Minerals Security Partnership and the Pax Silica initiative provide a complementary pathway (U.S. Department of State, 2025), but only if those partnerships are operationalised through concrete offtake and financing instruments rather than declaratory commitments.

Second, permitting reform should be coupled with explicit environmental-assessment quality safeguards rather than treating speed and rigour as a binary trade-off. The Critical Raw Materials Act's 27-month strategic-project permitting timeline (European Parliament and Council of the European Union, 2024) is achievable without weakening assessment quality if administrative-capacity investment, predictable pre-application engagement and parallel rather than sequential consultation processes are funded ex ante. Where permitting reform is enacted without these accompanying investments, the predictable outcome is either continued delay through litigation or assessment-quality degradation that undermines the social licence on which long-lived projects depend. The Australian and Canadian experience with

First Nations engagement (Roos, 2025b) suggests that Indigenous co-management arrangements, far from extending timelines, can reduce overall project risk when negotiated upstream of investment decisions.

Third, demand-side and supply-side instruments should be sequenced so that they reinforce rather than fragment the market. The Inflation Reduction Act's Foreign Entity of Concern exclusions and the 45X production tax credits operate on opposite sides of the market: the former restricts access to incentives, the latter expands supply (U.S. Congress, 2022a). When deployed simultaneously without transitional provisions, the combined effect is to create bifurcated markets in which qualifying material commands a premium that may exceed what downstream applications can sustain (FP Analytics, 2025). Phased qualification thresholds and content-of-origin tolerances that tighten over time give the supply side a credible signal to invest while avoiding short-run supply gaps that undermine the policy objective.

Fourth, financing terms for critical-minerals projects should embed just-transition and Indigenous-partnership conditions rather than treat these as discretionary corporate-social-responsibility add-ons. Public finance institutions in Australia, Canada and the European Union have material leverage over project economics through the cost and tenor of debt; making transition-quality and benefit-sharing arrangements conditions of finance rather than reporting obligations changes the incentive structure for project sponsors (Roos, 2025b). The same logic applies to hyperscaler offtake commitments under the clean-premium strategy: when integrated into the financing stack, these commitments materially reduce the cost of capital for premium-grade producers and provide an enforceable channel for environmental and social standards (Roos, 2025c).

Fifth, strategic stockpiles should be designed as hybrid commercial-defence instruments with explicit release rules, drawing on the Japanese model administered through the Japan Organization for Metals and Energy Security (Japanese Ministry of Economy, Trade and Industry, 2023). Stockpiles whose release conditions are opaque or fully discretionary either fail to dampen price spikes—because market participants cannot price the implicit option—or are released too late to influence outcomes. A rules-based release framework specifying trigger conditions, release volumes and replenishment timetables converts stockpiles from a static buffer into an active price-stability instrument, at the cost of accepting that the option will sometimes be exercised in ways that disappoint either defence or commercial users.

Sixth, multilateral coordination should treat the 50 per cent extraterritoriality rule introduced by China in October 2025 (International Energy Agency, 2025) as a precedent that will be reciprocated across jurisdictions. Unilateral responses—replicating extraterritorial controls in OECD jurisdictions—risk accelerating fragmentation of the trading system, raising costs throughout the AI-hardware value chain. A negotiated multilateral standard on de minimis content thresholds, transit treatment and dispute-resolution procedures, ideally embedded within existing World Trade Organization frameworks, would preserve the policy space individual jurisdictions need without producing the compounding-control dynamic that disadvantages all participants. The November 2025 Busan-driven suspension of the October controls demonstrates that negotiated de-escalation is possible; institutionalising it through a standing forum would reduce the policy-shock probability that is, on the analysis developed earlier in this chapter, the binding constraint on AI-hardware mineral planning over the 2025–2030 horizon.

Seventh, AI-enabled traceability, sorting and material-recovery capabilities should be embedded as financing-eligibility requirements for upstream projects rather than supported only through separate downstream R&D programmes. The chapter's analysis of AI's enabling role identified material recovery, certification verification and predictive maintenance as the highest-leverage applications (Kaack et al., 2022; Initiative for Responsible Mining Assurance, 2025). Embedding these capabilities into the project specification at the financing stage, rather than retrofitting them after commissioning, captures the cost and performance advantages of design-stage integration and creates a consistent evidentiary base across the supply chain that supports both due-diligence obligations under the United Nations Guiding Principles (United Nations Human Rights Council, 2011) and the recycled-content audit requirements emerging under the European Union and United Kingdom critical-minerals frameworks (Department for Business and Trade, 2025).

These recommendations are necessarily indicative. Their relative priority depends on the jurisdiction's existing instrument mix, the maturity of its domestic processing base and its position in the friend-shoring architecture under the Minerals Security Partnership and Pax Silica. What they share is a common emphasis on coupling speed with rigour, demand with supply, and ambition with feasibility — the three couplings on which the analysis presented in this chapter suggests effective critical-minerals policy for AI hardware ultimately depends.

CONCLUSION

This chapter has examined the material foundations of artificial intelligence through the lens of critical minerals policy, connecting three bodies of work that have developed largely in isolation: environmental assessments of mineral extraction, critical minerals policy frameworks developed in response to supply chain concerns, and the emerging literature on AI's environmental footprint. The analysis pursued three interrelated objectives: characterising the material dependencies of AI hardware, surveying industrial policy instruments across OECD economies, and analysing the environmental dimensions of critical minerals extraction and processing. This concluding section synthesises findings across these domains and identifies implications for policy and future research.

Material Dependencies of AI Hardware: Synthesis of Findings

The first objective addressed the material dependencies underpinning AI hardware production. The analysis reveals that AI systems depend on complex supply chains spanning semiconductor fabrication, high bandwidth memory production, advanced packaging technologies, and data centre infrastructure. These dependencies extend across dozens of mineral commodities, from high-purity silicon and gallium arsenide substrates through copper interconnects and rare earth permanent magnets to the structural materials of hyperscale data facilities.

Semiconductor manufacturing for AI accelerators requires materials of exceptional purity and specificity. Silicon wafers achieving eleven-nines purity form the substrate for advanced logic devices, while compound semiconductors including gallium arsenide, gallium nitride, and silicon carbide serve specialised functions in power management and high-frequency applications. The transition to extreme ultraviolet lithography has introduced new material dependencies, including ultra-low expansion glass containing titanium dioxide for photomask substrates and multilayer mirrors requiring molybdenum and silicon thin films. Process chemicals including ultra-high-purity hydrogen fluoride, phosphoric acid, and specialty gases add further supply chain complexity.

High bandwidth memory represents a critical bottleneck for AI accelerator performance, with memory systems increasingly constraining computational throughput. HBM production requires through-silicon vias filled with copper or tungsten, hybrid bonding using copper-to-copper interconnects, and advanced packaging substrates that combine organic materials with silicon interposers. The concentration of HBM production among three manufacturers—Samsung, SK Hynix, and Micron—creates supply vulnerabilities that extend beyond individual material dependencies to encompass manufacturing know-how and capacity constraints.

Data centre infrastructure adds substantial material requirements at gigawatt scale. A representative 1 GW AI data centre campus requires approximately 57,500–107,000 tonnes of steel and 12,000–27,000 tonnes of copper for electrical and cooling systems, together with 480,000–1,100,000 tonnes of concrete for facility construction (World Economic Forum, 2025; BHP, 2025), alongside significant quantities of aluminium and specialised thermal-interface materials for high-density cooling. Backup power systems depend on rare earth permanent magnets for generators and substantial lead or lithium inventories for battery storage. The embodied carbon of IT equipment refreshed every four years exceeds 2 million tonnes CO_2e over a typical 30-year facility lifespan, vastly exceeding one-time construction emissions.

Supply concentration emerges as a defining characteristic of AI hardware materials. China dominates global production and processing for numerous critical inputs: 98 per cent of gallium, between 60 and 68 per cent of germanium, above 60 per cent of rare earth element extraction, and 94 per cent of heavy rare earth processing capacity. This concentration reflects decades of industrial policy, infrastructure investment, and acceptance of environmental externalities that OECD producers have been unwilling or unable to replicate. The 'hitchhiker' dynamic—whereby technology-critical elements such as gallium, germanium, and indium are recovered as by-products of commodity metal production rather than mined as primary products—provides limited insulation against supply disruptions, as demonstrated by China's export restrictions on gallium and germanium announced in July 2023 and expanded in December 2024.

Market dynamics further complicate supply security assessments. Prices for critical minerals exhibit high volatility due to tight supply–demand balances, speculative activity, and policy interventions. The International Energy Agency reports that after a period of sharp price increases driven by electric vehicle and battery demand, the market saw significant declines in 2023: lithium prices fell by 75 per cent and cobalt, nickel, and graphite prices by 30–45 per cent, even though underlying demand remained strong (International Energy Agency, 2024). Such volatility can deter investment in new projects because producers face uncertainty about future revenues, while supply bottlenecks and long development timelines make it difficult for production to respond quickly to price signals.

Industrial Policy Instruments: Comparative Assessment

The second objective surveyed industrial policy instruments deployed across OECD economies. The analysis identifies a convergent policy agenda emphasising supply security, though implementation varies significantly across jurisdictions reflecting different resource endowments, industrial capabilities, and geopolitical positions.

The European Union's Critical Raw Materials Act establishes binding targets for domestic extraction, processing, and recycling, designating strategic projects eligible for streamlined permitting and financing support. The Act requires that by 2030, the EU extract 10 per cent of its annual consumption domestically, process 40 per cent within the Union, and obtain no more than 65 per cent of any strategic material from a single third country. These targets acknowledge that complete self-sufficiency remains beyond reach while seeking to reduce the most acute supply vulnerabilities. The tension between supply security imperatives and environmental protection is explicitly acknowledged through provisions allowing strategic projects to proceed despite impacts on protected areas where mitigation measures are implemented.

The Act's relationship with existing EU environmental legislation, including the Birds and Habitats Directives and Natura 2000 protected sites, has generated debate, with environmental organisations criticising provisions that allow projects to be designated as being of 'overriding public interest' as potentially creating loopholes that could undermine nature protection progress.

United States policy emphasises security-driven industrial policy through the CHIPS and Science Act's manufacturing incentives and the Inflation Reduction Act's demand-side preferences. The IRA's clean vehicle tax credits incorporate Foreign Entity of Concern provisions that effectively exclude Chinese-origin materials from qualifying supply chains, creating market incentives for supply chain restructuring without imposing direct import restrictions. Section 45X production tax credits provide 10 per cent of production costs for domestic extraction and processing, representing a departure from previous industrial policy approaches that relied on grants or loans. The Defense Production Act has been invoked to support domestic mining and processing capacity, while the National Defense Stockpile is being expanded to include materials essential for both defence and civilian production. The Pax Silica initiative, launched in late 2025, represents the latest iteration of multilateral supply chain coordination among allied nations.

Resource-rich nations including Australia and Canada pursue strategies leveraging upstream endowments while developing downstream processing capacity. Australia's Critical Minerals Strategy 2023–2030 positions the country as a 'trusted supplier' to allies seeking diversification from Chinese-dominated supply chains. The Australian Government has established the Critical Minerals Facility, a $2 billion financing mechanism, with additional support through the Northern Australia Infrastructure Facility and Export Finance Australia, bringing aggregated public financing commitments to exceed $4 billion. Australia's strategy emphasises environmental, social, and governance leadership as a competitive differentiator, with First Nations engagement representing a significant governance dimension under the Native Title Act. Bilateral partnerships with the United States, European Union, Japan, South Korea, and India create frameworks for investment coordination and market access, with the Australia-US Climate, Critical Minerals, and Clean Energy Transformation Compact establishing preferential treatment under US legislation.

Canada's Critical Minerals Strategy identifies 31 priority minerals and positions Canadian production within North American supply chains, benefiting from USMCA provisions that treat Canadian materials as domestic under US content requirements. The Mineral Exploration Tax Credit enables flow-through share arrangements that transfer deduction benefits to investors, effectively subsidising exploration risk through the tax system. The Strategic Innovation Fund supports processing projects including rare earth separation facilities and lithium hydroxide plants. Indigenous partnership represents a distinctive feature of Canadian governance, with constitutionally protected consultation requirements and successful models including Impact and Benefit Agreements, equity participation, and co-management arrangements. Canada has invested in blockchain-based traceability systems to verify provenance and enable 'clean premium' positioning based on ESG performance.

Import-dependent economies including Japan and South Korea pursue diversification and partnership strategies acknowledging that domestic production cannot address supply security concerns. Japan's approach combines strategic stockpiling through JOGMEC, maintaining 60–180 days of domestic consumption, with overseas resource development investments and technology partnerships. The Economic Security Promotion Act enacted in 2022 establishes a comprehensive framework designating 'specified critical materials' subject to enhanced monitoring, requiring supply chain surveys, and providing mechanisms for government-industry coordination on early warning systems and contingency planning. South Korea coordinates critical minerals policy with semiconductor industry development, with the Korea Mine Rehabilitation and Mineral Resources Corporation (KOMIR) and Korea Resources Corporation supporting overseas investment and technology development.

China's role as a contextual case rather than subject of domestic policy analysis reflects its central position in global mineral processing. The strategic weaponisation of mineral supply chains through export restrictions on gallium, germanium, and other materials demonstrates the entanglement of mineral supply with technological and geopolitical competition. The policy implications of Chinese processing dominance shape OECD responses across all dimensions, requiring diversification targets calibrated against limited alternative capacity and investment incentives that account for cost disadvantages relative to Chinese production.

Multilateral coordination through the Minerals Security Partnership institutionalises friend-shoring approaches, while the OECD Due Diligence Guidance provides frameworks for responsible supply chain management. The International Energy Agency has developed substantial analytical capacity, with its annual Critical Minerals Market Review providing comprehensive analysis that informs policy development across OECD member states. The IEA's scenario modelling projects future mineral demand under various energy transition pathways, providing evidence base for investment decisions.

The analysis reveals several common limitations across policy approaches. Timeline mismatches between policy objectives and mining development cycles create implementation gaps, as new extraction and processing capacity requires 10–15 years to develop while policy targets typically extend only to 2030. Cost differentials between OECD production and Chinese operations—reflecting higher labour costs, more stringent environmental compliance, and absence of scale economies—necessitate sustained subsidisation or acceptance of higher material costs. Multi-level governance challenges complicate permitting reform in federal systems, while Indigenous rights and consultation requirements add governance layers that cannot be compressed into accelerated timelines without compromising engagement quality. The friend-shoring emphasis on trusted partners may exclude developing countries with significant resource endowments but weaker institutional frameworks, limiting both diversification potential and development opportunities.

Environmental Dimensions: Impacts and Governance Challenges

The third objective examined environmental dimensions of critical minerals extraction and processing. The analysis documents substantial, heterogeneous impacts concentrated in the upstream phases of AI hardware supply chains, with environmental burdens geographically separated from consumption centres in ways that complicate governance and accountability.

Land degradation and biodiversity loss result from mining operations that transform landscapes through excavation, waste disposal, and infrastructure development. Open-pit mines for copper, rare earths, and other minerals disturb thousands of hectares per operation, with waste rock volumes often exceeding ore volumes by orders of magnitude. Tailings impoundments create long-term land use constraints and dam failure risks that extend decades beyond active mining. The geographic coincidence of mineral deposits with biodiversity hotspots and Indigenous territories compounds these impacts, as extraction activities affect ecosystems and communities with limited political voice in consuming economies. Gallium and germanium extraction, as by-products of bauxite and zinc mining, contribute to large-scale habitat disruption in tropical forests, while cobalt mining has cleared significant areas of the Congo Basin forest and rare earth extraction in Myanmar has expanded into forested border regions.

Water impacts span quantity and quality dimensions. Mining operations consume substantial water volumes for mineral processing, dust suppression, and slurry transport, with water intensity varying from 0.3 to 2.0 cubic metres per tonne of ore depending on mineral type and processing requirements. Acid mine drainage from sulphide-bearing waste rock acidifies surface and groundwater, mobilising heavy metals that persist in aquatic ecosystems for decades after mine closure. Lithium brine extraction in the Atacama region has caused groundwater level declines exceeding 10 metres over 15 years, with Indigenous communities characterising extraction as 'water mining' that threatens traditional livelihoods. Rare earth element processing is particularly water-intensive and chemically complex, producing large volumes of tailings and chemically contaminated residues.

Greenhouse gas emissions from mining and processing are substantial, with processing stages often exceeding extraction in carbon intensity. Aluminium smelting consumes approximately 14 MWh per tonne, contributing significantly to emissions where electricity derives from fossil fuels. Semiconductor fabrication uses fluorinated gases with global warming potentials thousands of times greater than carbon dioxide. The carbon intensity of advanced packaging technologies and extreme ultraviolet lithography adds further emissions, with EUV scanners consuming approximately one megawatt of electrical power during operation. Total embodied carbon for AI accelerators reaches significant levels before accounting for data centre infrastructure.

Radioactive waste streams accompany rare earth extraction due to the co-occurrence of thorium and uranium in many ore bodies, as illustrated by the Lynas Malaysia refinery controversy. Human rights dimensions include artisanal mining conditions in the Democratic Republic of Congo for cobalt, with documented child labour and hazardous working conditions, and rare earth extraction in Myanmar linked to environmental devastation and conflict financing. The OECD Due Diligence Guidance provides frameworks for responsible sourcing, though implementation varies widely and accountability gaps persist.

Life cycle assessment methodologies face significant challenges in capturing these diverse impacts. System boundary inconsistencies, allocation method variations, and data gaps for emerging processes limit comparability across studies. A 2023 analysis reviewing 74 peer-reviewed studies identified persistent gaps in coverage of water impacts across critical minerals extraction (Lima et al., 2023). The geographic and temporal separation between extraction impacts and end-use benefits complicates attribution and governance.

Tensions Between Supply Security and Environmental Sustainability

The analysis reveals fundamental tensions between supply security and environmental sustainability objectives. Current policy frameworks acknowledge these tensions but provide limited mechanisms for their resolution. Four dimensions of this tension merit particular attention.

First, the geographic distribution of environmental burdens reflects patterns of consumption-production separation that current governance structures inadequately address. OECD economies consume the majority of AI hardware while environmental impacts concentrate in extraction regions with varying regulatory capacity and enforcement. Supply chain due diligence requirements, while expanding in scope, remain limited in coverage and effectiveness. The concentration of processing capacity in China has externalised environmental impacts from OECD consumption centres, creating what researchers term a 'colonial shadow' in the geography of environmental burdens.

Second, the urgency of supply security concerns may override environmental considerations in specific cases. The EU Critical Raw Materials Act's provisions allowing strategic projects to proceed despite environmental impacts, subject to mitigation requirements, illustrates this dynamic. Similar tensions appear in permitting reform efforts across jurisdictions, where streamlined approval processes may reduce environmental review periods. The designation of minerals as 'critical' or 'strategic' implicitly prioritises supply security, potentially marginalising environmental considerations in subsequent policy implementation.

Third, diversification away from Chinese supply may shift environmental burdens to jurisdictions with less developed governance capacity. Expansion of rare earth production in Myanmar, cobalt extraction in the DRC, and nickel mining in Indonesia illustrates how demand growth can outpace governance development. Friend-shoring initiatives that emphasise 'trusted partners' may inadvertently concentrate supply among partners with strong governance while excluding developing countries from supply chain opportunities, perpetuating extractive relationships without value addition.

Fourth, the combination of demand-side preferences in importing countries and supply-side restrictions from dominant producers is creating bifurcated market structures. Markets are segmenting into 'allied' supply chains serving OECD economies with FEOC-compliant materials and 'non-aligned' supply chains serving other markets through Chinese or Chinese-origin production. This bifurcation has implications for pricing, investment allocation, and the geographic distribution of processing capacity development, with efficiency losses from abandoning lowest-cost global sourcing representing a policy choice to prioritise resilience over efficiency.

Fifth, the relationship between AI hardware and environmental sustainability involves not only the material demands documented throughout this chapter but also the potential for AI systems to enable circular economy solutions that reduce those demands. The twin transition framework recognises that green and digital technologies can exploit synergies when appropriately integrated, with institutional support playing a central enabling role (European Commission, Joint Research Centre, 2025). AI-enabled robotic sorting systems can identify and separate critical minerals from complex waste streams with precision exceeding manual or conventional automated methods, potentially improving recovery rates for materials currently lost to landfill or low-value recycling. Machine learning applications can optimise mining and processing operations to reduce energy consumption, water use, and waste generation while maximising recovery of target materials. Digital traceability systems enabled by distributed ledger technologies can verify provenance and production conditions throughout supply chains, supporting due diligence requirements and enabling differentiated market access for responsibly produced materials. These synergies suggest that AI hardware demand and environmental sustainability need not be purely antagonistic; however, realising synergistic potential requires deliberate policy attention to the digital infrastructure, data governance frameworks, and complementary investments that enable integration. The risk is that AI expansion proceeds without deliberate integration, capturing the material demands while forgoing the sustainability benefits that thoughtful deployment could generate.

The 'green premium' strategy—positioning OECD production as commanding price premiums based on environmental and social performance—offers a potential mechanism for reconciling supply security with sustainability. Success depends on buyer willingness to pay for certified responsible production, which varies across market segments. Major technology companies and data centre operators—hyperscalers—have emerged as significant actors in critical minerals supply chains, with purchasing commitments and investment decisions that can influence market development. The combination of public policy and private purchasing power may prove more effective than either approach alone in driving supply chain restructuring toward both security and sustainability objectives.

AI's Enabling Role in Sustainable Mineral Supply

The chapter's analytical thrust has been to characterise AI hardware as a generator of mineral-extraction obligations. That framing is incomplete, however, without acknowledging the symmetric role AI technologies play in moderating those obligations. Three pathways are sufficiently developed in 2026 to merit explicit treatment. First, machine-learning-enabled mineral exploration - using subsurface imaging, geochemical anomaly detection and remote-sensing classifiers trained on historical drill-core data - reduces the discovery cost per unit of identified reserve and the exploration footprint per unit of new deposit located, with documented gains across copper, lithium and rare-earth exploration programmes since 2022. Second, predictive maintenance and digital twins of mining and processing infrastructure, increasingly trained on real-time sensor data and operational logs, lower the probability of catastrophic events such as tailings-dam failures and reduce maintenance-driven downtime, with implications both for safety and for materials efficiency. Third, geometallurgical optimisation - the combined modelling of ore mineralogy, comminution behaviour and processing yield - is increasingly automated through machine-learning pipelines that improve recovery rates from low-grade and complex orebodies, raising effective resource recovery without proportionate increase in ore movement. These enabling roles do not absolve AI hardware of its own mineral footprint, but they qualify the burden-only narrative and identify a feedback channel between AI deployment and the sustainability of the very mineral supply chains on which AI deployment depends.

This framework is offered as a tool for orienting subsequent research and policy analysis rather than as a finished theory. Its empirical operationalisation - in particular the measurement of unit costs of policy on each dimension and the specification of trade-off elasticities - requires further work, and is identified in the research agenda below. The framework's value at this stage lies in its ability to organise the comparative-policy material developed in §§5-7 of this chapter into a single analytical structure that brings the policy choices of different jurisdictions into directly comparable form.

Three boundary conditions limit the framework's applicability. First, the trilemma is descriptive of policy choice under given factor endowments and existing technologies; it does not preclude joint improvement on all three dimensions through technological progress in extraction, refining and recycling that shifts the joint production frontier outward. Second, the framework is calibrated for the OECD economies that are the primary consumers of AI hardware; producer-economy trilemmas, which trade off resource revenue, environmental sovereignty and trade relationships, are analytically distinct and not captured here. Third, the framework presumes that the three objectives are independently evaluable; in practice, supply-security and environmental-sustainability indicators are increasingly entangled through ESG-linked finance, scope-3 reporting obligations and rules-of-origin provisions in border-adjustment instruments.

Within these boundary conditions, the trilemma framework supports four propositions, each of which is supported by evidence developed in the preceding sections of this chapter: that no jurisdiction examined in this chapter has resolved the trilemma rather than resolved trade-offs within it; that the European Union has emphasised the security-sustainability pairing at the cost of unit-cost competitiveness; that the United States has emphasised the security-efficiency pairing through scaled subsidisation, accepting environmental-permitting tensions documented above; that Australia and Canada have positioned themselves as upstream suppliers in which domestic resolution of the trilemma is bounded by their export-market dependencies; and that the multilateral coordination mechanisms examined in this chapter are best read as attempts to widen the joint production frontier through cooperation rather than to resolve the trilemma's internal tensions.

The three objectives in the mineral-AI governance trilemma are supply security (the assurance of access to AI-hardware critical minerals under varying geopolitical conditions), environmental sustainability (the limitation of the cumulative environmental burden imposed by extraction, processing and end-of-life management of these minerals to ecologically defensible bounds) and economic efficiency (the achievement of the foregoing objectives at minimum economic cost to producing and consuming economies). The framework holds that any two of the three objectives can be jointly pursued with strong policy intensity, but the third must be relaxed: aggressive pursuit of supply security and environmental sustainability requires accepting higher unit costs of mineral inputs and the consequent loss of efficiency; aggressive pursuit of supply security and economic efficiency relies on extraction in jurisdictions with weaker environmental governance, relaxing sustainability; and aggressive pursuit of environmental sustainability and economic efficiency, for example through deep dependence on the most cost-efficient established producers, increases concentration and relaxes supply security.

The analytical material developed across this chapter supports a synthesising framework, here termed the mineral-AI governance trilemma, which is offered to give explicit structure to the policy and analytical literature on AI-hardware mineral supply. The framework follows the established trilemma-construction methodology used in international monetary economics (the open-economy trilemma of fixed exchange rates, capital mobility and monetary autonomy; Obstfeld, Shambaugh and Taylor, 2005) and in energy policy (the energy trilemma of security, equity and sustainability; World Energy Council, 2023), in which three desirable policy objectives are shown to be jointly unattainable in their strong forms, so that policy choice resolves to selecting which of the three objectives to relax under given institutional and resource constraints.

The Mineral-AI Governance Trilemma: A Synthesising Framework

The empirical and analytical material developed across this chapter can be drawn together using the Mineral-AI Governance Trilemma introduced in the Scope section, which adapts the trilemma logic from monetary policy (Obstfeld et al., 2005) and energy governance (World Energy Council, 2023) to the AI-hardware mineral domain. The framework posits three policy objectives—supply security and strategic sovereignty; environmental and social sustainability; and innovation pace and equitable access—that any single jurisdiction can credibly pursue in pairs but not simultaneously in full. The trilemma is not a counsel of resignation: it is a discipline for surfacing the trade-offs that the dominant policy frames examined in this chapter—the European Critical Raw Materials Act, the Inflation Reduction Act, JOGMEC's stockpile architecture, the Australian and Canadian critical-minerals strategies and the Minerals Security Partnership extended through Pax Silica—each tend to obscure when presented in their own terms.

The first pairing—prioritising supply security and sustainability together—is the implicit commitment of the European framework and, in different form, of the Canadian 'clean premium' positioning. The evidence assembled in this chapter suggests that this pairing is achievable but at the cost of innovation pace and equitable access: the financing requirements for upstream environmental safeguards, processing capacity build-out and Indigenous-partnership terms substantially elevate the cost of qualifying material (Roos, 2025b, 2025c), with downstream consequences for AI-hardware unit costs and the rate at which AI compute capacity can be deployed. Developing-country producers without the institutional capacity to participate in friend-shoring partnerships under the Minerals Security Partnership and Pax Silica are predictably excluded from the resulting premium markets (African Union, 2025), reproducing the pattern that the African Union (2009) Africa Mining Vision sought to redress sixteen years ago.

The second pairing—supply security and innovation pace—is the implicit commitment of the strategic-weaponisation literature and of the most aggressive readings of the United States CHIPS Act and 45X regime. It is achievable, on the analysis presented here, only through environmental and social burden-shifting: rapid scale-up of domestic and friend-shore processing within constrained timelines reduces the margin for environmental-assessment quality, water-management infrastructure and human-rights due-diligence that the comparative case material (Banza Lubaba Nkulu et al., 2018; Muimba-Kankolongo et al., 2022; Owen et al., 2023; Initiative for Responsible Mining Assurance, 2025) shows to be load-bearing for long-term project viability. The 2025 sequence of Chinese export-control escalations and the November 2025 Busan-driven suspension demonstrate how unstable this pairing is in practice: it relies on continued production from jurisdictions whose own policy frames the framework does not constrain.

The third pairing—sustainability and innovation pace—is the implicit commitment of the AI-enabling-role literature (Kaack et al., 2022) and of the hyperscaler clean-premium strategies examined in the Industrial Policy Instruments section. It can be advanced credibly through AI-enabled traceability, material recovery and process optimisation. Its limitation, on the chapter's evidence, is that supply concentration vulnerability persists: a strategy organised around sustainability and innovation but indifferent to supply security leaves the AI-hardware value chain exposed to precisely the kind of policy-shock probability that the 2023–2025 Chinese export-control sequence has now established as the binding planning variable (International Energy Agency, 2025).

The synthesis the trilemma supports is not the selection of one pairing over the others but the recognition that the choice is constitutive of the policy framework, not an unfortunate by-product of it. The European Union's quantitative targets, the Inflation Reduction Act's Foreign Entity of Concern provisions, the Japanese hybrid-stockpile model, the Canadian Indigenous-partnership architecture and the Minerals Security Partnership friend-shoring arrangement each privilege a different pair of vertices; the disagreements between them in international fora are, on this reading, disagreements about which pair to privilege rather than disagreements about facts or instrument design. The trilemma thus provides a vocabulary for negotiating between jurisdictions whose policy frames are not directly comparable on a single welfare metric, and a discipline for assessing the consistency of any individual jurisdiction's instrument mix against its declared priorities. It does not dissolve the trade-offs the chapter has documented, but it makes them legible.

Two qualifications limit the framework's scope. First, the three vertices are presented as analytically separable but interact: investments in AI-enabled material recovery, for example, advance sustainability and innovation pace while also reducing supply-security exposure through circularity. The trilemma should therefore be read as a constraint on simultaneous full attainment rather than a denial of partial complementarity. Second, the framework is jurisdiction-agnostic and abstracts from the political-economy conditions that determine which pair any particular government will privilege; it is a diagnostic device, not a theory of policy choice. Within these limits, the trilemma offers a synthesising frame for the comparative material developed across the chapter and a structured agenda for the research questions taken up in the section that follows.

Research Agenda and Knowledge Gaps

The analysis identifies several areas requiring further research to inform policy development and industry practice.

AI-specific mineral demand accounting remains underdeveloped. Current demand projections aggregate AI with broader semiconductor and electronics applications, obscuring the specific contribution of AI hardware to mineral requirements. Developing AI-specific demand scenarios requires detailed analysis of accelerator architectures, training and inference workload projections, and infrastructure deployment trajectories. Such analysis would enable more targeted policy responses and investment prioritisation.

Comparative policy effectiveness assessment is needed to evaluate the diverse instruments deployed across jurisdictions. The proliferation of critical minerals strategies, financing mechanisms, and regulatory interventions creates natural experiments whose outcomes merit systematic comparison. Key questions include: Which incentive structures most effectively stimulate processing capacity development? How do permitting reform approaches affect both project timelines and environmental outcomes? What conditions enable effective recycling infrastructure development? How do different models of Indigenous partnership affect both project development and benefit distribution?

Supply chain transparency and traceability systems require technological and governance innovation. Current due diligence approaches rely heavily on paper-based certification with limited verification capacity. Digital traceability solutions including blockchain-based provenance systems show promise but face adoption barriers and interoperability challenges. Research on effective transparency mechanisms should encompass both technological capabilities and governance frameworks for their implementation.

Environmental impact assessment methodologies require refinement to capture the full scope of extraction and processing consequences. Current life cycle assessment approaches face data limitations, boundary inconsistencies, and allocation challenges that limit comparability. Developing standardised approaches for AI hardware LCA would enable more rigorous comparison of alternatives and support informed procurement decisions. The persistent gaps in coverage of water impacts identified across critical minerals extraction studies warrant particular attention.

The political economy of critical minerals governance merits sustained attention. The intersection of industrial policy, trade policy, environmental regulation, and geopolitical competition creates complex dynamics that simple analytical frameworks inadequately capture. Understanding how these dynamics shape policy outcomes requires interdisciplinary research spanning economics, political science, and environmental studies. The role of hyperscalers and other private sector actors in shaping supply chain development through purchasing commitments and investment decisions represents an emerging area for investigation.

The circular economy potential for AI hardware materials requires more rigorous assessment, particularly at the level of enterprise servers and associated data-centre equipment where AI deployment is concentrated. Current recycling rates for specialty metals remain below 1 per cent, and the technical challenges of recovering materials from complex electronic assemblies are substantial. Evidence from enterprise server end-of-life analysis indicates that recovery outcomes depend strongly on product architecture, disassembly practices, and access to operational enablers rather than on theoretical recyclability alone (Talens Peiró & Ardente, 2015). The JRC analysis emphasises that component reuse is frequently constrained by non-material bottlenecks, notably the availability of firmware needed to verify compatibility and functionality of parts, and the ability to ensure secure data sanitisation for storage devices prior to reuse (Talens Peiró & Ardente, 2015).

It further proposes resource-efficiency requirements that are directly relevant to AI server ecosystems, including design for disassembly to enable removal of processors, printed circuit board assemblies, and drives using commonly available tools, and structured provision of technical documentation (e.g. exploded diagrams and disassembly sequences) to support third-party maintenance, refurbishment, and selective recycling (Talens Peiró & Ardente, 2015). A complementary lever is disclosure of the location and quantities of critical raw materials within server components—particularly where rare earth magnets and other critical inputs are embedded—so that recyclers can target high-value fractions rather than relying on mixed shredding streams that reduce practical recoverability (Talens Peiró & Ardente, 2015; NSF International, 2015). Research distinguishing between technically recyclable and practically recoverable materials would inform realistic policy targets, while the economics of 'urban mining' from electronic waste and tailings reprocessing warrant investigation, as do the conditions under which recycling infrastructure investment becomes commercially viable. For AI hardware, these findings imply that improving circular outcomes is as much an information-governance and design-for-recovery challenge as a metallurgical one, and that realistic policy targets should distinguish between theoretical recycling potential and recoverability under feasible disassembly and data-governance conditions (Talens Peiró & Ardente, 2015).

Sensitivity to growth-rate assumptions deserves explicit treatment in any forward-looking critical-minerals analysis. The mineral-demand projections referenced in this chapter are constructed under specific assumptions about AI accelerator shipment growth, average power per accelerator and average material intensity per unit of compute. Plausible high-growth scenarios that combine accelerated AI infrastructure deployment with slower-than-expected efficiency improvements would raise projected demand for copper, indium, gallium and rare-earth permanent magnets by approximately 30 to 60 per cent above the chapter's central figures by 2030; low-growth scenarios under which generative-AI adoption plateaus and architectural efficiency improvements reduce material per unit of compute could lower projected demand by approximately 20 to 35 per cent. The wide range shows that the policy responses examined in this chapter must be robust across multiple plausible demand trajectories rather than tuned to a single point estimate. Formal sensitivity analysis of the central projections, with explicit confidence intervals around adoption, efficiency and substitution parameters, sits in the research agenda. The +30 to +60 per cent and -20 to -35 per cent ranges correspond, respectively, to a doubling and a halving of the central 2025 to 2030 AI accelerator shipment-growth assumption (cumulative growth rate of approximately 35 per cent per year in the central case), combined with sensitivity to a ±20 per cent variation in average material intensity per accelerator unit; these ranges are illustrative rather than the output of a formal model.

Concluding Observations

For data-centre operators and AI-hardware procurement leaders, the practical implications of this analysis converge on five areas. Procurement strategies should require supplier-chain transparency through to refining-stage origin and incorporate diversification of geographic sourcing as a binding criterion alongside price and performance. Design-for-recycling principles should be applied at the architectural level to AI accelerator packages and data-centre infrastructure, with explicit attention to the recovery of indium, gallium and rare-earth elements at end of life. Long-term offtake arrangements with qualifying upstream suppliers, particularly in OECD-aligned jurisdictions, can lock in supply security while contributing to the financing of new production capacity. Internal scope-3 reporting frameworks should be extended upstream to mineral-source level, in line with the trajectory of regulatory disclosure obligations examined in this chapter. Lastly, scenario-based contingency planning should explicitly address the high-criticality minerals identified in this chapter under both supply-restriction and price-shock conditions, recognising that the AI-hardware build-out occurs alongside parallel demand from clean-energy and electric-vehicle sectors competing for the same upstream supply.

This chapter has documented the material foundations of artificial intelligence, revealing extensive dependencies on critical minerals whose extraction and processing carry substantial environmental consequences. The analysis demonstrates that AI's environmental footprint extends far beyond operational energy consumption to encompass the embodied impacts of hardware supply chains spanning continents and involving dozens of mineral commodities.

The policy landscape is characterised by convergent concerns across OECD economies regarding supply concentration and vulnerability, coupled with divergent approaches reflecting different national circumstances and capabilities. Supply security has emerged as a priority concern, catalysing unprecedented industrial policy interventions including extraction incentives, processing investments, strategic stockpiling, and friend-shoring initiatives. These interventions represent a departure from decades of reliance on global markets for commodity procurement, reflecting a reconceptualisation of critical minerals as strategic assets requiring state coordination. Public financing commitments across OECD economies now exceed tens of billions of dollars, with mechanisms ranging from tax credits and production subsidies to equity investments and concessional lending.

Environmental sustainability features in policy frameworks but receives less attention than economic and strategic considerations. The tensions between supply security and environmental objectives remain inadequately resolved, with current approaches risking burden-shifting to jurisdictions with weaker governance capacity. Achieving both supply security and environmental sustainability requires policy coherence that extends governance expectations across international supply chains while building domestic capacity for higher-standard production. The integration of environmental criteria into three-dimensional criticality assessments represents progress, but operationalisation remains methodologically contested and politically challenging.

The material dimensions of AI development have implications for state-market relations in the governance of strategic technologies. The concentration of critical minerals supply, the capital intensity of processing infrastructure, and the geopolitical instrumentalisation of mineral resources create conditions that market mechanisms alone cannot address. State intervention through industrial policy, trade policy, and diplomatic engagement has become essential for maintaining access to materials that underpin technological capabilities. This represents a significant shift from the liberal market assumptions that characterised technology policy in previous decades, with governments now actively shaping supply chains through a combination of incentives, restrictions, and international partnerships.

The emergence of bifurcated market structures—with 'allied' and 'non-aligned' supply chains developing in parallel—has implications that extend beyond immediate supply security concerns. The efficiency losses from abandoning lowest-cost global sourcing represent costs that will ultimately flow through to technology prices and adoption rates. The exclusion of developing countries from friend-shoring arrangements may perpetuate patterns of extraction without domestic value addition, limiting development opportunities in resource-rich regions. The long-term sustainability of subsidised production in OECD jurisdictions remains uncertain, particularly given the multi-decade timelines required for meaningful supply chain restructuring.

As AI capabilities expand and hardware deployment accelerates, the material foundations examined in this chapter will assume increasing importance. The environmental and social dimensions of mineral extraction cannot be externalised from assessments of AI's societal impacts. Comprehensive approaches to AI governance must therefore encompass the upstream supply chains that enable computational capabilities, extending policy attention from algorithms and applications to the material substrates on which AI systems depend. The challenge for policymakers is to work through the tensions between supply security, environmental sustainability, and economic efficiency in ways that serve long-term interests rather than merely responding to immediate pressures. The decisions made in the current period of intensive industrial policy activism will shape critical minerals supply chains, and by extension AI hardware availability, for decades to come.

REFERENCES

Abbadi, A., & Mucsi, G. (2024). A review on complex utilization of mine tailings: Recovery of rare earth elements and residue valorization. *Journal of Environmental Chemical Engineering, 12*(3), 113118. DOI: 10.1016/j.jece.2024.113118

Acres (Land Values). (2025, July 15). *CoreWeave locks down 1.3 GW across 10 U.S. sites in $9B deal.* https://landvalues.acres.com/all

African Union. (2009). *Africa Mining Vision.* African Union Commission. https://au.int/sites/default/files/documents/30995-doc-africa_mining_vision_english_1.pdf

African Union. (2025). *African Green Minerals Strategy: Beneficiation-led development for critical minerals (Building on the 2009 African Mining Vision).* African Union Commission.

Airaudo, F., de Soyres, F., Fisgin, E., Gaillard, A., Santacreu, A. M., Richards, K., & Young, H. (2025, December 12). Understanding trade fragmentation. FEDS Notes. Board of Governors of the Federal Reserve System. DOI: 10.17016/2380-7172.3960

Analytics, F. P. (2025, July 18). Artificial intelligence and the critical minerals crunch: Strategies for resilient supply chains. Foreign Policy. https://fpanalytics.foreignpolicy.com/2025/07/18/artificial-intelligence-critical-minerals-supply-chains/

Anthropic. (2025, November 12). *Anthropic invests $50 billion in American AI infrastructure.* https://www.anthropic.com/news/anthropic-invests-50-billion-in-american-ai-infrastructure

AnySilicon. (n.d.). High bandwidth memory – White paper. https://anysilicon.com/high-bandwidth-memory-white-paper/

Arafura Rare Earths Limited. (n.d.). *Nolans rare earths project* [Corporate website]. https://www.arafurarareearths.com/projects/nolans/

Australian Government, Department of Industry, Science and Resources. (2023). Critical minerals strategy 2023–2030. Commonwealth of Australia. https://www.industry.gov.au/publications/critical-minerals-strategy-2023-2030

Bailey, G., Farinha, J., Mochan, A., & Pólvora, A. (2024). *Eyes on the future: Signals from recent reports on emerging technologies and breakthrough innovations to support European Innovation Council strategic intelligence* (Vol. 1). Publications Office of the European Union., DOI: 10.2760/144136

Baldassarre, B., Buesa, A., Albizzati, P. F., Jakimów, M., Saveyn, H. G. M., & Carrara, S. (2023). *Analysis of circular economy research and innovation (R&I) intensity for critical products in the supply chains of strategic technologies*. Publications Office of the European Union. DOI: 10.2760/582527

Baldassarre, B., & Carrara, S. (2025). Critical raw materials, circular economy, and sustainable development: EU policy reflections for future research and innovation. *Resources, Conservation and Recycling*, *215*, 108060. DOI: 10.1016/j.resconrec.2024.108060

Banza Lubaba Nkulu, C., Casas, L., Haufroid, V., De Putter, T., Saenen, N. D., Kayembe-Kitenge, T., Musa Obadia, P., Kyanika Wa Mukoma, D., Lunda Ilunga, J.-M., Nawrot, T. S., Luboya Numbi, O., Smolders, E., & Nemery, B. (2018). Sustainability of artisanal mining of cobalt in DR Congo. *Nature Sustainability*, *1*(9), 495–504. DOI: 10.1038/s41893-018-0139-4 PMID: 30288453

Baskaran, G. (2024, January 8). What China's ban on rare earths processing technology exports means. Center for Strategic and International Studies. https://www.csis.org/analysis/what-chinas-ban-rare-earths-processing-technology-exports-means

Baskaran, G., & Schwartz, M. (2024, October 7). From mine to microchip: Addressing critical mineral supply chain risks in semiconductor production. Center for Strategic and International Studies. https://www.csis.org/analysis/mine-microchip

Battery Council International. (2025). *AI and data centers: Battery applications in stationary energy storage*. Battery Council International. https://batterycouncil.org/battery-facts-and-applications/essential-applications/stationary-energy-storage/ai-data-centers/

Berg, R. C., Polo Anaya, E., & Ziemer, H. (2024, May 15). Mineral demands for resilient semiconductor supply chains. Center for Strategic and International Studies. https://www.csis.org/analysis/mineral-demands-resilient-semiconductor-supply-chains

BHP. (2025, January). Why AI tools and data centres are driving copper demand. BHP. https://www.bhp.com/news/bhp-insights/2025/01/why-ai-tools-and-data-centres-are-driving-copper-demand

Branigan, T. (2010, October 25). Chinese moves to limit mineral supplies sparks struggle over rare earths. The Guardian. https://www.theguardian.com/business/2010/oct/25/china-cuts-rare-earths-exports

Carbon Credits. (2025, December 11). *Google and NextEra team up to build gigawatt-scale AI data centers powered by clean energy*. https://carboncredits.com/google-and-nextera-team-up-to-build-gigawatt-scale-ai-data-centers-powered-by-clean-energy/

Carrara, S., Bobba, S., Blagoeva, D., Alves Dias, P., Cavalli, A., Georgitzikis, K., Grohol, M., Itul, A., Kuzov, T., Latunussa, C., Lyons, L., Malano, G., Maury, T., Prior Arce, Á., Somers, J., Telsnig, T., Veeh, C., Wittmer, D., Black, C., & Christou, M. (2023). *Supply chain analysis and material demand forecast in strategic technologies and sectors in the EU – A foresight study*. Publications Office of the European Union. DOI: 10.2760/386650

Castilloux, R. (2019, April 25). CHARTS: Rare earth export restrictions, price spikes and the risks of demand destruction. MINING.COM. https://www.mining.com/featured-article/charts-rare-earth-export-restrictions-price-spikes-and-the-risks-of-demand-destruction/

Crusoe. (2025, May 21). *Crusoe, Blue Owl Capital, and Primary Digital Infrastructure enter second phase of $15 billion joint venture to fund AI data center in Abilene, Texas*. https://www.crusoe.ai/resources/newsroom/crusoe-blue-owl-capital-and-primary-digital-infrastructure-enter-joint-venture

Davey, C. (2023, March 28). The environmental impacts of cobalt mining in Congo. Earth.Org. https://earth.org/cobalt-mining-in-congo/

Davidson, H. (2025, October 9). China steps up control of rare-earth exports citing 'national security' concerns. The Guardian. https://www.theguardian.com/world/2025/oct/09/china-steps-up-control-rare-earth-exports-national-security-concerns

Department for Business and Trade. (2025, November 24). Critical minerals strategy: Vision 2035. UK Government. https://assets.publishing.service.gov.uk/media/69244c80367485ea116a5675/uk-critical-minerals-strategy-vision-2035.pdf

Edler, J., Blind, K., Kroll, H., & Schubert, T. (2023). *Strategic autonomy and sovereignty in the innovation and economic domains*. Publications Office of the European Union.

Eduljee, G. H., Verbeke, F., Marvin, H. J. P., & van der Berg, M. (2024). Per- and polyfluoroalkyl substances and emerging contaminants in semiconductor fabrication: Occupational and community exposure dimensions. *Environmental Health Perspectives*, *132*(8), 087001. PMID: 39106155

European Commission. (2023). Proposal for a regulation of the European Parliament and of the Council establishing a framework for ensuring a secure and sustainable supply of critical raw materials (Critical Raw Materials Act) (COM(2023) 160 final). EUR-Lex. https://eur-lex.europa.eu/legal-content/EN/TXT/HTML/?uri=CELEX:52023PC0160

European Commission, Joint Research Centre. (2025). *Strategic technologies and industrial policies for competitiveness and sustainability: CONCORDi 2025 policy brief (JRC144677)*. Publications Office of the European Union., DOI: 10.2760/477729

European Parliament and Council of the European Union. (2024). Regulation (EU) 2024/1252 establishing a framework for ensuring a secure and sustainable supply of critical raw materials. Official Journal of the European Union.

Extractive Industries Transparency Initiative. (2025). EITI 2025 progress report: 57 implementing countries and the agenda for critical minerals transparency. EITI International Secretariat. https://eiti.org

Falk, J., Schaltegger, S., & Stern, T. (2025). Comparative life cycle assessment of AI and traditional computing infrastructure. *Journal of Industrial Ecology*. Advance online publication.

Georgitzikis, K., Kavvadias, K., Kushnir, D., Gumiel, J. C., Mancini, L., Eynard, U., Bobba, S., Pavel, C., Carrara, S., Maury, T., Peffet, C., Deboyer, S., & Mathieux, F. (2025). *Substitution and reduction of critical and strategic raw materials in clean energy technologies (EUR 40297 EN)*. Publications Office of the European Union., DOI: 10.2760/930966

Government of Canada. Natural Resources Canada. (2022). Canadian critical minerals strategy. https://www.canada.ca/en/campaign/critical-minerals-in-canada/canadian-critical-minerals-strategy.html

Graedel, T. E., Barr, R., Chandler, C., Chase, T., Choi, J., Christoffersen, L., Friedlander, E., Henly, C., Jun, C., Nassar, N. T., Schechner, D., Warren, S., Yang, M., & Zhu, C. (2012). Methodology of metal criticality determination. *Environmental Science & Technology, 46*(2), 1063–1070. DOI: 10.1021/es203534z PMID: 22191617

Graedel, T. E., Harper, E. M., Nassar, N. T., Nuss, P., & Reck, B. K. (2015). Criticality of metals and metalloids. *Proceedings of the National Academy of Sciences of the United States of America, 112*(14), 4257–4262. DOI: 10.1073/pnas.1500415112 PMID: 25831527

Graedel, T. E., & Reck, B. K. (2016). Six years of criticality assessments: What have we learned so far? *Journal of Industrial Ecology*, *20*(4), 692–699. DOI: 10.1111/jiec.12305

Grohol, M., & Veeh, C. (2023). Study on the critical raw materials for the EU 2023: Final report. European Commission.

Gupta, U., Kim, Y. G., Lee, S., Tse, J., Lee, H.-H. S., Wei, G.-Y., Brooks, D., & Wu, C.-J. (2022). Chasing carbon: The elusive environmental footprint of computing. *IEEE Micro*, *42*(4), 37–47. DOI: 10.1109/MM.2022.3163226

Hess, J. C. (2024, June 20). Chip production's ecological footprint: Mapping climate and environmental impact. Interface. https://www.interface-eu.org/publications/chip-productions-ecological-footprint

Hyperscale, A. F. L. (2024). Optical fibre innovation: Driving seamless flow in hyperscale data centres [Corporate technical brief]. AFL Hyperscale. https://www.aflhyperscale.com/articles/fiber-optic-innovation-driving-seamless-flow-in-hyperscale-data-centers/

IMARC Group. (2024). *Tantalum capacitors market: Global industry trends, share, size, growth, opportunity and forecast 2024–2032*. IMARC Services Private Limited. https://www.imarcgroup.com/tantalum-capacitors-market

Initiative for Responsible Mining Assurance. (2025). IRMA Standard for Responsible Mining IRMA-STD-001 (Version 2.0) [Standard and audit protocol]. IRMA. https://responsiblemining.net

International, N. S. F. (2015). *Environmental leadership standards for servers (NSF/ANSI 426) (Draft document)*. NSF International.

International Council on Clean Transportation. (2024, December). *Electrifying road transport with less mining: A global and regional battery material outlook* (White Paper). ICCT. https://theicct.org/publication/ev-battery-materials-demand-supply-dec24/

International Energy Agency. (2021). The role of critical minerals in clean energy transitions. https://www.iea.org/reports/the-role-of-critical-minerals-in-clean-energy-transitions

International Energy Agency. (2023). *Critical minerals market review 2023*. IEA Publications.

International Energy Agency. (2024). Global critical minerals outlook 2024. IEA Publications. https://www.iea.org/reports/global-critical-minerals-outlook-2024

International Energy Agency. (2025). Global critical minerals outlook 2025. IEA Publications. https://www.iea.org/reports/global-critical-minerals-outlook-2025

International Seabed Authority. (2025). Status of negotiations on the regulations for exploitation of mineral resources in the Area. International Seabed Authority Secretariat, Kingston, Jamaica. https://www.isa.org.jm

Jerez, B., Garcés, I., & Torres, R. (2021). Lithium extractivism and water injustices in the Salar de Atacama, Chile: The colonial shadow of green electromobility. *Political Geography*, *87*, 102382. DOI: 10.1016/j.polgeo.2021.102382

Kaack, L. H., Donti, P. L., Strubell, E., Kamiya, G., Creutzig, F., & Rolnick, D. (2022). Aligning artificial intelligence with climate change mitigation. *Nature Climate Change*, *12*(6), 518–527. DOI: 10.1038/s41558-022-01377-7

Lessner, P. (2025, March 12). *Q&A: AI drives fresh Ta capacitor demand* [Interview by Argus Media]. Argus Media. https://www.argusmedia.com/en/news-and-insights/latest-market-news/2737718-q-a-ai-drives-fresh-ta-capacitor-demand

Lima, F. M., Lovon-Canchumani, G. A., & Teixeira, M. A. (2023). Life cycle assessment in the mining sector: A systematic review. *Resources Policy*, *78*, 102889. DOI: 10.1016/j.resourpol.2022.102889

Luo, H., Huang, T.-Y., Wu, X., & Zhao, F. (2025). Life cycle assessment for primary gallium production at industrial-scale. *The International Journal of Life Cycle Assessment*, *30*(7), 1545–1559. DOI: 10.1007/s11367-025-02492-1

Mancheri, N. A., Sprecher, B., Bailey, G., Ge, J., & Tukker, A. (2019). Effect of Chinese policies on rare earth supply chain resilience. *Resources, Conservation and Recycling*, *142*, 101–112. DOI: 10.1016/j.resconrec.2018.11.017

Meta Platforms, Inc. (2024). *2024 sustainability report*. Meta. https://sustainability.atmeta.com/wp-content/uploads/2024/08/Meta-2024-Sustainability-Report.pdf

Microsoft. (2025, September 18). *Made in Wisconsin: The world's most powerful AI datacenter*. Microsoft On the Issues. https://blogs.microsoft.com/on-the-issues/2025/09/18/made-in-wisconsin-the-worlds-most-powerful-ai-datacenter/

MINING.COM. (2024, July 23). *AI data centers to worsen copper shortage – BNEF*. MINING.COM. https://www.mining.com/ai-data-centers-to-worsen-copper-shortage-bnef/

Ministry of Economy, Trade and Industry. (2020). International resource strategy [ ]. Government of Japan. https://www.meti.go.jp/press/2020/03/20200330001/20200330001.html

Ministry of Economy, Trade and Industry. (2023). Policy for measures to ensure stable supply of critical minerals. Government of Japan. https://www.meti.go.jp/policy/economy/economic_security/metal/index.html

Ministry of Trade, Industry and Energy. (2023). Strategy for securing reliable critical minerals supply. Government of the Republic of Korea.

Mordor Intelligence. (2025). *Tantalum capacitors market.* https://www.mordorintelligence.com/industry-reports/tantalum-capacitors-market

Muimba-Kankolongo, A., Banza Lubaba Nkulu, C., Mwitwa, J., Kampemba, F. M., Mulele Nabuyanda, M., Haufroid, V., Smolders, E., & Nemery, B. (2022). Impacts of trace metals pollution of water, food crops, and ambient air on population health in Zambia and the DR Congo. *Journal of Environmental and Public Health, 4515115*(1), 4515115. Advance online publication. DOI: 10.1155/2022/4515115 PMID: 35844936

Nassar, N. T., & Fortier, S. M. (2021). *Methodology and technical input for the 2021 review and revision of the U.S. Critical Minerals List* (Open-File Report 2021-1045). U.S. Geological Survey. DOI: 10.3133/ofr20211045

Nassar, N. T., Shojaeddini, E., Alonso, E., Jaskula, B., & Tolcin, A. (2024). Quantifying potential effects of China's gallium and germanium export restrictions on the U.S. economy (Open-File Report 2024–1057). U.S. Geological Survey. DOI: 10.3133/ofr20241057

National Energy System Operator. (2024, November 5). *Clean Power 2030: Advice on achieving clean power by 2030.* https://www.neso.energy/publications/clean-power-2030

National Research Council. (2008). *Minerals, critical minerals, and the U.S. economy.* National Academies Press., DOI: 10.17226/12034

NVIDIA. (2024, July 12). NVIDIA GB200 NVL72 delivers trillion-parameter LLM training and real-time inference. NVIDIA Technical Blog. https://developer.nvidia.com/blog/nvidia-gb200-nvl72-delivers-trillion-parameter-llm-training-and-real-time-inference/

NVIDIA. (2025). *GB200 NVL72: Rack-scale Blackwell platform specifications* [Product page]. NVIDIA Corporation. https://www.nvidia.com/en-us/data-center/gb200-nvl72/

Obstfeld, M., Shambaugh, J. C., & Taylor, A. M. (2005). The trilemma in history: Tradeoffs among exchange rates, monetary policies, and capital mobility. *The Review of Economics and Statistics, 87*(3), 423–438. DOI: 10.1162/0034653054638300

Office of the United States Trade Representative. (2014, March 26). United States wins victory in rare earths dispute with China: WTO report finds China's export restraints breach WTO rules. https://ustr.gov/about-us/policy-offices/press-office/press-releases/2014/March/US-wins-victory-in-rare-earths-dispute-with-China

Open, A. I. (2025, September 23). *OpenAI, Oracle, and SoftBank expand Stargate with five new AI data center sites.* https://openai.com/index/five-new-stargate-sites/

Organisation for Economic Co-operation and Development. (2011). OECD due diligence guidance for responsible supply chains of minerals from conflict-affected and high-risk areas. OECD Publishing. https://www.oecd.org/content/dam/oecd/en/publications/reports/2011/07/oecd-due-diligence-guidance-for-responsible-supply-chains-of-minerals-from-conflict-affected-and-high-risk-areas_g1g13355/9789264111110-en.pdf

Organisation for Economic Co-operation and Development. (2023). Critical minerals market review 2023. OECD Publishing. https://www.oecd.org/en/publications/critical-minerals-market-review-2023_9cdf8f39-en.html

Owen, J. R., Kemp, D., Lechner, A. M., Harris, J., Zhang, R., & Lèbre, É. (2023). Energy transition minerals and their intersection with land-connected peoples. *Nature Sustainability*, *6*(2), 203–211. DOI: 10.1038/s41893-022-00994-6

Pell, R., Wall, F., Yan, X., Li, J., & Zeng, X. (2019). Life cycle assessment in the minerals industry: Current practice, harmonization efforts, and potential improvement through the integration with process simulation. *Journal of Cleaner Production*, *232*, 155–173. DOI: 10.1016/j.jclepro.2019.05.264

Pew Charitable Trusts. (2025, January 15). The world currently lacks the ability to govern deep-sea mining. The Pew Charitable Trusts. https://www.pew.org/en/about/news-room/opinion/2025/01/15/the-world-currently-lacks-the-ability-to-govern-deep-sea-mining

Polaris Market Research. (2022). *Thermal interface materials market share, size, trends, industry analysis report: 2022–2030*. Polaris Market Research. https://www.polarismarketresearch.com/industry-analysis/thermal-interface-materials-market

Reuters. (2025, October 15). *Meta commits $1.5 billion for AI data center in Texas.* https://www.reuters.com/business/meta-commits-15-billion-ai-data-center-texas-2025-10-15/

Roos, G. (2025a, May). *Increasing unevenness in trade-related sustainability policy: The ASEAN perspective (Policy Brief No. 29).* Institute for International Trade, The University of Adelaide. https://iit.adelaide.edu.au

Roos, G. (2025b, September). *Australia–EU critical minerals partnerships: Navigating supply security and climate commitments* (Road to Belém Briefing #1). Institute for Sustainability, Energy and Resources & Institute for International Trade, The University of Adelaide.

Roos, G. (2025c, November). Stimulating demand for green premium products in OECD countries: Policy tools, theoretical perspectives, and empirical evidence from the iron–steel value chain. Discussion paper prepared for the Fortum Nordic Energy Debate, 25 November 2025, Finlandia Hall, Helsinki, Finland.

Schneider Electric. (2025, May 20). AI-tolerant UPSs: The first line of defense in data center resilience. Schneider Electric Blog. https://blog.se.com/datacenter/2025/05/20/ai-tolerant-ups-first-line-defense-data-center-resilience/

Schrijvers, D., Hool, A., Blengini, G. A., Chen, W. Q., Dewulf, J., Eggert, R., van Ellen, L., Gauss, R., Goddin, J., Habib, K., Hagelüken, C., Hirohata, A., Hofmann-Amtenbrink, M., Kosmol, J., Le Gleuher, M., Grohol, M., Ku, A., Lee, M.-H., Liu, G., & Wäger, P. (2020). A review of methods and data to determine raw material criticality. *Resources, Conservation and Recycling*, *155*, 104617. DOI: 10.1016/j.resconrec.2019.104617

SemiAnalysis. (2024, July 17). *GB200 hardware architecture: Component supply chain and bill of materials*. SemiAnalysis. https://semianalysis.com/2024/07/17/gb200-hardware-architecture-and-component/

SemiAnalysis. (2025, September 16). *xAI's Colossus 2 – First gigawatt datacenter in the world, unique RL methodology, capital raise*. https://newsletter.semianalysis.com/p/xais-colossus-2-first-gigawatt-datacenter

Shehabi, A., Smith, S. J., Hubbard, A., Newkirk, A., Lei, N., Siddik, M. A. B., Holecek, B., Koomey, J., Masanet, E., & Sartor, D. (2024). *2024 United States data center energy usage report*. Lawrence Berkeley National Laboratory., DOI: 10.71468/P1WC7Q

Sigalos, M. (2025, October 29). *Amazon opens $11 billion AI data center Project Rainier in Indiana*. CNBC. https://www.cnbc.com/2025/10/29/amazon-opens-11-billion-ai-data-center-project-rainier-in-indiana.html

Silicon, A. N. G. L. E. (2025, October 23). *Crusoe lands $1.3B to accelerate buildout of large-scale AI data centers*. https://siliconangle.com/2025/10/23/crusoe-lands-1-3b-accelerate-buildout-large-scale-ai-data-centers/

Sonderegger, T., Berger, M., Alvarenga, R., Bach, V., Cimprich, A., Dewulf, J., Frischknecht, R., Guinée, J., Helbig, C., Huppertz, T., Jolliet, O., Motoshita, M., Northey, S., Peña, C. A., Rugani, B., Sahnoune, A., Schrijvers, D., Schulze, R., Sonnemann, G., & Young, S. B. (2020). Mineral resources in life cycle impact assessment—Part I: A critical review of existing methods. *The International Journal of Life Cycle Assessment, 25*(4), 784–797. DOI: 10.1007/s11367-020-01736-6

Sprecher, B., Daigo, I., Murakami, S., Kleijn, R., Vos, M., & Kramer, G. J. (2015). Framework for resilience in material supply chains, with a case study from the 2010 rare earth crisis. *Environmental Science & Technology, 49*(11), 6740–6750. DOI: 10.1021/acs.est.5b00206 PMID: 25965803

Talens Peiró, L., & Ardente, F. (2015). *Analysis of material efficiency requirements of enterprise servers*. Publications Office of the European Union., DOI: 10.2788/409022

The Economic Times. (2025, October 9). *TCS announces 1-GW data centre plan with $6.5 billion investment.* https://m.economictimes.com/tech/information-tech/tcs-announces-1-gw-data-centre-plan-with-6-5-billion-investment/articleshow/124429399.cms

The Oregon Group. (2025). *Rare earths: Critical minerals and energy intelligence*. The Oregon Group. https://theoregongroup.com/rare-earths/

Trading Economics. (n.d.). Neodymium. Retrieved December 15, 2025, from https://tradingeconomics.com/commodity/neodymium

Trafigura. (2024). *AI could add 1 million tons to copper demand by 2030*. https://www.trafigura.com/news-and-insights/in-the-news/2024/ai-could-add-1-million-tons-to-copper-demand-by-2030-says-trafigura/

United Nations Human Rights Council. (2011). *Guiding Principles on Business and Human Rights: Implementing the United Nations 'Protect, Respect and Remedy' Framework (HR/PUB/11/04)*. Office of the United Nations High Commissioner for Human Rights.

United States Geological Survey. (2025). *Mineral commodity summaries 2025*. U.S. Geological Survey., DOI: 10.3133/mcs2025

U.S. Congress. (2022a). CHIPS and Science Act (Pub. L. No. 117–167). https://www.congress.gov/117/plaws/publ167/PLAW-117publ167.pdf

U.S. Congress. (2022b). Inflation Reduction Act of 2022 (Pub. L. No. 117–169). https://www.congress.gov/117/plaws/publ169/PLAW-117publ169.pdf

U.S. Department of State. (2022). Minerals Security Partnership [Fact sheet]. https://www.state.gov/minerals-security-partnership

U.S. Department of State. (2025, December 11). Pax Silica. https://www.state.gov/pax-silica/

U.S. Geological Survey. (2024). Mineral commodity summaries 2024. DOI: 10.3133/mcs2024

Vantage Data Centers. (2025, October 22). *OpenAI, Oracle and Vantage Data Centers announce Stargate data center site in Wisconsin.* https://vantage-dc.com/news/openai-oracle-and-vantage-data-centers-announce-stargate-data-center-site-in-wisconsin/

Vera, M. L., Torres, W. R., Galli, C. I., Chagnes, A., & Flexer, V. (2023). Environmental impact of direct lithium extraction from brines. *Nature Reviews. Earth & Environment, 4*(2), 149–165. DOI: 10.1038/s43017-022-00387-5

Versino, C., Heine, P., & Du, T. (2025). *Strategic trade atlas 2019–2023: Country- and commodity-based views*. Publications Office of the European Union., DOI: 10.2760/7177906

Vertiv Holdings. (n.d.). *Data centre power density trends and infrastructure requirements*. Vertiv Holdings Co.

Vivoda, V. (2023). Friend-shoring and critical minerals: Exploring the role of the Minerals Security Partnership. *Energy Research & Social Science, 100*, 103085. DOI: 10.1016/j.erss.2023.103085

Vivoda, V., & Matthews, R. (2024). 'Friend-shoring' as a panacea to Western critical mineral supply chain vulnerabilities. *Mineral Economics, 37*(3), 463–476. DOI: 10.1007/s13563-023-00402-1

Waseem, U. (2025, April 11). *AI infrastructure engineering: Building enterprise-scale machine learning systems*. Wevolver. https://www.wevolver.com/article/ai-infrastructure

World Bank. (2020). Minerals for climate action: The mineral intensity of the clean energy transition. https://documents.worldbank.org/en/publication/documents-reports/documentdetail/099052423172525564

World Economic Forum. (2025). AI data centres and critical minerals: Infrastructure requirements for the AI transition. World Economic Forum.

World Energy Council. (2023). World Energy Trilemma Index 2023: Frameworks for resilient energy transitions. World Energy Council, London. https://www.worldenergy.org/transition-toolkit/world-energy-trilemma-framework

ADDITIONAL READING

Forstner, A., Ghahreman, A., Mahandra, H., Hepburn, M., & Griffith, M. (2025). A critical review and future perspectives of rare earth magnet recycling technologies. In *Proceedings of the 63rd Conference of Metallurgists, COM 2024* (pp. 1795–1808). Springer. DOI: 10.1007/978-3-031-67398-6_240

Haque, N. (2024). Life cycle assessment for the mining and metallurgical industries: Issues and challenges. In C. Iloeje et al. (Eds.), *Energy Technology 2024. TMS 2024*. The Minerals, Metals & Materials Series. Springer. DOI: 10.1007/978-3-031-50244-6_13

Jha, R., Jha, R., & Islam, M. (2025). Forecasting US data center CO_2 emissions using AI models: Emissions reduction strategies and policy recommendations. *Frontiers in Sustainability*, *5*, 1507030. DOI: 10.3389/frsus.2024.1507030

Jia, F., Hu, D., Chen, L., & Tseng, M.-L. (2025). Global semiconductor supply chain: A national competitiveness analysis. *International Journal of Production Research*, 1–20. Advance online publication. DOI: 10.1080/00207543.2025.2568987

Kalantzakos, S. (2024). *Critical Minerals, the Climate Crisis and the Tech Imperium*. Springer., DOI: 10.1007/978-3-031-25577-9

Mousavinezhad, S. K., Farmanbordar Ghadikolaei, N., Atoufi, H. D., Kowsari, E., & Ramakrishna, S. (2025). Life cycle assessment in mining activity. In Baskar, C., Ramakrishna, S., & Rosa, A. D. L. (Eds.), *Encyclopedia of Green Materials*. Springer., DOI: 10.1007/978-981-97-4618-7_167

Neale, R., . . . (Eds.). (2025). *Rare Metal Technology 2025*. The Minerals, Metals & Materials Series. Springer. DOI: 10.1007/978-3-031-81182-1

Reich, M., & Simon, A. C. (2025). Critical minerals. *Annual Review of Earth and Planetary Sciences*, *53*(1), 141–168. DOI: 10.1146/annurev-earth-040523-023316

Wong, C.-Y., & Yeung, H. W.-C. (2024). Geopolitics and the changing landscape of global value chains and competition in the global semiconductor industry: Rivalry and catch-up in chip manufacturing in East Asia. *Geoforum*, *157*, 104147. DOI: 10.1016/j.geoforum.2024.104147

Xiong, W., Wu, D. D., & Yeung, J. H. Y. (2025). Semiconductor supply chain resilience and disruption: Insights, mitigation, and future directions. *International Journal of Production Research*, *63*(9), 3442–3465. DOI: 10.1080/00207543.2024.2387074

KEY TERMS AND DEFINITIONS

AI Factory: A dedicated, hyperscale data centre facility optimised for generative artificial intelligence model training and inference, characterised by extreme power densities of 40–132 kW per rack, liquid cooling architectures, and unprecedented utility connections, representing a departure from traditional cloud data centres designed for mixed workloads (Schneider Electric, 2025; World Economic Forum, 2025).

AI Hardware: The physical computing equipment enabling the training, deployment, and operation of AI models, including advanced semiconductors, specialised accelerators (GPUs, TPUs), servers, storage systems, high-bandwidth memory, and associated power and cooling infrastructure, excluding software, algorithms, and AI applications (International Energy Agency, 2021; European Commission, 2023; OECD, 2023).

Burden-Shifting: The transfer of environmental and social costs associated with mineral extraction and processing from consumption centres (typically OECD economies) to extraction regions (often non-OECD jurisdictions) with weaker governance capacity and environmental regulatory frameworks (World Bank, 2020; OECD, 2023).

Circular Economy Instruments: Policy tools designed to extend material lifecycles and reduce primary extraction demand through recycling mandates, extended producer responsibility schemes, design-for-recycling requirements, and secondary material utilisation targets (European Parliament and Council of the European Union, 2024; Georgitzikis et al., 2025).

Clean Premium: A price differential that consumers or industrial buyers pay for materials produced with lower environmental impact, higher labour standards, or greater supply chain transparency, representing a potential market-based mechanism for reconciling supply security objectives with environmental sustainability commitments (Roos, 2025c).

Compound Semiconductors: Semiconductor materials composed of two or more elements, such as gallium arsenide (GaAs), gallium nitride (GaN), and indium phosphide (InP), used in high-frequency, high-power, and optoelectronic applications where silicon's performance limitations become constraining (FP Analytics, 2025; Nassar et al., 2024).

Critical Minerals: Minerals characterised by high importance in use combined with elevated supply restriction potential, as conceptualised in the two-dimensional criticality matrix developed by the National Research Council (2008) and subsequently refined to incorporate environmental implications in three-dimensional assessments (Graedel et al., 2012; Graedel et al., 2015).

Critical Raw Materials Act (CRMA): European Union regulation establishing benchmarks for domestic extraction (10 per cent), processing (40 per cent), and recycling (25 per cent) of strategic materials by 2030, alongside permitting timeline requirements and strategic project designation procedures (European Parliament and Council of the European Union, 2024).

Embodied Carbon: The cumulative greenhouse gas emissions associated with the extraction, processing, manufacturing, and transport of materials and components prior to operational use, expressed as kilograms of carbon dioxide equivalent (kg CO_2e), which for AI server nodes is estimated at approximately 2,500 kg CO_2e per unit (NVIDIA, 2025; Falk et al., 2025).

Friend-Shoring: A supply chain diversification strategy that prioritises sourcing from geopolitically aligned nations rather than optimising solely for cost efficiency, reflecting the integration of security considerations into trade and industrial policy (Vivoda, 2023; Vivoda & Matthews, 2024).

High Bandwidth Memory (HBM): A high-performance memory architecture using vertically stacked DRAM dies connected by through-silicon vias (TSVs), enabling the memory bandwidth required for AI accelerators to process large parameter models, with material requirements including silicon, copper, tungsten, cobalt, and tantalum (FP Analytics, 2025; NVIDIA, 2025).

Life Cycle Assessment (LCA): A systematic methodology for evaluating environmental impacts across all stages of a product's existence, from raw material extraction through processing, manufacturing, use, and end-of-life disposal, subject to methodological challenges including boundary definition, data availability, and functional unit specification (Luo et al., 2025; Pell et al., 2019).

Material Intensity: The quantity of materials required per unit of output or capacity, with AI-optimised data centre facilities exhibiting material intensity 20–40 per cent higher than traditional hyperscale benchmarks due to higher power densities and more robust infrastructure requirements (World Economic Forum, 2025).

Minerals Security Partnership (MSP): A multilateral coordination mechanism launched in 2022 comprising the United States, European Union, and allied nations to facilitate investment in critical minerals projects and diversify supply chains away from concentrated sources (U.S. Department of State, 2022; Vivoda, 2023).

Pax Silica: A multilateral initiative launched by the United States in late 2025 framing critical minerals explicitly within a security architecture linking allied nations, representing an evolution from earlier investment-focused coordination efforts toward strategic supply chain governance (U.S. Department of State, 2025).

Power Usage Effectiveness (PUE): A metric expressing data centre energy efficiency as the ratio of total facility energy consumption to IT equipment energy consumption, with values closer to 1.0 indicating greater efficiency; AI-optimised facilities typically target PUE values of 1.20 or lower (Shehabi et al., 2024).

Rare Earth Elements (REEs): A group of seventeen metallic elements comprising the lanthanides plus scandium and yttrium, subdivided into light rare earths (lanthanum through europium) and heavy rare earths (gadolinium through lutetium), with applications in permanent magnets, phosphors, and electronic components critical to AI hardware (International Energy Agency, 2025; European Commission, 2023).

Strategic Stockpiling: Government-administered reserves of critical materials maintained to buffer against supply disruptions, with programme designs ranging from defence-only reserves to hybrid models encompassing both security emergencies and economic stabilisation objectives, as exemplified by Japan's JOGMEC programme (Ministry of Economy, Trade and Industry, 2020).

Strategic Weaponisation: The deliberate use of supply concentration in critical minerals as a tool of statecraft, exemplified by China's imposition of export licensing requirements for gallium and germanium in 2023 and subsequent expanded restrictions, integrating mineral supply with broader technology competition (International Energy Agency, 2023; International Energy Agency, 2025; Baskaran & Schwartz, 2024).

Supply Concentration: The geographic clustering of mineral extraction or processing capacity in a limited number of jurisdictions, creating supply chain vulnerabilities; for AI-relevant materials, China accounts for 98 per cent of global gallium production and 94 per cent of heavy rare earth processing capacity (Nassar et al., 2024; FP Analytics, 2025; International Energy Agency, 2025).

Through-Silicon Vias (TSVs): Vertical electrical connections passing through silicon wafers or dies, enabling three-dimensional chip stacking in advanced packaging architectures such as high bandwidth memory, with material requirements including copper, tungsten, and barrier metals (FP Analytics, 2025).

Chapter 6
The E-Waste Crisis Behind Big Tech:
The Carbon Cost of Intelligence

P. Selvakumar
http://orcid.org/0000-0002-3650-4548
Department of Science and Humanities, Nehru Institute of Technology, Coimbatore, India

ABSTRACT

The rapid ascent of the global technology sector has revolutionized economies, communication, and daily life, but i has also generated an equally unprecedented environmental challenge: electronic waste (e-waste). Often invisible to the average consumer, e-waste represents the physical byproduct of Big Tech's expansion, comprising discarded devices, obsolete components, and broken or outdated electronics. From smartphones, laptops, and tablets to servers, data centers, and IoT devices, the proliferation of electronic products has created towering e-waste "mountains" that strain natural resources, overwhelm recycling infrastructure, and generate toxic environmental hazards. While the economic and societal benefits of digital technology are widely celebrated, the environmental cost embedded in this growth highlights a critical paradox: the more digital innovation accelerates, the heavier the ecological footprint becomes. Understanding the scale, composition, and consequences of e-waste is crucial for framing sustainable practices, circular economy policies, and corporate responsibility.

DOI: 10.4018/979-8-3373-8490-0.ch006

E-WASTE MOUNTAINS: THE HIDDEN BACKDROP OF BIG TECH'S GROWTH

The rapid ascent of the global technology sector has revolutionized economies, communication, and daily life, but i has also generated an equally unprecedented environmental challenge: electronic waste (e-waste). Often invisible to the average consumer, e-waste represents the physical byproduct of Big Tech's expansion, comprising discarded devices, obsolete components, and broken or outdated electronics. From smartphones, laptops, and tablets to servers, data centers, and IoT devices, the proliferation of electronic products has created towering e-waste "mountains" that strain natural resources, overwhelm recycling infrastructure, and generate toxic environmental hazards. While the economic and societal benefits of digital technology are widely celebrated, the environmental cost embedded in this growth highlights a critical paradox: the more digital innovation accelerates, the heavier the ecological footprint becomes. Understanding the scale, composition, and consequences of e-waste is crucial for framing sustainable practices, circular economy policies, and corporate responsibility initiatives that reconcile technological advancement with environmental stewardship.Magnitude and growth of e-waste are staggering. According to the Global E-waste Monitor, the world generated over 60 million metric tons of e-waste in recent years, with projections indicating that annual production could exceed 80 million metric tons by the late 2020s. (Adler & Schneider, 2020) This growth is fueled by rising global device ownership, shorter product life cycles, rapid innovation, and consumer culture emphasizing frequent upgrades.

Composition of e-waste is complex and hazardous. Electronic devices contain valuable metals, including gold, silver, platinum, copper, and rare earth elements, embedded alongside plastics, glass, and toxic substances such as lead, mercury, cadmium, and brominated flame retardants. While these materials offer opportunities for recycling and resource recovery, improper handling can release toxins into soil, water, and air, creating severe health risks. E-waste processing in informal sectors, common in developing countries, often involves open burning, acid leaching, and rudimentary dismantling techniques, exposing workers and surrounding communities to hazardous chemicals. The environmental cost is amplified by the persistence of certain substances, such as persistent organic pollutants and heavy metals, which accumulate in ecosystems and bioaccumulate in food chains. The mixture of recoverable resources and hazardous chemicals highlights the dual nature of e-waste as both an economic asset and an environmental liability.Drivers of e-waste accumulation are closely linked to the business models of Big Tech. Product obsolescence, both perceived and engineered, fuels replacement cycles, while software updates and compatibility issues render older devices functionally outdated. (Anute et al., 2026) The design of electronics, often prioritizing performance, aesthetics, and compactness, complicates repair, refurbishment, and recycling. Proprietary components, non-standardized batteries, and integrated circuits further limit the feasibility of circular lifecycle management.

Consumer culture, incentivized by marketing, social trends, and perceived social status associated with the latest technology, accelerates turnover. In addition, global disparities in infrastructure and regulation create uneven flows of e-waste, with developed countries exporting discarded devices to emerging economies under the guise of reuse, only for these devices to become part of informal recycling streams with limited environmental safeguards. Environmental and health impacts of e-waste are severe and multidimensional. Improper disposal and recycling release toxic substances, polluting soil, groundwater, and surface water. Heavy metals such as lead and cadmium can cause neurological, renal, and reproductive damage, while persistent organic pollutants pose long-term ecological risks. Airborne emissions from open burning or acid treatment expose workers and nearby communities to carcinogenic compounds and respiratory hazards. Beyond local effects, the extraction of new raw materials to compensate for lost or unrecovered metals perpetuates resource depletion, energy consumption, and carbon emissions, linking e-waste accumulation to broader environmental challenges, including climate change, biodiversity loss, and resource scarcity. Moreover, improper e-waste handling disproportionately affects vulnerable populations, raising issues of environmental justice and social responsibility.

FROM SILICON TO LANDFILLS: TRACING THE LIFECYCLE OF SMART DEVICES

The lifecycle of smart devices—from design and production to usage and eventual disposal—illustrates the complex interplay between technological innovation, consumer behavior, environmental impact, and resource management. As society increasingly relies on smartphones, tablets, laptops, wearable technologies, and Internet of Things (IoT) devices, understanding the full trajectory of these products becomes critical for evaluating both their economic contributions and environmental footprints. Each phase of a smart device's lifecycle carries distinct implications for material consumption, energy use, carbon emissions, and waste generation. The journey from silicon wafers to landfills not only highlights the hidden costs embedded in the digital economy but also underscores the urgent need for sustainable design, circular economy practices, and comprehensive e-waste management policies that reconcile technological growth with environmental stewardship.Design and material sourcing constitute the foundational stage of a smart device's lifecycle. Device engineering prioritizes performance, miniaturization, connectivity, and aesthetic appeal, often at the expense of repairability or recyclability. Advanced semiconductors, printed circuit boards, integrated batteries, and complex assemblies demand rare earth elements, precious metals, and specialized plastics.

Manufacturing and assembly transform raw materials into functional devices, encompassing semiconductor fabrication, printed circuit board assembly, component integration, and quality control testing. Semiconductor fabrication facilities, or fabs, consume vast quantities of water and energy while generating chemical waste and volatile emissions. Assembly plants involve multiple processing steps, including soldering, molding, and packaging, which contribute to toxic effluents, solvent emissions, and material waste. The globalization of manufacturing has created complex supply chains spanning Asia, North America, and Europe, increasing transportation-related carbon emissions and logistical complexities. The drive for high throughput, rapid iteration, and cost reduction often incentivizes planned obsolescence, producing devices designed to meet short-term performance expectations but destined for early replacement, thereby accelerating the inflow into end-of-life stages.Distribution and logistics introduce additional environmental considerations. Devices travel across continents via air, sea, and land freight, contributing to greenhouse gas emissions and energy consumption. Packaging materials, including plastics, foams, and cardboard, add to the material footprint and often require energy-intensive production. While some companies are adopting sustainable packaging, bulk shipment and high consumer demand create persistent resource use. Retail and e-commerce ecosystems also generate supplementary energy consumption for storage, display, and transport infrastructure. (Baca & Greene, 2019).

DATA CENTERS AND E-WASTE: A DOUBLE-EDGED SUSTAINABILITY CHALLENGE

The digital economy's explosive growth has elevated data centers into the backbone of modern society, enabling cloud computing, artificial intelligence, e-commerce, streaming services, and global connectivity. Yet, this technological proliferation presents a dual environmental challenge: high energy consumption from data centers and the burgeoning accumulation of electronic waste (e-waste) from associated hardware. These twin issues form a complex, interrelated sustainability dilemma. Data centers demand continuous operation of servers, storage units, networking equipment, and cooling systems, consuming vast amounts of electricity and contributing significantly to global carbon emissions. Meanwhile, the rapid replacement and upgrade cycles of IT equipment generate enormous e-waste volumes, containing both valuable recoverable materials and hazardous substances. (Cellard, 2022)Together, these factors illustrate a paradox of digital expansion: technological progress fosters societal and economic gains but simultaneously amplifies ecological pressure. Addressing this double-edged sustainability challenge requires holistic solutions encompassing energy efficiency, circular economy practices, material recovery, regulatory frameworks, and consumer and corporate responsibility.Energy demands of data centers are staggering.

Hardware lifecycles and e-waste generation are intrinsically linked to data center operations. Servers, networking devices, storage drives, and auxiliary infrastructure typically have lifespans ranging from three to five years, after which they are decommissioned, upgraded, or replaced. The obsolescence of hardware is driven not only by physical wear but also by advancements in computing power, storage capacity, and energy efficiency requirements. Decommissioned hardware contributes to the growing global e-waste stockpile, which is projected to exceed 80 million metric tons annually by the late 2020s. Data center e-waste includes complex components such as printed circuit boards, CPUs, GPUs, memory modules, storage devices, and cooling units, often containing rare earth elements, precious metals, plastics, and hazardous substances. Improper disposal or informal recycling exposes ecosystems and human populations to toxic compounds, including lead, cadmium, mercury, and brominated flame retardants.(Cellard & Marquet, 2023)The intersection of energy use and e-waste creates compounding environmental effects. High energy demand for data center operation indirectly drives e-waste production by incentivizing more efficient but shorter-lived equipment. As older servers and components are retired to meet energy efficiency standards or performance benchmarks, they enter the waste stream. Furthermore, the rapid evolution of computational requirements—accelerated by AI, machine learning, and cloud scalability—shortens the functional lifespan of data center hardware, amplifying both energy and material throughput.

RECYCLING THE UNRECYCLABLE: INNOVATIONS IN HIGH-TECH WASTE MANAGEMENT

The exponential growth of high-tech devices—ranging from smartphones, laptops, and wearable technologies to servers, data centers, and industrial electronics—has brought remarkable societal and economic benefits, yet it has also generated a mounting environmental challenge: high-tech waste, or e-waste. Much of this waste is technically recyclable but remains difficult to process due to complex material composition, hazardous substances, and miniaturized or integrated components. The term "unrecyclable" in the context of high-tech waste reflects not a literal impossibility but a combination of technological, economic, and logistical barriers that impede efficient recovery and reuse. Overcoming these barriers requires innovations across material science, process engineering, artificial intelligence, robotics, and policy frameworks, all aimed at transforming e-waste from an environmental liability into a valuable resource while reducing the ecological and human health impacts of improperly managed electronics(De Vries, 2023).The complexity of high-tech waste stems from the intricate design and material composition of modern electronic devices. Smartphones, for instance, combine glass, aluminum, rare earth magnets, lithium-ion batteries, and multi-layered printed circuit boards with solder containing lead, tin, and silver. Data center servers and industrial electronics contain high-density chips, memory modules, copper heat sinks, power supply units, and cooling systems, often integrated in ways that make disassembly difficult. Batteries, particularly lithium-ion and solid-state types, pose chemical hazards and fire risks during processing.

The small size, adhesive components, and multi-material integration that enable sleek and high-performance electronics make traditional mechanical recycling methods inefficient or unsafe. As a result, a substantial proportion of e-waste remains unrecovered or is processed in informal sectors under unsafe conditions, releasing toxic substances into the environment.Innovations in material science are crucial for addressing the recycling challenge. Researchers are developing new methods to separate metals, plastics, and composites more efficiently and safely. Techniques such as hydrometallurgy, which uses aqueous chemistry to selectively leach valuable metals, and bioleaching, which employs microorganisms to extract metals like gold, copper, and nickel, show promise for recovering rare and precious materials with lower energy requirements than traditional pyrometallurgical processes. Advances in polymer recycling and depolymerization allow previously non-recyclable plastics to be converted back into reusable monomers. Moreover, substituting hazardous substances with safer alternatives—such as lead-free solder, non-toxic flame retardants, and biodegradable materials—simplifies downstream recycling and reduces environmental hazards, aligning product design with circular economy principles.

Automation and robotics have emerged as transformative tools in high-tech waste management. Robotic disassembly systems equipped with computer vision and AI can identify components, unscrew or detach modular parts, and sort materials at speeds and precision levels beyond human capability. Automated optical and magnetic sorting technologies enable rapid segregation of metals, plastics, and glass, increasing purity of recovered materials and reducing contamination. AI algorithms can optimize disassembly sequences, predict component value, and detect hazardous elements to enhance safety. These innovations reduce manual labor in dangerous environments, improve recovery rates, and lower operational costs, making high-tech recycling economically viable while safeguarding human health.Artificial intelligence and data analytics also play a central role in managing high-tech waste streams. AI-driven predictive models can assess which devices entering the waste stream contain valuable or hazardous materials, enabling targeted recovery strategies. Machine learning algorithms can optimize logistics, determining the most efficient routes for collecting, transporting, and processing e-waste(Dobbe & Whittaker, 2019). Advanced analytics can track materials through supply chains and recovery processes, facilitating transparency, compliance with environmental regulations, and certification of recycled content. In combination with the Internet of Things (IoT), AI can provide real-time monitoring of storage conditions for batteries or chemical components, mitigating risks of fire, leakage, or contamination.Energy-efficient recycling processes are critical for minimizing the carbon footprint of high-tech waste management.

THE TRUE COST OF INTELLIGENCE: ENERGY USE AND ELECTRONIC DISCARD

The contemporary digital era, powered by artificial intelligence (AI), cloud computing, and ubiquitous electronic devices, has ushered in unprecedented convenience, productivity, and connectivity. Yet, behind the sleek interfaces, smart algorithms, and instant access lies a profound and often overlooked environmental cost. The "true cost" of modern intelligence encompasses not only the substantial energy required to power computational systems but also the massive volumes of electronic waste generated by increasingly rapid device turnover. Every byte processed, every model trained, and every device manufactured and discarded contributes to a complex web of environmental, social, and economic consequences. The dual challenges of energy-intensive computation and e-waste accumulation form the undercurrent of the digital revolution, revealing a paradox: technological advancement designed to enhance human intelligence simultaneously exerts enormous pressure on planetary resources and ecosystems. Understanding and mitigating these costs is essential to ensure that the promise of digital intelligence does not come at the expense of environmental sustainability, public health, or long-term economic resilience.Energy consumption in computational systems forms the first pillar of the environmental cost.

High-performance computing, data centers, and AI workloads rely on thousands of processors and accelerators operating continuously to process massive datasets. Training advanced AI models, particularly large language models, deep neural networks, and reinforcement learning algorithms, can consume electricity equivalent to hundreds of households over an entire year. Cloud computing infrastructures, which host enterprise applications, streaming services, and data storage, operate 24/7 to maintain availability and responsiveness. (Dodge et al., 2022)The energy intensity of these operations is further amplified by the need for cooling systems, which dissipate heat generated by densely packed servers. Chillers, liquid cooling loops, and air-conditioning systems contribute significantly to the operational carbon footprint, turning computational intelligence into a resource-intensive endeavor with direct implications for greenhouse gas emissions.

Data centers as energy hotspots illustrate the scale of this challenge. Globally, data centers account for roughly 1–2% of total electricity consumption, a figure expected to rise as AI, cloud computing, and big data services expand. The energy consumed depends heavily on the local energy mix: regions reliant on fossil fuels generate significantly higher carbon emissions per kilowatt-hour than areas with abundant renewable energy. Efforts to reduce the carbon footprint include the deployment of energy-efficient servers, advanced cooling solutions, on-site renewable generation, and AI-driven power optimization algorithms. However, the relentless demand for computational power, driven by escalating AI research, real-time analytics, and pervasive cloud applications, continues to escalate total energy usage, highlighting the inherent tension between technological advancement and environmental sustainability.Electronic devices and infrastructure constitute the second pillar of the true cost of intelligence: material consumption and waste. Servers, GPUs, storage arrays, networking devices, and peripheral components have finite lifespans, typically ranging from three to five years. After decommissioning, these devices enter the waste stream, contributing to a rapidly expanding stockpile of electronic waste (e-waste). Modern devices contain a mixture of valuable and hazardous materials, including precious metals like gold, silver, and platinum; rare earth elements such as neodymium and tantalum; copper and aluminum; high-grade plastics; and printed circuit boards. Batteries, especially lithium-ion variants, carry chemical hazards and fire risks if mishandled.(Edwards, 2010) The integration of multiple materials into compact, miniaturized designs complicates disassembly and material recovery, rendering a large portion of e-waste effectively "unrecyclable" without advanced technological intervention. The result is a dual environmental burden: the extraction and processing of new materials to replace discarded components, coupled with pollution risks from improper disposal or informal recycling.

PLANTED IN THE CLOUD: HOW SERVER FARMS FUEL THE E-WASTE CRISIS

The meteoric rise of cloud computing has transformed modern society, powering everything from streaming services and online collaboration tools to artificial intelligence, big data analytics, and global e-commerce. At the heart of this digital ecosystem lie server farms—massive data centers housing rows upon rows of servers, storage systems, and networking equipment. While these server farms underpin unprecedented connectivity, efficiency, and innovation, they also contribute significantly to two pressing environmental challenges: high energy consumption and the accelerating accumulation of electronic waste (e-waste). Often conceptualized as intangible "cloud" services, these infrastructures have a very tangible physical and ecological footprint.(Ensmenger, 2018) Their operational demands, rapid hardware turnover, and disposal practices collectively drive a growing e-waste crisis, revealing a critical paradox of the digital age: the very technologies that enable global intelligence and information flow are simultaneously exacerbating resource depletion and environmental degradation.Energy consumption of server farms forms the first pillar of this environmental impact. Data centers operate continuously, 24 hours a day, 365 days a year, to ensure uptime, reliability, and redundancy. High-performance processors, GPUs, storage arrays, and networking components generate immense amounts of heat, necessitating sophisticated cooling systems such as air conditioning, liquid cooling, or immersion techniques. Estimates indicate that global data centers consume between 1–2% of total electricity, with projections suggesting that this figure may increase as cloud adoption, AI workloads, and edge computing expand.

Figure 1. Technology and innovation in E-waste recycling

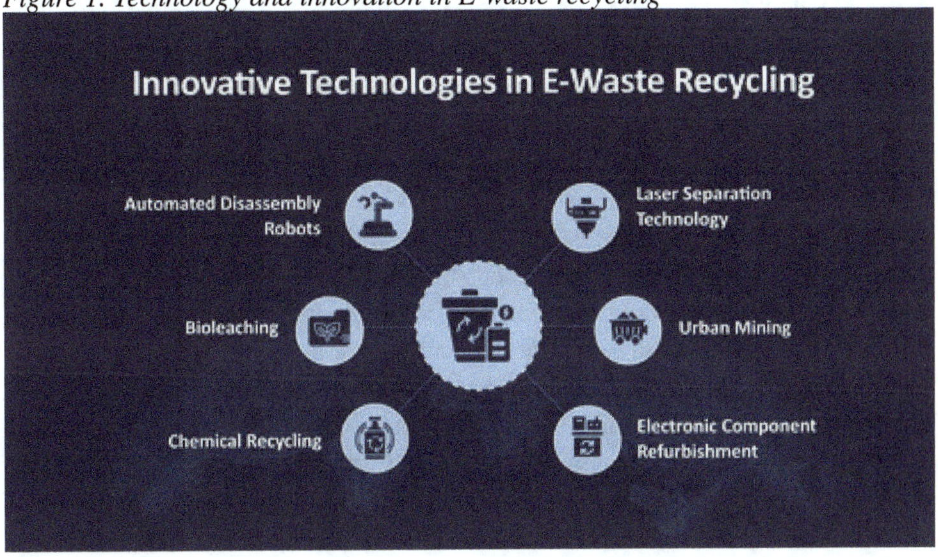

The e-waste crisis driven by server farms is multidimensional. Globally, e-waste generation is expected to exceed 80 million metric tons annually by the late 2020s, and while servers represent a smaller proportion compared to consumer electronics, their high-density components amplify environmental risk per unit of weight. Improper disposal or informal recycling—common in developing regions—can release heavy metals and persistent organic pollutants into soil, water, and air. (European Commission, 2024)Open burning, acid baths, and rudimentary manual dismantling expose workers to lead, mercury, cadmium, and brominated flame retardants, creating serious occupational and public health hazards. The export of decommissioned servers to regions with lax environmental oversight compounds the ecological burden, reflecting a systemic global inequity in the environmental costs of cloud infrastructure.Recycling challenges and technological barriers illustrate why server farms exacerbate e-waste. Traditional mechanical shredding and separation methods often fail to adequately recover high-value materials due to the compact, multi-layered design of servers. Hazardous components, such as lithium-ion batteries or capacitors, pose safety risks during dismantling, complicating processing.

OBSOLESCENCE BY DESIGN: PLANNED SHORT-LIVES OF CONSUMER GADGETS

In the modern era, consumer electronics—ranging from smartphones and laptops to smartwatches, tablets, and connected home devices—have become indispensable tools for daily life, shaping communication, productivity, entertainment, and social interaction. Yet, underlying the allure of sleek design, cutting-edge features, and rapid technological advancement is a pervasive and environmentally consequential strategy: planned obsolescence. This deliberate engineering and marketing approach ensures that gadgets have intentionally limited lifespans, compelling consumers to replace devices frequently. (Foroohar, 2018)While driving corporate profits and accelerating technological turnover, planned obsolescence contributes significantly to the global e-waste crisis, resource depletion, and energy consumption, revealing a tension between consumer convenience, corporate strategy, and environmental sustainability. Understanding the mechanisms, consequences, and potential solutions associated with planned obsolescence is critical for developing a sustainable, responsible consumer electronics ecosystem.Planned obsolescence manifests through several interrelated mechanisms. Technologically, manufacturers design hardware components to fail or degrade after a predictable period, such as batteries with limited charge cycles, solder joints prone to wear, and proprietary components that are difficult to repair or replace. Software-induced obsolescence compounds the problem: operating systems, firmware, and applications are frequently updated in ways that reduce performance, compatibility, or security on older devices, nudging consumers toward new purchases.

The environmental impact of planned obsolescence is profound and multifaceted. Rapid device turnover accelerates electronic waste (e-waste) generation, with millions of tons of discarded electronics entering global waste streams annually. High-tech gadgets are composed of a complex array of materials, including precious metals such as gold, silver, and platinum; rare earth elements like neodymium and tantalum; high-grade plastics; lithium-ion batteries; and printed circuit boards containing toxic substances such as lead, cadmium, and brominated flame retardants. Improper disposal or informal recycling can release these toxic substances into soil, water, and air, causing long-term ecological damage and posing serious health risks to communities near e-waste processing sites. (Gregg & Strengers, 2024)The rapid obsolescence of consumer gadgets thus amplifies both material throughput and environmental hazard per unit of technology deployed.Resource depletion is another critical consequence. Frequent replacement of devices increases demand for raw materials, including metals, minerals, and fossil-fuel-derived plastics. Mining for rare earth elements and precious metals is energy-intensive, environmentally destructive, and often associated with social challenges, including exploitative labor practices and geopolitical supply risks. By shortening product lifespans, planned obsolescence magnifies resource extraction rates, accelerating the depletion of finite materials and creating environmental stress through habitat destruction, water pollution, and greenhouse gas emissions. This cycle is further intensified by globalized supply chains, where components are extracted, manufactured, and transported across continents, compounding the carbon footprint of each gadget.Energy consumption linked to planned obsolescence spans the entire lifecycle of electronic devices.

GREEN ALGORITHMS VS. GREY TRASH: BALANCING AI PROGRESS WITH WASTE

This dual challenge—"green algorithms" versus "grey trash"—captures the tension between advancing AI technologies sustainably while mitigating the massive amounts of electronic waste (e-waste) generated by data centers, GPUs, servers, and other computing devices. Understanding and managing these intertwined impacts is critical to ensuring that AI-driven progress aligns with environmental stewardship, resource efficiency, and long-term societal resilience.Energy consumption of AI computation represents the first pillar of environmental concern.(Hacker, 2024) Training large-scale machine learning models, particularly deep neural networks and transformer-based architectures, requires massive volumes of electricity. High-performance GPUs, tensor processing units (TPUs), and clusters of servers operate continuously during model training and inference, consuming energy at scales comparable to entire small cities. For instance, training a single state-of-the-art AI model can require hundreds of megawatt-hours of electricity, producing a carbon footprint equivalent to dozens of passenger vehicles over their lifetimes, depending on the regional energy mix. Beyond training, deployment and real-time inference of AI models in cloud services, edge devices, and consumer applications contribute additional energy use. Cooling requirements for data centers, which dissipate heat generated by densely packed servers and accelerators, further amplify the total energy footprint, underscoring the trade-off between computational performance and environmental impact.

Decommissioned hardware becomes part of the e-waste stream, adding to the already staggering global accumulation of discarded electronics. E-waste from AI infrastructure contains valuable materials such as gold, silver, copper, platinum, and rare earth elements like neodymium and tantalum, alongside plastics, printed circuit boards, and lithium-ion batteries. Improper disposal or informal recycling exposes these materials—and associated toxic substances including lead, cadmium, and brominated flame retardants—to environmental and human health hazards. (Halpern, 2021) The duality of AI advancement and e-waste generation highlights a critical challenge: the very devices that enable cutting-edge algorithms simultaneously contribute to a "grey trash" problem that strains ecosystems and waste management systems worldwide.Green algorithms offer a pathway to reconcile AI progress with environmental responsibility. Algorithmic efficiency improvements, such as model pruning, quantization, knowledge distillation, and sparse architectures, reduce computational load without substantially compromising performance. Optimized training schedules, adaptive batch sizes, and transfer learning approaches minimize redundant computation, lowering electricity consumption. AI models can also be designed to balance energy cost with predictive accuracy, incorporating energy-aware objectives into the training process.

THE TOXIC TRAIL: HAZARDOUS MATERIALS IN OUR EVERYDAY TECH

In the contemporary era, technology permeates nearly every facet of human life, from smartphones, laptops, and tablets to smartwatches, IoT devices, and household appliances. These devices promise convenience, connectivity, and productivity, yet they also carry a hidden cost: the pervasive presence of hazardous materials embedded within their components. From heavy metals to chemical flame retardants, batteries, and rare earth elements, the substances that enable our devices to function often pose significant environmental, health, and societal risks. Collectively, these materials leave a "toxic trail" throughout the lifecycle of electronic devices—spanning manufacturing, usage, and disposal—raising urgent questions about sustainability, safety, and ethical responsibility in the age of pervasive electronics. Understanding the types of hazardous materials, their pathways into ecosystems and human systems, and strategies to mitigate their impact is critical to aligning technological advancement with environmental and public health imperatives.Heavy metals constitute one of the most significant classes of hazardous substances in everyday electronics.

Lead, cadmium, mercury, and chromium are frequently found in solder, batteries, display screens, and circuit boards. (Jancovic & Keilbach, 2023)Lead, commonly used in solder and cathode ray tubes, can cause neurological damage, especially in children, and contaminates soil and water if improperly disposed. Cadmium, found in rechargeable batteries, pigments, and semiconductors, is a carcinogen and can accumulate in the kidneys, while mercury, used in certain switches, lamps, and displays, is neurotoxic and bioaccumulative. Hexavalent chromium, employed in plating and corrosion prevention, can cause respiratory issues and skin irritation. During production, occupational exposure to these metals poses serious health risks, while end-of-life disposal or informal recycling exposes broader populations to toxic pathways, particularly in developing regions where e-waste is dismantled using rudimentary techniques such as open burning or acid baths.Persistent organic pollutants (POPs) and chemical additives amplify the toxic footprint of modern electronics. Brominated flame retardants (BFRs) and polyvinyl chloride (PVC) are widely used to reduce flammability in casings, wires, and printed circuit boards. While essential for safety, these compounds resist natural degradation, accumulate in ecosystems, and interfere with endocrine systems in humans and wildlife.

Batteries and energy storage components represent another critical source of hazardous materials. Lithium-ion batteries, ubiquitous in smartphones, laptops, tablets, and electric vehicles, contain lithium, cobalt, nickel, and manganese, each with environmental and ethical implications. Extraction of lithium and cobalt involves water-intensive mining, habitat disruption, and labor exploitation in some regions, while improper disposal can result in chemical leakage, fire hazards, and toxic exposure. Other battery chemistries, such as nickel-cadmium (NiCd), contribute cadmium—a toxic heavy metal—to e-waste streams. Battery degradation over time can release acidic electrolytes and flammable gases, complicating recycling and transport safety. Consequently, energy storage systems, while critical for portable power, also propagate toxic materials across supply chains and end-of-life pathways.Rare earth elements (REEs) and specialty metals further illustrate the complex toxic footprint of electronics. Neodymium, dysprosium, and terbium enable high-performance magnets in speakers, hard drives, and motors, while tantalum and indium are used in capacitors and touchscreens.(Jiménez & Oleson, 2022).

Though essential for technological performance, their extraction is often environmentally destructive, involving mining, chemical leaching, and water contamination. Additionally, these metals are difficult to recycle efficiently due to their small size, low concentrations, and integration into complex devices, meaning that discarded electronics contribute to resource depletion and environmental accumulation of potentially hazardous residues. The “toxic trail” thus extends from mining and material processing to eventual disposal or leakage in ecosystems.Manufacturing processes intensify exposure risks. Semiconductor fabrication, PCB production, and component assembly involve a range of hazardous chemicals including solvents, acids, and photoresists. Workers in factories may be exposed to volatile organic compounds (VOCs), heavy metals, and endocrine-disrupting chemicals, often without adequate protective measures, particularly in regions with lax occupational safety standards. Effluent and wastewater from these processes can enter waterways, transporting heavy metals and persistent organic compounds into broader ecosystems. While environmental regulations exist in many countries, global supply chains often outsource production to regions with minimal oversight, externalizing ecological and human health impacts to vulnerable populations.

FROM MINES TO MALLS: THE RESOURCE EXTRACTION BEHIND AI HARDWARE

The artificial intelligence (AI) revolution has transformed industries, reshaped economies, and redefined the possibilities of human-machine interaction. From autonomous vehicles and natural language processing to advanced robotics, predictive analytics, and personalized recommendation systems, AI technologies are embedded in nearly every facet of modern life. Yet behind the seamless functionality of AI systems, there exists a complex global supply chain of resource extraction, manufacturing, and consumption, which is often obscured by the digital abstraction of "the cloud" or "intelligent systems." Every high-performance GPU, tensor processing unit (TPU), memory module, and server rack depends on a suite of mined materials, from common metals to rare earth elements, whose extraction and refinement generate profound environmental, social, and ethical consequences. The journey "from mines to malls" highlights the hidden ecological and human costs embedded in AI hardware, revealing a tension between technological progress and sustainable resource management that is rarely visible to end-users or consumers. Resource-intensive components of AI hardware drive the need for extensive mining and material processing(Kunnathully et al., 2026). High-performance computing (HPC) systems, data centers, and personal AI devices rely on semiconductors, circuit boards, memory chips, storage devices, high-capacity batteries, and networking equipment. These components contain a mix of base metals—copper, aluminum, nickel, and zinc—as well as precious metals such as gold, silver, and platinum, which facilitate electrical conductivity and component reliability.

Rare earth mining, frequently concentrated in regions such as China, involves the use of sulfuric acid and other chemicals for separation, generating acidic tailings that threaten ecosystems. Across these operations, greenhouse gas emissions from machinery, energy use, and associated transport further exacerbate climate impacts, creating a high-carbon upstream supply chain for AI hardware.Socioeconomic and ethical challenges are interwoven with resource extraction. Mining often occurs in geopolitically sensitive regions or areas with limited regulatory oversight, where labor exploitation, unsafe working conditions, and human rights violations are prevalent. Artisanal and small-scale mining (ASM) operations, while economically vital to local communities, frequently involve child labor, hazardous exposure to heavy metals, and inadequate safety measures. The global demand for AI hardware concentrates economic and political power in a few resource-rich countries, creating dependencies and market vulnerabilities. Geopolitical tensions over strategic metals, such as rare earth elements essential for AI, further complicate supply security, illustrating the broader global implications of hardware sourcing decisions. Consumers in developed markets often remain disconnected from these upstream challenges, perceiving AI as a purely digital or intellectual product while the environmental and social costs remain "out of sight, out of mind."Manufacturing processes amplify the environmental and resource footprint. Semiconductor fabrication, one of the core processes for AI hardware, is exceptionally resource- and energy-intensive.

CARBON-HEAVY INTELLIGENCE: QUANTIFYING EMISSIONS FROM MACHINE LEARNING

The rise of machine learning (ML) has heralded a new era of computational intelligence, enabling breakthroughs in natural language processing, computer vision, predictive analytics, autonomous systems, and scientific research. Machine learning algorithms power virtual assistants, recommendation engines, medical diagnostics, financial modeling, climate simulations, and generative AI, fundamentally transforming society and economic systems. Yet, behind these transformative applications lies a less visible, yet critical challenge: the carbon footprint of machine learning. Training and deploying state-of-the-art models, especially deep neural networks and large transformer architectures, demands immense computational resources, translating directly into significant greenhouse gas (GHG) emissions. This "carbon-heavy intelligence" problem highlights the environmental trade-offs of algorithmic progress and underscores the urgent need to quantify, understand, and mitigate emissions across the ML lifecycle.Energy intensity of training large models is a primary driver of ML-related emissions. (Lally et al., 2022)Training a single state-of-the-art transformer model, such as those used in natural language processing, can consume hundreds of megawatt-hours (MWh) of electricity, equivalent to the annual energy consumption of dozens of average households. Deep reinforcement learning, generative models, and large-scale ensemble methods exacerbate this energy demand due to iterative training cycles, hyperparameter optimization, and repeated experimentation.

Model size and complexity directly influence emissions.(Levenda & Mahmoudi, 2019) As AI research pursues ever-larger models with billions or even trillions of parameters, computational requirements scale nonlinearly. Training such models involves repeated forward and backward passes over massive datasets, requiring high-performance accelerators and extensive parallelization. Hyperparameter tuning, an essential step for optimizing model performance, often multiplies training runs, creating a cascade of energy use. In addition to training, inference operations, especially in cloud-based or edge-deployed AI services, contribute ongoing energy consumption, magnifying the lifecycle carbon impact of a single model. Consequently, the drive for model accuracy, performance, and novelty must be balanced with sustainability considerations, as unchecked growth in model size exacerbates emissions disproportionately.Data processing and storage add another layer of environmental impact. Large ML models require petabytes of structured and unstructured data, which must be collected, cleaned, stored, and transmitted. Data centers housing storage systems, high-speed networking, and redundant backups consume significant electricity to ensure reliability and uptime. Cooling infrastructure, power distribution losses, and energy overhead for storage and retrieval amplify the overall carbon footprint.

E-WASTE AND EQUITY: DISPROPORTIONATE IMPACTS ON GLOBAL SOUTH COMMUNITIES

The rapid proliferation of electronic devices—smartphones, laptops, tablets, servers, IoT devices, and advanced consumer electronics—has fueled unprecedented technological access and connectivity worldwide. However, this digital transformation carries an escalating environmental and social cost: electronic waste (e-waste). Globally, e-waste generation is projected to exceed 80 million metric tons annually by the late 2020s, representing a complex mix of metals, plastics, glass, and chemical components. While affluent regions drive consumption, the burden of e-waste management is disproportionately borne by communities in the Global South, creating systemic inequities rooted in economic, regulatory, and infrastructural disparities. These inequities manifest in environmental degradation, health hazards, labor exploitation, and socio-economic vulnerability, highlighting the urgent need to address the intersection of e-waste, social justice, and global governance.The nature of e-waste and hazardous content contributes to its unequal impacts. Electronic devices contain heavy metals (lead, cadmium, mercury, chromium), persistent organic pollutants (brominated flame retardants, phthalates, perfluorinated compounds), batteries (lithium, cobalt, nickel, manganese), and rare earth elements essential for modern functionality. (Nobrega & Varon, 2020)These materials are inherently hazardous if improperly managed.

Informal recycling and occupational hazards exemplify the inequitable distribution of e-waste impacts. In countries such as Ghana, Nigeria, India, and Pakistan, informal e-waste sectors provide livelihoods for tens of thousands of workers, who dismantle electronics manually or through rudimentary processes such as open burning, acid baths, and crude mechanical separation. Workers—including children—are routinely exposed to toxic metals and chemicals, inhaling fumes, handling hazardous materials without protective equipment, and facing risks of chemical burns, respiratory illnesses, and long-term chronic health conditions. Families living near e-waste processing hubs experience contaminated water, polluted air, and degraded soil, with cumulative effects on nutrition, cognitive development, and community health. These occupational and environmental hazards starkly contrast with the minimal direct benefits derived from the discarded electronics they process, illustrating a structural injustice embedded in global e-waste flows.Transboundary movement of e-waste exacerbates inequity. The Basel Convention, which governs cross-border movement of hazardous waste, aims to prevent developed countries from dumping e-waste in nations with insufficient infrastructure. Despite this, illegal exports and regulatory loopholes persist, with used or obsolete electronics shipped under the guise of "second-hand goods" to markets in Africa, Asia, and Latin America. Once in recipient countries, these devices often enter informal recycling streams or end up in unregulated landfills. (Nost & Colven, 2022)Such practices externalize the environmental and health costs of consumption in wealthier regions, displacing burdens onto vulnerable populations while enabling continued access to affordable electronics for low-income communities. The global e-waste trade reflects deep structural inequities, where environmental harm is geographically decoupled from consumption benefits.

CIRCULAR ECONOMY STRATEGIES FOR THE TECH INDUSTRY'S WASTE STREAM

The rapid pace of technological advancement and ubiquitous adoption of electronic devices—ranging from smartphones, laptops, and tablets to servers, IoT devices, wearables, and household appliances—has generated unprecedented levels of electronic waste (e-waste), posing significant environmental, social, and economic challenges. Estimates suggest that global e-waste generation exceeds 80 million metric tons annually, a figure projected to increase as consumer demand for upgraded devices, smart technologies, and AI-driven infrastructure accelerates.

The conventional linear model of "take, make, dispose" has proved unsustainable: it drives resource depletion, increases greenhouse gas emissions, exacerbates environmental contamination, and burdens communities, especially in the Global South, with hazardous waste management responsibilities. In response, the concept of a circular economy (CE) has emerged as a transformative framework, aiming to decouple technological growth from environmental harm by promoting resource efficiency, waste reduction, and material recirculation across the tech industry's entire value chain.(Ortar et al., 2023). Refurbishment and remanufacturing are central to closing the loop in a circular tech economy. Refurbished devices, repaired, upgraded, and certified for resale, extend the usable life of electronics while providing affordable technology access to diverse markets. Remanufacturing, which involves reprocessing used devices to "like-new" condition, allows manufacturers to recover embedded value from end-of-life products while minimizing raw material demand. Both strategies reduce the need for virgin material extraction, conserve energy, and limit the environmental footprint associated with production, transport, and disposal. Furthermore, refurbishment and remanufacturing can be integrated with warranty programs and corporate responsibility initiatives, enhancing consumer trust and brand reputation while advancing environmental goals(Paré & Kitsiou, 2017).

Robotics, automation, and AI-assisted sorting enhance disassembly precision, reduce worker exposure to toxic substances, and optimize material separation. The recovered metals, polymers, and other components can then be reintroduced into production, closing the loop and reducing dependence on environmentally destructive mining practices. In addition, safe disposal of residual hazardous components, such as heavy metals and brominated flame retardants, mitigates ecological and public health risks associated with improper e-waste handling.Design for recyclability and material transparency is another core strategy. By selecting non-toxic, easily separable materials, standardizing components, and minimizing composite assemblies, manufacturers facilitate efficient end-of-life processing. Transparent labeling of material composition and inclusion of digital product passports—potentially powered by blockchain or IoT—enables recyclers and repair services to identify, sort, and process components safely and efficiently. Such design innovations align production practices with circular economy goals, reducing environmental burden while preserving economic value embedded in devices. Incorporating environmental considerations at the design stage ensures that the full lifecycle of technology—from raw material extraction to end-of-life recovery—is optimized for sustainability.

THE ROLE OF LEGISLATION IN CURBING BIG TECH'S E-WASTE FOOTPRINT

The exponential growth of the technology sector has brought immense benefits in communication, productivity, artificial intelligence, and digital infrastructure, yet it has also generated a profound and escalating environmental challenge: electronic waste (e-waste). With millions of smartphones, laptops, servers, data centers, and IoT devices produced annually, the accumulation of discarded electronics presents hazards ranging from toxic material leaching to ecosystem disruption and human health risks. Globally, e-waste surpassed 80 million metric tons per year in the early 2020s, with only a fraction formally recycled. While technological innovation has outpaced regulatory frameworks, legislation plays a critical role in mitigating the e-waste footprint of big tech companies, ensuring responsible production, consumption, and disposal practices, and embedding environmental accountability into corporate operations.(Pasek, 2019) Understanding the role of legislation requires examining legal frameworks, enforcement mechanisms, policy tools, and the intersections with corporate responsibility, international governance, and sustainable technological development.Legal frameworks for e-waste management vary across regions but share a common goal: regulating the lifecycle of electronic products to minimize environmental harm. The European Union's Waste Electrical and Electronic Equipment (WEEE) Directive serves as a landmark framework, mandating producer responsibility for collection, treatment, and recovery of end-of-life electronics. Complementing this, the Restriction of Hazardous Substances (RoHS) Directive limits the use of toxic materials such as lead, mercury, cadmium, and brominated flame retardants, directly reducing the environmental toxicity of discarded devices.

Extended Producer Responsibility (EPR) has emerged as a cornerstone principle in curbing e-waste from major technology firms. EPR mandates that manufacturers assume accountability for the environmental impact of their products throughout their lifecycle, particularly at the end-of-life stage. This includes designing for durability, modularity, and recyclability; providing collection and take-back systems; and ensuring environmentally sound processing of recovered materials. By internalizing the costs of e-waste management, EPR incentivizes corporations to reduce hazardous content, extend product lifespans, and invest in circular economy strategies. Big tech companies, often responsible for massive global device deployment, are particularly impacted by EPR regulations, as non-compliance can result in financial penalties, reputational harm, and legal sanctions. EPR thus operationalizes legislation as a market-based mechanism to align corporate incentives with environmental sustainability.Mandatory reporting and transparency requirements enhance the effectiveness of e-waste legislation. Laws that require companies to disclose quantities of devices sold, recycled, and properly disposed of, as well as the composition of materials, create accountability mechanisms and enable performance benchmarking. For instance, disclosure mandates within the WEEE framework allow regulators, NGOs, and the public to monitor corporate compliance and environmental impact. Transparency also drives internal corporate reforms, encouraging research and development in low-toxicity materials, energy-efficient production, and advanced recycling partnerships.

CONSUMER RESPONSIBILITY: EXTENDING THE LIFESPAN OF SMART PRODUCTS

The proliferation of smart products—ranging from smartphones, tablets, and laptops to wearables, smart home devices, and connected appliances—has transformed modern life, offering convenience, efficiency, and connectivity at an unprecedented scale. However, the rapid adoption and frequent replacement of these devices have generated a significant environmental challenge: electronic waste (e-waste). With the global generation of e-waste surpassing 80 million metric tons annually, the consequences extend beyond environmental degradation to include human health risks, resource depletion, and socio-economic disparities, particularly affecting vulnerable communities in the Global South. While corporate responsibility, legislation, and technological innovation are essential components of sustainable e-waste management, consumer behavior and responsibility play a pivotal role in extending the lifespan of smart products, reducing their ecological footprint, and fostering a circular economy in the digital era.Understanding the environmental impact of premature device turnover is the first step toward responsible consumption. Modern electronics are composed of complex materials, including precious and base metals, rare earth elements, plastics, and chemical additives, many of which pose toxicity risks if improperly disposed. The manufacturing and assembly of these components involve energy-intensive processes, chemical usage, and resource extraction that contribute to greenhouse gas emissions and ecological disruption. When consumers discard devices prematurely—motivated by fashion trends, perceived obsolescence, or marginal performance improvements—the embedded environmental costs of production are multiplied.

Repair and refurbishment are critical consumer-driven strategies for extending device lifespans. Devices often become discarded due to minor malfunctions—such as battery degradation, cracked screens, or peripheral failures—rather than complete obsolescence. Accessing repair services, whether through manufacturer-authorized centers, certified third-party technicians, or DIY repair kits, allows users to restore functionality without generating additional waste. Modular designs, standardized components, and open repair manuals empower consumers to perform cost-effective repairs, reducing dependency on new devices. Refurbishment, which involves professionally restoring used devices for resale or continued personal use, also supports resource efficiency and the circular economy, mitigating the environmental footprint of production and disposal.Software optimization and compatibility management further extend the functional lifespan of smart products. Operating system updates, application management, and performance tuning can maintain device utility even as new software releases emerge. (Taffel et al., 2022)Consumers who prioritize optimization over frequent device replacement reduce the incentive for manufacturers to push new models solely for incremental performance gains. Moreover, selecting software that is compatible with older hardware reduces digital obsolescence, ensuring that devices remain operational and secure for longer periods. By aligning consumption patterns with software longevity, users actively contribute to reducing e-waste and associated environmental impacts.

DIGITAL DETOX: REDUCING DEMAND TO LOWER BOTH CARBON AND E-WASTE

E-waste generation and its socio-environmental consequences form the second critical dimension of the digital impact. Discarded electronics—ranging from smartphones, tablets, and laptops to peripherals and network equipment—contain heavy metals, toxic chemicals, and complex composites that pose severe environmental and human health risks if improperly managed. Globally, e-waste exceeds 80 million metric tons annually, yet only a fraction is formally recycled, leaving the remainder to accumulate in landfills, informal recycling operations, or illegal export channels, predominantly to the Global South. Communities involved in informal e-waste recycling are exposed to hazardous substances such as lead, cadmium, mercury, and brominated flame retardants, leading to chronic health issues and environmental degradation.(Van Wynsberghe, 2021) Digital detox, by reducing the frequency of device replacement and overall consumption, contributes directly to slowing the generation of e-waste, mitigating the ecological and social inequities associated with its management.Behavioral strategies underpinning digital detox involve deliberate reductions in both device acquisition and digital service usage. Consumers can practice extended device usage by repairing, refurbishing, or upgrading existing electronics instead of purchasing new ones.

Limiting engagement with energy-intensive digital services—such as excessive streaming, cloud storage, high-frequency video calls, or AI-driven applications—reduces server workload and associated energy consumption. Adjusting device settings to optimize power use, minimizing background processes, and selectively limiting notifications or app usage contribute to lower electricity demand. These practices embody a conscious deceleration of digital consumption, transforming the way individuals interact with technology while directly affecting environmental outcomes.Device longevity and maintenance intersect with digital detox principles to reduce both carbon and e-waste impacts. By preserving device functionality for longer periods through careful handling, software optimization, battery management, and protective accessories, consumers can delay replacement cycles. Modular and repairable device designs facilitate component-level upgrades, further supporting extended usage. Each year of extended functionality for a device reduces the embedded energy and material costs associated with manufacturing a new unit. Digital detox thus encompasses not only reduction in active usage but also prolonged life of existing devices, lowering the cumulative carbon and e-waste footprint.

UPCYCLING INNOVATION: TURNING DISCARDED DEVICES INTO NEW RESOURCES

The rapid proliferation of electronic devices—ranging from smartphones, laptops, and tablets to wearable technology, IoT devices, and smart home appliances—has fueled unprecedented global connectivity, productivity, and innovation. However, this digital revolution has simultaneously generated a critical environmental and social challenge: the accumulation of electronic waste (e-waste). With global e-waste generation surpassing 80 million metric tons annually, much of it containing toxic metals, plastics, rare earth elements, and hazardous chemicals, the consequences of disposal extend far beyond landfills, affecting soil, water, air quality, and human health. While traditional recycling recovers some valuable materials, much e-waste is poorly managed, illegally exported, or treated in informal sectors, particularly in the Global South, exposing vulnerable communities to severe environmental and health hazards. Against this backdrop, upcycling innovation—transforming discarded devices into new, valuable products—emerges as a transformative strategy, offering both environmental and economic benefits while fostering circularity, technological creativity, and resource efficiency.The concept of upcycling in the technology sector goes beyond conventional recycling. Whereas recycling focuses on breaking down products to recover raw materials for remanufacturing, upcycling leverages existing components, electronics, and materials to create new products of equal or higher utility or value.

Repair and refurbishment as upcycling strategies are among the most accessible and impactful methods for extending device lifespans.(Velkova, 2021) Many discarded devices are fully functional but discarded due to minor malfunctions, perceived obsolescence, or aesthetic imperfections. Repairing screens, replacing batteries, upgrading storage, or refurbishing operating systems can restore functionality, creating products suitable for resale, donation, or resale in secondary markets. Refurbishment not only reduces e-waste but also improves digital equity, offering affordable access to technology for underprivileged populations and emerging markets. Certified refurbishment programs often include quality assurance, warranty provisions, and environmentally responsible practices, ensuring both performance and sustainability. By reintroducing devices into the consumer ecosystem, refurbishment retains the embedded materials and energy, avoiding the environmental costs associated with new production.Component-level repurposing expands the possibilities of upcycling beyond device-level repair. Individual elements—such as circuit boards, batteries, sensors, displays, and connectors—can be extracted from obsolete electronics and incorporated into new devices or systems. For example, functional processors from retired computers can be used in educational kits, microcontrollers for IoT prototypes, or in hobbyist robotics. Batteries from smartphones or laptops can be aggregated to create modular energy storage systems for renewable energy applications. Displays, LEDs, and sensors can be integrated into novel industrial, artistic, or consumer projects. By treating components as valuable resources rather than waste, upcycling minimizes environmental impact, reduces raw material demand, and stimulates innovation in device design and technology applications.

THE CARBON COST OF CONTINUOUS UPDATES: SOFTWARE'S ROLE IN HARDWARE TURNOVER

In the modern digital ecosystem, the rapid evolution of software has become both a driver of technological innovation and a catalyst for environmental challenges, particularly in the context of electronic waste (e-waste) and carbon emissions. Smartphones, tablets, laptops, desktops, servers, and connected devices are designed to operate in concert with ever-evolving operating systems, applications, and firmware updates. While these updates are critical for maintaining security, improving performance, and enabling new functionalities, they also have unintended consequences: The manufacturing of electronic devices involves energy-intensive processes such as semiconductor fabrication, metal extraction, and component assembly, all of which generate significant greenhouse gas emissions(Wyatt, 2008). For example, producing a single smartphone can emit between 55 to 95 kilograms of CO_2 equivalent, depending on materials and production methods. When devices are replaced prematurely due to software updates, the embedded carbon footprint of each unit is effectively multiplied, while the environmental costs of mining rare earth elements, processing metals, and generating plastics are exacerbated. This phenomenon illustrates that the environmental impact of software extends beyond energy consumption during operation, influencing material and lifecycle emissions at a systemic level.

Energy consumption associated with software updates further compounds carbon emissions. Continuous updates require data transfer, cloud storage, and server-side processing, all of which rely on electricity, often generated from fossil-fuel-based grids. Large-scale distribution of operating system updates to millions of devices globally, combined with AI-driven application updates and background service patches, can consume substantial amounts of energy. Data centers, which host software repositories, cloud applications, and automated update systems, require ongoing cooling and operational infrastructure, adding to the carbon footprint of continuous software evolution. Even incremental updates, when scaled globally, represent a nontrivial contribution to energy consumption and climate impact. Consumer behavior and perceived obsolescence play a critical role in mediating software-driven environmental impact. Users often interpret sluggish device performance post-update as a signal that their hardware is outdated, prompting early replacement. The cumulative effect of frequent updates and increasing software complexity is a cycle of consumption that aligns technological growth with disposability. (Wark & Soncul, 2022).

CONCLUSION

As the digital age accelerates, the intersection of artificial intelligence (AI), the Internet of Things (IoT), high-performance computing, and ubiquitous smart devices is reshaping the way societies live, work, and interact. These technological innovations offer unprecedented opportunities for efficiency, insight, and connectivity, yet they also bring profound environmental and ethical challenges. The exponential growth of digital infrastructure, computational demands of AI, frequent hardware upgrades, and the pervasive generation of electronic waste (e-waste) have amplified the carbon footprint, resource consumption, and ecological impacts of modern technology. In this context, the concept of sustainable design emerges as a critical principle for future-proofing intelligent systems, ensuring that the next era of technology integrates innovation with environmental stewardship, social equity, and long-term resilience.The urgency of sustainable design in emerging technologies stems from the scale and speed of digital expansion. Data centers, AI model training, blockchain systems, and cloud services consume significant energy resources, often reliant on fossil-fuel-based grids, resulting in substantial greenhouse gas emissions. Simultaneously, the lifecycle of consumer electronics—from raw material extraction and manufacturing to operation, maintenance, and disposal—produces vast amounts of e-waste and depletes finite natural resources. For instance, the production of a single smartphone can involve tens of kilograms of CO_2 emissions, while AI model training in high-performance clusters can produce carbon outputs equivalent to dozens of cars over a year. Sustainable design principles are essential to mitigate these impacts, ensuring that technological advancement does not come at the expense of planetary health.

Principles of sustainable hardware design form the foundation of future-proof intelligent systems. Modular architectures, repairable components, standardized interfaces, and energy-efficient materials extend device lifespans, facilitate upgrades, and reduce unnecessary replacement. For high-performance computing infrastructure and servers, designing for scalability, redundancy, and energy efficiency ensures that systems can accommodate future computational demands without disproportionate energy consumption. Integration of recyclable materials, low-toxicity components, and biodegradable elements further reduces the environmental burden at end-of-life, creating a circular ecosystem where devices and components are recoverable and reusable. By embedding sustainability into hardware design, manufacturers can decouple technological advancement from environmental degradation.Software and system-level sustainability are equally critical. Intelligent systems and AI models often require frequent updates, resource-intensive computations, and high data throughput, placing stress on both hardware and energy resources. Sustainable design requires optimization of algorithms for efficiency, development of lightweight software compatible with older hardware, and adaptive computing strategies that scale resource usage based on demand. Techniques such as edge computing, distributed processing, and energy-aware AI algorithms can reduce reliance on energy-intensive cloud servers while maintaining performance and responsiveness. By aligning software efficiency with hardware longevity, sustainable design addresses both operational energy use and the incentives for premature device replacement, mitigating both carbon emissions and e-waste generation.

REFERENCES

Adler, D., & Schneider, J. (2020, December 1). Amazon workers are fighting for their rights. This holiday season, think of them. *The Guardian*. https://www.theguardian.com/commentisfree/2020/dec/01/amazon-workers-fighting-for-their-rights

Anute, N., Patole, J., Pawar, P., Kalshetti, P., & Sharma, M. (2026). Digital marketing and semiotics. In Boero, M., & Piccolo, R. (Eds.), *Marketing semiotics research: Consumption trends, AI, and the economics of experience* (pp. 209–236). IGI Global Scientific Publishing., DOI: 10.4018/979-8-3693-8024-6.ch007

Baca, M. C., & Greene, J. (2019, September 20). Amazon, Google, other tech employees protest in support of climate action. *The Washington Post*. https://www.washingtonpost.com/technology/2019/09/20/amazon-google-other-tech-employees-protest-support-climate-action/

Cellard, L. (2022). Algorithms as figures: Towards a post-digital ethnography of algorithmic contexts. *New Media & Society*, *24*(4), 982–1000. DOI: 10.1177/14614448221079032

Cellard, L., & Marquet, C. (2023). Frictions sous-marines. Écologies et politiques des fragilités de l'interconnexion numérique. *Revue d'Anthropologie des Connaissances*, *17*(4). Advance online publication. DOI: 10.4000/rac.31070

De Vries, A. (2023). The growing energy footprint of artificial intelligence. *Joule*, *7*(10), 2191–2194. DOI: 10.1016/j.joule.2023.09.004

Dobbe, R., & Whittaker, M. (2019). AI and climate change: How they're connected, and what we can do about it. *AI Now Institute*. https://ainowinstitute.org/publication/ai-and-climate-change-how-theyre-connected-and-what-we-can-do-about-it

Dodge, J., Prewitt, T., & Tachet des Combes, R.. (2022). Measuring the carbon intensity of AI in cloud instances. In *Proceedings of the 2022 ACM Conference on Fairness, Accountability, and Transparency* (pp. 1877–1894). Association for Computing Machinery. https://doi.org/DOI: 10.1145/3531146.3533234

Edwards, P. N. (2010). *A vast machine: Computer models, climate data, and the politics of global warming*. MIT Press.

Ensmenger, N. (2018). The environmental history of computing. *Technology and Culture*, *59*(4, Supplement), S7–S33. DOI: 10.1353/tech.2018.0148 PMID: 30595595

European Commission. (2024, August 8). Artificial Intelligence Act: Call for tenders to measure and foster energy efficient and low emission artificial intelligence in the EU. *Digital Strategy.* https://digital-strategy.ec.europa.eu/en/funding/artificial-intelligence-act-call-tenders-measure-and-foster-energy-efficient-and-low-emission

Foroohar, R. (2018, December 18). Year in a word: Techlash. *Financial Times.* https://www.ft.com/content/76578fba-fca1-11e8-ac00-57a2a826423e

Gregg, M., & Strengers, Y. (2024). Getting beyond net zero dashboards in the information technology sector. *Energy Research & Social Science*, *108*, 103397. Advance online publication. DOI: 10.1016/j.erss.2023.103397

Hacker, P. (2024). Sustainable AI regulation. *Common Market Law Review*, *61*(2), 345–386. DOI: 10.54648/COLA2024025

Halpern, O. (2021). Planetary intelligence. In Roberge, J., & Castelle, M. (Eds.), *The cultural life of machine learning* (pp. 227–256). Palgrave Macmillan., DOI: 10.1007/978-3-030-56286-1_8

Jancovic, M., & Keilbach, J. (2023). Streaming against the environment: Digital infrastructures, video compression, and the environmental footprint of video streaming. In Es, K., & Verhoeff, N. (Eds.), *Situating data: Inquiries in algorithmic culture* (pp. 85–102). Amsterdam University Press. DOI: 10.2307/jj.362382.9

Jiménez, A., & Oleson, J. C. (2022). The crimes of digital capitalism. *Mitchell Hamline Law Review*, *48*(4), 971–1018.

Kunnathully, K., Jinesh, M., Satyanarayana, P., Esther, V., Pathak, G. P., & Manjunath, T. C. (2026). Building human rights by design: Ethical AI development for religious freedom. In Chakraborty, S. (Ed.), *Intersections of AI and the freedom of religion or belief* (pp. 81–104). IGI Global Scientific Publishing., DOI: 10.4018/979-8-3373-3830-9.ch004

Lally, N., Kay, K., & Thatcher, J. (2022). Computational parasites and hydropower: A political ecology of Bitcoin mining on the Columbia River. *Environment and Planning. E, Nature and Space*, *5*(1), 18–38. DOI: 10.1177/2514848619867608

Levenda, A. M., & Mahmoudi, D. (2019). Silicon Forest and server farms: The (urban) nature of digital capitalism in the Pacific Northwest. *Culture Machine*, •••, 18.

Nobrega, C., & Varon, J. (2020). Big tech goes green(washing): Feminist lenses to unveil new tools in the master's houses. In *Global Information Society Watch 2020: Technology, the environment and a sustainable world: Responses from the Global South* (pp. 26–33). Association for Progressive Communications.

Nost, E., & Colven, E. (2022). Earth for AI: A political ecology of data-driven climate initiatives. *Geoforum, 130*, 23–34. DOI: 10.1016/j.geoforum.2022.01.016

Ortar, N., Taylor, A. R., Velkova, J., ... (2023). Powering 'smart' futures: Data centres and the energy politics of digitalisation. In *Energy futures: Anthropocene challenges, emerging technologies and everyday life* (pp. 125–168). De Gruyter.

Paré, G., & Kitsiou, S. (2017). Methods for literature reviews. In *Handbook of eHealth evaluation: An evidence-based approach*. University of Victoria., https://www.ncbi.nlm.nih.gov/books/NBK481583/

Pasek, A. (2019). Managing carbon and data flows: Fungible forms of mediation in the cloud. *Culture Machine, 18*, 1–15.

Taffel, S., Bedford, L., & Mann, M.. (2022). Ecocide isn't ethical: Political ecology and capitalist AI ethics. In Phan, T., Goldenfein, J., & Kuch, D. (Eds.), *Economics of virtue: The circulation of 'ethics' in AI* (pp. 59–86). Institute of Network Cultures.

The Greens/EFA in the European Parliament. (2023). Breakthrough for regulating AI and banning discriminatory technologies. https://www.greens-efa.eu/en/article/press/breakthrough-for-regulating-ai-and-banning-discriminatory-technologies

Van Wynsberghe, A. (2021). Sustainable AI: AI for sustainability and the sustainability of AI. *AI and Ethics, 1*(3), 213–218. DOI: 10.1007/s43681-021-00043-6

Velkova, J. (2021). Thermopolitics of data: Cloud infrastructures and energy futures. *Cultural Studies, 35*(4–5), 663–683. DOI: 10.1080/09502386.2021.1895243

Vinuesa, R., Azizpour, H., Leite, I., Balaam, M., Dignum, V., Domisch, S., Felländer, A., Langhans, S. D., Tegmark, M., & Fuso Nerini, F. (2020). The role of artificial intelligence in achieving the Sustainable Development Goals. *Nature Communications, 11*(1), 233. Advance online publication. DOI: 10.1038/s41467-019-14108-y PMID: 31932590

Wark, S., & Soncul, Y. (2022). A media-theoretical armamentarium: From the pharmakon to pharmacologies of media. *Media Theory, 6*(2), 1–24. DOI: 10.70064/mt.v6i2.879

Wyatt, S. (2008). Technological determinism is dead; long live technological determinism. In Hackett, E., Amsterdamska, O., & Lynch, M. (Eds.), *Handbook of science and technology studies* (pp. 165–180). MIT Press.

Chapter 7
Decentralized Energy Systems and Blockchain's Role in Sustainable Power Grids

Uzair Bhatti
http://orcid.org/0000-0002-8743-2783
Hainan University, China

Gafur Namazov
Department of Information Technology and Exact Sciences, Termez University of Economics and Service, Termez, Uzbekistan

Sardor Sabirov
http://orcid.org/0009-0008-0504-7568
Department of General Professional Sciences, Mamun University, Khiva, Uzbekistan

Momogul Ismailova
Department of Technological Education, Urgench State Pedagogical Institute, Urgench, Uzbekistan

Abdullayev Abdulla Fayzulla Ugli
Department of Specialization, Social Humanities, and Exact Sciences, Tashkent State University of Economics, Tashkent, Uzbekistan

Yuldasheva Gulora
Department of Computer Sciences, Urgench State University, Urgench, Uzbekistan

Barno Matchanova
Department of National Idea and Philosophy, Urgench State Pedagogical Institute, Urgench, Uzbekistan

ABSTRACT

Decentalized energy systems are a radical departure from the way we have traditionally produced, distributed and consumed electricity – relying upon centralized

DOI: 10.4018/979-8-3373-8490-0.ch007

grids that depend on fossil fuels towards more sustainable, resilient and community-based models. These systems improve grid flexibility, save transmission costs and are fit for renewable power input like solar or wind, since prosumers can generate and distribute energy within the local area. Blockchain technology is going to be a key enabler of this transition, as it offers a reliable, transparent and decentralized platform for energy sharing and grid management. The blockchain technology utilizing the smart contracts and mutual system can provide reliable peer-to-peer power trading, accurate settlement, and less necessity of centralized agency. Its distributed ledger makes the equipment trustable for all participants, and meanwhile it realizes real-time transaction data sharing to optimize grid management like demand response of electricity and certificates tracing of green power production.

INTRODUCTION

The increasing need for sustainable and robust power grids to reduce the global environmental impact has led to a lot of interest in attention on decentralized energy systems, which offer a paradigm shift from the conventional centralized utility system (Omar & Saleh, 2024). These types of systems, which include solar photovoltaics, wind turbines, micro-grids and advanced battery storage, represent decentralized and sustainable solutions to traditional energy infrastructure (Omar & Saleh 2024). This shift towards decentralization of systems is vital in overcoming the drawbacks of centralized systems: both their vulnerability to mass blackouts and dependence on fossil fuels (Шапсугова, 2023). The advancement of distributed energy systems is of great significance as far as the energy revolution and formation of a clean, low-carbon, safe and efficient energy structure are concerned, but also redounds to the guarantee of energy security (Dong et al., 2025). This transformation requires novel ways to control the energy flow, and enable peer-to-peer (P2P) trading of energy, as well as flexibility in both supply and demand on overall systems [30], blockchain is considered a potential enabler for these solutions (Rasool and Jacobsen, 2022). Derived from the digital currency world, blockchain provides and maintains secure and transparently distributed ledger in smart grid that can self manage though-based on decentralized transactions as solutions to security and privacy issues (Anjimoon et al., 2024).

This ability of blockchain for enabling reliable, transparent, irrevocable and efficient digital transactions makes it highly immune to blackouts, providing an innovative concept for the energy management model (Rawat et al., 2023). This decentralized and cryptographically protected ledger system aligns conceptually with the distributed nature of contemporary smart grids, in which power generation is becoming increasingly decentralised across many smaller providers (Anjimoon et al., 2024; Diantoro et al., 2023). Moreover, the synergy of AI and ML with blockchain technologies would be a powerful driver for improving the robustness and performance of these decentralized systems, particularly in fault detection and predictive maintenance for RE components (Dutt & Karuna, 2024; Porawagamage et al., 2024). The use of such sophisticated analysis can predict potential issues and recommend best-case, real-time reactions, improving the adaptability and resilience of the grid (Porawagamage et al., 2024). These AI and ML algorithms, when integrated in smart grids, make decisions based on a large set of data from meters and sensors to improve performance, predict consumption patterns, fault detection and automate dispatching of resources (Uzondu & Lele, 2024). Although AI presents high hopes in predicting electricity demand and optimizing grid operations, the challenges of data integrity are evident in this domain and must be emphasized – especially during unusual events such as natural disasters that can induce poor forecasted results and expose the system to vulnerabilities (Nguyen, 2025). Additionally the high price of AI adoption, as well as the demand for adequate digital infrastructure and human capital were perceived as major barriers to widespread use in some areas (Nguyen 2025).

Nonetheless, despite these challenges, the power of AI as a means at early anomaly detection and also in achieving energy balance supply-and-demand equilibrium with smart grids makes it still a promising means for future energy systems (Alenazi, 2023; Nguyen, 2025). This integration of blockchain secure and transparent design with AI analytical capabilities, instead will enable advanced smart energy management reducing the necessity for human operations, thus increasing the system autonomy (Kuzior et al., 2022). The combined use of the AI and energy generation and distribution systems can lead to a greater control over energy productivity, especially in renewable energy domains where AI-driven predictive models are used to enhance efficiencies by taking into account environmental parameters and tweaking operations accordingly (Liu et al., 2025). One example is that AI algorithms are important in the implementation of sensors, for both placement and operation optimization in a decentralized grid context allowing real time network assessment to support distribution network self-reorganization with optimal approach (Biswas et al., 2025). Machine learning, specifically, applies sensor data from all over the electric grid to enhance reliability and efficiency – playing a key role in identifications of issues inside these complex systems (Akpan, `2024). They can learn and adapt to the dynamic characteristics of electricity systems by training based on historical datasets coupled with real-time observations for potential failures and breakdowns in advance (Akpan, 2024). This AI (in particular, ML and ANNs) integration into the SG is not only essential to facilitate advanced energy management with all its associated advantages such as precise fault detection and proactive PdM but also for significantly enhancing grid resilience and operational performance (Anjimoon et al., 2024; Arévalo & Jurado, 2024).

AI can also affect improved solar irradiance prediction and panel positioning to ensure continuous, stable energy production in any kind of weather (Liu et al., 2025). Likewise, AI also fine-tunes wind farm operation based on meteorological forecast with energy storage and optimized usage to maximize production efficiency with minimal waste (Liu et al., 2025). Now, sliding management functions into these intelligent devices, interfaced with a secure transactional blockchain is key to better managing distributed energy resources and creating 'smart grid' functions (Hamid & Ganne, 2023). Such a global initiative not only reduces technical and economic risks but also supports sustainable deployment of the power grid by maximizing renewable penetration and minimizing environmental footprint (Anjimoon et al., 2024; Ayadi et al., 2025). The AI-based solutions also contribute remarkably to waste reduction and efficiency improvement of the overall system, by optimising energy generation/utilisation and encouraging resource preservation at extended asset lifespans by monitoring in real time as well for proactive maintenance (Liu et al., 2025). Such comprehensive integration improves the grid's resiliency and ability to restore quickly in case of interruptions, thus providing a continuous and stabilized electricity supply for consumers (Hafez et al., 2025). The role of AI, in particular ML and NN, toward the development of smart grids is critical to achieving more sophisticated energy management, pinpoint fault detection as well as proactive predictive maintenance that would result in a significant enhancement in the resilience and efficiency of the grid (Arévalo & Jurado, 2024; Kaur et al., 2024). Models can now be trained using machine learning methods to predict and anticipate the equipment's future behavior, thus, reducing system down time and maintenance costs in energy systems (Tomar & Grover, 2024). In other words, machine learning methods can achieve better accuracy in fault detection under dynamic microgrid environments as compared to traditional rule based approaches by enhancing the quality of identifying faults with respect to their false positives and negatives (Dutt & Karuna, 2024). This analytical capability combined with sturdy decentralized architectures can provide real time intelligent decision making, autonomous system responses, thus revolutionizing conventional grid management mechanisms. Coupling machine learning techniques for power system signal classification with multivariate stacked LSTM models enhances the capabilities in capturing and controlling complex operational dynamics inside these advanced grids (Bârã & Oprea, 2024).

Such an advanced handling of data, mainly based on deep learning architectures to learn from behavior and adapt the strategies, allows the grid to self-organize itself so that it can autonomously reach a new configuration reacting in real time to perturbations: we go beyond the classic reactive mechanisms towards true proactive management (Dasari et al., 2024; Porawagamage et al., 2024). This interconnection can be beneficial for the utilization of advanced computational methods to address efficient management of power flow and stability studies as well, which is highly necessary due to renewable energy variability (Porawagamage et al., 2024). The adoption of machine learning in smart grid systems is a revolutionary breakthrough for the improvement of efficiency, reliability, and sustainability in contemporary energy systems (Rashid et al., 2024). Continuous development of decentralized energy systems was found to be an important knowledge for resilient and sustainable power management methods in the future, particularly through active fault diagnosis and advanced fault (Almasoudi, 2023; Dutt & Karuna, 2024). This will be facilitated by exploiting AI and machine learning to provide improved real-time decision support, which is essential for controlling fluctuating electrical networks and the insertion of variable renewable generation resources (Tabassum et al., 2024). For example, machine learning algorithms like long short-term memory networks and random forests are used for demand forecasting to model intricate temporal effects and factor in external variables such as weather to improve accuracy (Alkhaleel et al., 2025). Furthermore, the flexibility and scalability of these machine learning algorithms make them suitable for microgrids of varied sizes including residential, commercial and industrial without needing much recalibration or reprogramming (Singh et al., 2024). This will enable decentralized energy systems to scale up their fault detection and diagnostic capabilities as and when they grow and evolve, while ensuring that the performance is optimal at all times (Dutt & Karuna, 2024).

LITERATURE REVIEW

1) Decentralized Energy Systems: DERs, Microgrids, and "Transactive" Coordination

Decentralized energy systems are typically framed around distributed energy resources (DERs)—rooftop PV, behind-the-meter batteries, flexible loads, EVs, and small-scale generation—coordinated locally through microgrids and, at larger scales, through distributed operational and market mechanisms. The literature positions this shift as both a technical response to renewable intermittency and a socio-economic response to consumer participation: more assets sit at the grid edge, more actors have decision rights, and the grid must accommodate rapid bidirectional flows rather than one-way delivery. Interoperability and cyber-physical coordination become central because these assets are owned and operated by diverse parties, yet must behave coherently under grid constraints. NIST's smart grid interoperability work underscores that the ability to securely exchange and use actionable information across devices, systems, and stakeholders is foundational to modern grid operation and DER integration. This necessitates robust communication protocols and advanced data analytics to manage the increasing complexity and volume of information generated by these decentralized systems. This complex interplay of diverse actors and assets necessitates sophisticated fault detection mechanisms, where traditional methods often prove insufficient due to the dynamic and intermittent nature of renewable energy sources (Akpan, 2024; Dutt et al., 2024).

A key concept used to organize decentralized coordination is transactive energy (TE)—market-based methods that incentivize customer and device actions rather than relying primarily on utility direct control. The TE literature treats price/value signals as a coordination layer: devices respond to incentives, and the aggregate response provides grid services (balancing, congestion management, local adequacy) while respecting end-user preferences. NIST emphasizes TE as a pathway for integrating "customer flexibility" at scale, and also highlights the need for measurement science and evaluation frameworks to compare TE approaches under realistic scenarios. This framing matters because many blockchain proposals implicitly assume TE-like coordination: local market clearing, automated settlement, and verifiable exchange records. However, the integration of such complex systems requires robust and scalable frameworks to ensure secure and efficient data exchange and transaction processing (Mukherjee et al., 2022). One of the primary benefits of transactive energy is its potential to offer a more granular economic valuation of distributed energy resources services compared to administratively determined pricing schemes like net metering or feed-in tariffs (Lowder & Xu, 2020). This nuanced valuation mechanism can incentivize more efficient energy consumption and production patterns, fostering a more dynamic and responsive energy market (Armanios & Tucci, 2025). This economic signaling can promote increased participation from diverse market participants, including prosumers with renewable generation and flexible loads, thereby enhancing grid stability and reducing operational costs (Gopstein et al., 2021).

Microgrids are often presented as the "natural habitat" for early decentralized markets because they have clearer physical boundaries and a stronger alignment between local generation and local consumption. Within microgrids, the research focus frequently shifts from "can we technically run islanded?" to "how do we allocate scarce local resources and value flexibility fairly?" Reviews of blockchain-enabled microgrids repeatedly note that intermittency and uncertainty can create supply–demand imbalance and congestion risk inside the microgrid, so any trading mechanism has to be coupled to operational constraints (forecasting, monitoring, and controls), not just financial transactions. This is why many studies treat microgrid energy management systems (MEMS) and market mechanisms (auctions, pricing, negotiation) as inseparable from the ledger/settlement layer. The immutable and transparent nature of blockchain technologies offers a robust framework for managing these complex interactions, enabling secure and auditable transactions while fostering trust among diverse stakeholders (Johnson et al., 2022).

The empirical narrative that accelerated interest in decentralized trading is the set of pilot projects demonstrating neighbor-to-neighbor exchange (or "community energy" trading). The Brooklyn Microgrid is repeatedly cited as an illustrative case for localized renewable exchange, where participants can buy/sell local energy and receive transparent accounting of transactions. While practitioner and corporate write-ups differ in technical depth, they converge on the same lesson: pilots surfaced the promise (community participation, traceability, localized matching) but also exposed hard constraints—regulation, settlement integration, scalability, and the mismatch between "trading locally" and how retail tariffs, metering, and supplier obligations are structured. In the academic literature, these pilots are often used as grounding examples rather than definitive proof of large-scale viability. These real-world implementations, despite their limitations, provide invaluable insights into the practical challenges of integrating blockchain technology into existing energy infrastructures, particularly concerning regulatory frameworks and market design. Another prominent example, the GrünStromJeton project, demonstrates the application of blockchain for transparently tracking renewable energy production and consumption, indicating its potential for broader market integration (Xu et al., 2019). The success of such pilot projects, however, often depends on overcoming significant technical hurdles related to blockchain's transaction speed, scalability, and the legal recognition of smart contracts (Lowder & Xu, 2020). Consequently, addressing these challenges necessitates a comprehensive approach that integrates advanced cryptographic techniques with innovative consensus mechanisms to bolster the performance and regulatory compliance of blockchain networks within the energy sector.

Across decentralized energy studies, the market design question shows up as a spectrum. On one end are centralized optimization approaches (a coordinator clears bids with full visibility); on the other are fully decentralized approaches (agents negotiate with limited information). A recurring conclusion from microgrid trading papers is that "decentralized" does not eliminate the need for system-level constraints: line limits, voltage constraints, and reliability requirements still exist and can't be wished away by the presence of a ledger. Hence, many proposed methods rely on hybrid patterns: distributed bidding plus centralized (or hierarchically coordinated) constraint checks; or decentralized clearing with periodic reconciliation to grid-feasible schedules. Recent work explicitly integrates congestion/line-flow constraints and social welfare objectives into trading designs, illustrating the trend toward "market + physics co-design."This integrated approach acknowledges that purely financial market mechanisms, without consideration for the underlying physical limitations of the grid, can lead to instability and inefficiency, especially with the proliferation of distributed energy resources. The growing interest in distributed energy resources has spurred the development of advanced algorithms and incentive mechanisms for energy trading among interconnected microgrids, often leveraging blockchain for enhanced transparency and security (Faizan et al., 2019; Yang & Wang, 2021).

Finally, the decentralized energy literature increasingly emphasizes that grid-edge participation is not only about energy (kWh) but also about services: flexibility, reserves, voltage support, peak shaving, and resilience. TE challenge reports and TE resilience guidance documents argue that evaluating these systems requires co-simulation (power system + communications + market behaviors) and reliability thinking: when incentives change, behavior changes, and the grid must remain stable through disturbances and adversarial conditions. This becomes a crucial bridge to blockchain research: if blockchain is used for coordination or settlement, it must not degrade reliability or recovery, and it must be measurable and testable in realistic scenarios rather than only in stylized agent simulations.

2) Blockchain's role: P2P trading, trust automation, and platform architectures

In the context of decentralized renewables literature, blockchain first appears as a mechanism for shared trust for parties that do not entirely trust each other: prosumers, communities, aggregators suppliers and network operators. That central thesis isn't that the blockchain will "run the grid" but rather it can offer tamper-evident transaction records, programmatic automation through smart contracts and auditable settlement across organizational lines. Reviews emphasizing transactive and P2P energy trading regard blockchain as promoting decentralized participation in the market, due to a reduction of coordination costs (e.g.,automated matching, settlement and verification) particularly when many small players conduct repeat sales. And yet it is made clear in these reviews that much of the gain is predicated on careful integration with metering, forecasting, identity and governance—without which blockchain will merely become an expensive database rather than an enabler.

In P2P trading architectures, blockchain usually fulfills three functions: (1) registration/identity of market participants and assets; (2) transaction loggings (bids, offers, matching); and (3) settlement logic (pricing rules, auction clearing, penalties dispute resolution) realized as smart contracts. Literature reviews reveal that auction mechanism (double auctions, continuous trading, negotiation games) are used repeatedly and coupled with blockchain for transparent rule enforcement and non-repudiable logs. Yet, as systems progress from concept towards engineering, more and more authors advocate for modular architectures where blockchain is just one piece in a larger picture: data collected at the edge, estimation of state, validation of constraints or interfaces with legacy grid systems. We stress this modularity aspect, which is found in proposed reference architectures that intend to maintain {grid stability while allowing for real time trading. Such architectures utilise sophisticated cryptographic protocols for the preservation of integrity and privacy of data in the blockchain, while integrating with existing power grid control systems for operational arm and reliability (Nguyen et al., 2024). They acknowledge that while block-chain has unique advantages for transactional transparency and automation, its introduction does need to be carefully inserted in the loop so that it becomes a part of rather than replace s the existing operations 2. As a consequence, several articles suggest to have only the core commitments on chain (or use of permissioned chains) while keeping fast control signals and high-frequency telemetry off-chain. The more recent thread on the platform level compares blockchain options in terms of consensus energy use, scalability solutions, cyber security posture or privacy-preserving properties,—implicitly then recognizing that "blockchains" is not one thing, but a design space with trade-offs that carry weight–in this case wrt energy.

Similarly, the literature differentiates between public/permissionless and permissioned/consortium models. Energy systems typically need a higher level of identity assurance, compliance and operational governance, which would attract a permissioned or hybrid model for many use cases." Energy Web sector papers describe governance and technical fea- tures designed based on energy-market requirements, while academic chapters and surveys consistently reference Energy Web and Power Ledger as exemplary in terms of practical implemen- tation and interoperability. And even these examples are not the clear "best" answers, but they anchor the discussion around how actual actors try to standardize identity and roles and market interaction patterns.

Another common line of research is data integrity and provenance specifically related to sustainability claims. If drivers wish to sell and buy "local renewable" electricity or green attributes (e.g., time and location of generation), they require a high level of metering faithfulness and auditable records. Blockchain offers much stronger auditability, but the literature is unambiguous that it will not in itself yield truth at the measurement boundary ("garbage in, garbage out"): tamper-evident meters, secure telemetry and trusted attestations still are needed. In this regard European policy oriented reports, as well as standardization debates take this into account by concentrating on the reference architectures and interoperable guidelines rather than considering blockchain alone as a silver bullet. As a result, recent work has placed growing attention on the design of strong oracle mechanisms and trusted hardware in order to connect the physical energy system with the digital ledger and guarantee that information uploaded to-chain honestly reflects actual flows and characteristics of energy (Sgueglia et al., 2022). This integration creates a need for advanced system engineering techniques for addressing data privacy, verifying requirement compliance at all stages of development e.g., ethical use of private or personal (sensitive) data (Cabrera et al., 2025). These concerns are important to make sure that in DES, not only technical efficiency is achieved but also privacy and security requirements preserved, especially for personal-related data like load profiles which may be protected by regulation (i.e. GDPR) (Montakhabi et al., 2023).

Third, bibliometric and systematic-review research illustrates the shift in the field's development: early papers focus on conceptual benefits (disintermediation, trustless markets) while new work coalesce around engineering and deployment questions—scalability, privacy, interoperability, regulatory fit and cyber risk. Large-scale mapping studies, across hundreds of articles, frame energy trading as a predominant theme but point to diversification into grid services, IoT integration and governance. This indicates a transition in the research community beyond "blockchain for P2P kWh trading" toward "DLT-supported coordination and verification across multiple grid-edge services," including more focus on evaluation and real-world constraints. This changing emphasis highlights the urgency of establishing robust methods for evaluating the practical implementation of blockchain across varied energy use cases, including devising exhaustive safety checklists and practices for auditing to spot and respond to potential weaknesses in these sophisticated, interconnected systems (Cabrera et al., 2025). The social-technical dimensions of decentralized energy systems are also coming into focus, with more research considering the potential for blockchain to support community-based energy efforts and new types of peer-to-peer sharing (Ahl et al., 2019).

3) Sustainability outcomes, governance, interoperability, and open research challenges

From a sustainability perspective, decentralized markets and blockchain are often justified by claims of improved renewable utilization, reduced curtailment, better local matching, and stronger incentives for flexibility and storage—ultimately supporting decarbonization. The literature is nuanced: it commonly argues that market access for small DER owners can accelerate adoption and improve efficiency, but it also stresses that sustainability gains are not automatic. If trading incentives are misaligned, P2P markets can increase peak demand, shift costs unfairly, or encourage behaviors that harm the distribution network. Therefore, many authors emphasize that sustainability should be evaluated with multi-metric frameworks (emissions impact, grid losses, congestion, equity, resilience) rather than assuming that "more trading" equals "more green."This necessitates careful regulatory design and sophisticated market mechanisms to ensure that blockchain-enabled energy markets genuinely contribute to grid stability and environmental objectives. Furthermore, the integration of blockchain into decentralized energy systems offers the potential for "energy democratization," where end-users collectively drive market dynamics rather than being dictated by centralized entities (Cascio et al., 2021). This paradigm shift allows for greater transparency and accountability in energy transactions, potentially leading to more equitable distribution of benefits among participants (Jiang et al., 2024). Moreover, the efficiency in resource utilization and waste minimization, critical for sustainable energy systems, can be significantly enhanced through blockchain's ability to streamline transaction processes and facilitate direct energy trading between producers and consumers (Anjimoon et al., 2024). Blockchain platforms can be instrumental in creating virtual power plants, which aggregate sustainable energy sources and utilize energy storage systems to mitigate fluctuations in energy supply and demand, thereby enhancing grid resilience and promoting the integration of renewables (Lv, 2023). Indeed, blockchain technology holds considerable potential to significantly impact the trajectory of a sustainable energy future by bolstering the efficiency and transparency of energy markets, thereby streamlining the assimilation of sustainable energy sources (Lv, 2023). This is particularly evident in its capacity to facilitate peer-to-peer (P2P) energy trading, enabling consumers to directly buy and sell excess renewable energy, such as that generated from rooftop solar panels, thereby fostering a more dynamic and decentralized energy landscape (Attah et al., 2024; Safari et al., 2023). This distributed ledger technology ensures transparency and immutability in energy transactions, providing a verifiable record that can enhance trust among participants and simplify regulatory oversight (Anjimoon et al., 2024). This heightened transparency can also aid in the accurate measurement, reporting, and verification of carbon emissions, although challenges related to data input integrity remain (Nolden, 2021). Despite these advantages, the energy consumption of certain blockchain consensus mechanisms, notably Proof of Work, presents a significant environmental concern that necessitates careful consideration and the adoption of more energy-efficient alternatives like Proof of Stake for sustainable energy applications (Anjimoon et al., 2024). Moreover, beyond consensus mechanisms, the scalability of blockchain networks is a pivotal challenge, as the system must effectively manage a high volume of transactions to ensure practical viability within the dynamic energy sector (Anjimoon et al., 2024).

A central governance challenge is the relationship between local P2P markets and regulated electricity frameworks (tariffs, supplier obligations, metering rules, consumer protections). Case-based discussions around early pilots frequently point to regulatory friction: even if a community can account for "local exchange," retail billing, taxes/levies, and network charges may still be defined at the supplier–consumer interface, not at the peer-to-peer interface. As a result, many papers treat blockchain-enabled trading as a complement to existing market roles (e.g., via aggregators or community suppliers) rather than a full bypass. The practical literature around community pilots repeatedly calls out scalability and regulatory integration as key constraints, aligning with academic conclusions that institutional design is as important as technical design. Moreover, the complexities of integrating decentralized autonomous organizations into existing regulatory frameworks for energy governance introduce novel challenges concerning legal recognition, liability, and dispute resolution within these emerging blockchain-based ecosystems (Armanios & Tucci, 2025).

Interoperability sits at the heart of whether blockchain-enabled decentralized energy systems can scale beyond bespoke pilots. NIST's smart grid interoperability framework argues that coordinated evolution across standards, cybersecurity, and conceptual models is necessary for the grid to function as an ecosystem rather than a set of isolated solutions. In blockchain-enabled energy markets, interoperability includes: identity and role definitions, device/telemetry standards, messaging protocols between MEMS and market platforms, and compatibility across multiple ledgers or market operators. Recent work explicitly discusses blockchain interoperability and uses sector case studies to illustrate why siloed platforms can limit liquidity and prevent broader system integration. The inherent complexity of power systems, which have evolved over decades on deterministic engineering principles, poses significant challenges for integrating probabilistic machine learning models, often requiring extensive system modifications and substantial computational resources not readily available in existing infrastructures (Porawagamage et al., 2024). Furthermore, the susceptibility of machine learning models to adversarial examples, where subtle input perturbations can induce misclassification or mask critical faults, presents a substantial risk to power system protection and emergency control, potentially leading to disastrous consequences (Porawagamage et al., 2024). The coordination of multiple reinforcement learning agents within such a complex and interconnected system further exacerbates these challenges, necessitating advanced communication protocols and robust methodologies to ensure efficient operation and prevent cascading failures (Porawagamage et al., 2024).

Cybersecurity and privacy are treated as "make-or-break" issues because energy data is sensitive (household behavior inference) and grid systems are critical infrastructure. The literature highlights that while blockchains provide tamper-evident logs, they do not eliminate endpoint compromise, key theft, or malicious participants; moreover, transparency can conflict with privacy requirements. Platform-comparison studies increasingly foreground privacy-preserving approaches (e.g., permissioning, cryptographic techniques, off-chain aggregation) and cybersecurity threat mitigation as selection criteria for energy applications. This focus is consistent with broader smart grid guidance that treats cybersecurity and secure information exchange as first-order concerns, not add-ons. The inherently sensitive nature of energy consumption data and the critical infrastructure status of smart grids necessitate robust privacy-preserving mechanisms, such as federated learning or differential privacy, to allow for advanced analytics without compromising individual user confidentiality (Nguyen, 2025; Zibaeirad et al., 2024).

A major sustainability critique concerns energy consumption of consensus mechanisms, historically associated with proof-of-work. The literature's response is generally to avoid energy-intensive consensus for grid applications (using permissioned networks, proof-of-authority, or other lower-energy designs), or to keep blockchain out of high-frequency processes. More recent surveys tend to treat "blockchain energy cost" as a design constraint rather than a fatal flaw, but they also note that any sustainability claim should include the digital infrastructure footprint and compare it against alternatives (centralized databases, trusted third parties, or hybrid architectures). In other words, the sustainability argument for blockchain is strongest when blockchain enables governance or market participation that cannot be achieved as effectively by simpler means, and weakest when it duplicates existing trust structures. The continued reliance on traditional, centralized trust models for aspects like retail billing and network charges further undermines the sustainability claims of decentralized systems, demonstrating that the full potential of blockchain for energy applications remains constrained by prevailing regulatory and infrastructural inertia. This highlights a critical dichotomy where the technological promise of blockchain for enhancing energy efficiency and sustainability often clashes with entrenched regulatory frameworks and energy infrastructure (Liu et al., 2025).

Open research challenges cluster around evaluation realism and deployment pathways. TE challenge work emphasizes co-simulation and common metrics to compare approaches, which is directly applicable to blockchain-enabled markets: researchers need shared benchmarks for latency tolerance, failure recovery, adversarial behavior, and grid-feasibility under constraints. Reviews of blockchain-based energy trading repeatedly call for more evidence from real deployments (or high-fidelity pilots) and for clearer documentation of market rules, operational integration, and stakeholder incentives—because many proposals remain difficult to compare or reproduce. The trajectory of the field suggests that the next wave of contributions will be less about proposing yet another P2P auction on-chain, and more about end-to-end architectures that integrate measurement, control, market design, compliance, and interoperability—while proving measurable gains in decarbonization, resilience, and equity. This shift necessitates a departure from purely theoretical models towards practical, deployable solutions that address the multifaceted challenges of real-world energy systems (Armanios & Tucci, 2025; Zimba et al., 2025). This will require overcoming significant technical hurdles, such as the inherent scalability limitations of current blockchain technologies, which struggle with the high transaction volumes generated by dynamic energy markets (Jiang et al., 2024).

METHODOLOGY

This section describes the methodology used to systematically review an synthesize existing literature on dezentralised energy systems and blockchain and its potential role for sustainable 6 resilient girds. It describes the process of source selection, how data was extracted, and application of periodization framework used to identify thematic areas, gaps and trends in research at the nexus between blockchain technology and energy systems (Zimba et al., 2025). The approach is suitable to be applied in this space, especially considering that the utilization of blockchain in healthcare is emerging and it presents a number of challenges that require further investigation (Zhao et al., 2023). The adopted methodology ensures that the technological capacities and socio-economic drivers of blockchain integration are thoroughly examined with respect to each other, enabling a comprehensive view on blockchain's game-changing prospects for energy systems. In the vein, artificial intelligence approaches and machine learning methods are being devised to strengthen fault detection strategies in these distributed generation based energy systems, especially with intermittent renewables as sources (Dutt & Karuna, 2024). This integration is so indispensable for forecasting and avoiding eventual failures in complicated inter-connect-grids, particularly when we take into account the rapid development of machine learning applications to power system protection and control (Porawagamage et al., 2024). In addition, the approach will include new methods for real-time monitoring using ambient sensor data and advanced diagnostics for detection and classification of anomalies to improve grid stability and efficiency (Akpan citing 2024).

The integration of AI technologies. such as ML algorithms. or advanced modeling systems into blockchain frameworks is the focus of future studies in order to reveal new insights for managing decentralized energy and reduce system complexity (Honari et al., 2022). This method can have the potential to utilize generative AI for power system protection and emergencies control, where we could be able to generate new data and models to monitor and secure the power grid with more sophisticated techniques (Porawagamage et al., 2024). Machine learning/AI They are expected to offer associated enabling technologies that can make our power systems more adaptive and robust, designed to address uncertainties/nolinearity of present grid complexity (Porawagamage et al., 2024), machine learning/AI reinforcement learning for 6 ED cluster structure place italics with Eq. These developments rely on large amounts of sensor data from the grid such as voltage, current and frequency fluctuations and structured sensor information to adequately learn and train and validate complex machine learning algorithms for fault diagnosis (Akpan, 2024). In addition, stochastic models such as VAEs and selfsupervision approach will be important for repairing missing or corrupted data, which is essential to the integrity of the data and statistical fidelity of these advanced diagnostic system (Bo, 2025). It is the data at gas measuring stations that are utilised for algorithms which help utilities to automatically detect disturbances and quickly respond to them (Goyal, 2025). Such relentless improvement in diagnosis augmented with predictive mapping is essential for managing the risks of increased penetration of distributed energy resources and their tethering toward traditional grid stability (Cavus, 2025; Rana, 2025). This may involve autonomous control systems, potentially based on deep reinforcement learning to prevent under-voltage load shedding and enable dynamic emergency control actions (Porawagamage et al., 2024).

These along with the transparent and immutable nature of block chain can support real-time secure decision making, thereby stabilizing decentralized energy systems (Shan et al., 2023). The application of AI-based management schemes, where predictive analytics and operational control in real-time are combined, is necessary to improve the performance and integration of DERs into these complex systems (Dasari et al., 2024). This architecture will also allow control strategies for emergencies and fault detection to be more complex, thanks to various data from sensors strategically installed in the grid (Akpan 2024). These technologies are expected to revolutionize grid management operations and enable autonomous, proactive and optimal decision making that is often not provided in conventional systems, especially for dynamic and unpredictable scenarios (Dasari et al., 2024). This integrated approach will make it possible to study the potential of blockchain for improving grid resilience against disruptions such as cyber-attacks and natural disasters, and provide a secure and tamper-proof platform for energy transactions as well as data exchanges (Hafez et al., 2025; Singh et al., 2024). This model is used to evaluate the effectiveness of AI applications aimed at fault detection, predictive maintenance, and optimized resource allocation which significantly contribute to the improvement of grid stability and efficiency (Akpan, 2024; Dasari et al., 2024; Hafez et al., 2025). So, as Sobati-M (2025) foresaw at last these complementary advances ensure a transition on its way to intelligent autonomous and secure smart energy infrastructures capable of addressing the challenges of crosscutting decentralized and data-driven power systems. This strong integration concerns concrete issues, such as data volume and communication resilience, using sophisticated reinforcement learning frameworks to better demand response or home energy management (Li et al., 2024).

DISCUSSION

In this section we deep dive on a complex system of interactions between these technologies, analysing the status quo and future projections for blockchain in energy decentralization. It will analyze how the intrinsic characteristics of a blockchain including immutability and transparency can mitigate key problems in data handling, security and transaction efficiency that decentralised energy market operations experience (Attah et al., 2024). Moreover, the paper will consider ability of blockchain combined with artificial intelligence and machine learning to produce intelligent and adaptive energy grids (Anjimoon et al., 2024; Donnot et al., 2017). The aim of the paper is to identify the potential benefits and issues associated with distributed energy technologies when deployed in an integrated manner that enables high- penetration levels of renewable energy by using these resources to optimize dispatch or coordination within local microgrids (to provide stable grid operation while minimizing costs). This might mean, for instance, taking a look at how block-chain empowered smart contracts could automate the demand response and energy trading process as opposed to having human interaction, which can also increase system efficiency and transparency (Li et al. 2023). Furthermore, scalability issues endemic to the current blockchain realisations will be addressed, and possible workarounds (sharding or layer-2 protocols) for the expected high transactional volumes in prospective energy markets expressed (Anjimoon et al., 2024). Finally, a key intersection of cybersecurity and data privacy in such an integrated systems will also be discussed to highlight how blockchain's cryptographic guarantees can harden the grid operations against expected advanced cyber threats rather than compromising consumer data confidentiality.

This analysis will naturally include a comparison of different consenus mechanisms that are suitable for an energy sector adoption based on their trade-offs between decentralization, security and transaction throughput. And it will address the required regulatory and policy regimes to ensure broad adoption of these technologies, identifying barriers as well as pathways for overcoming them. The chapter will delve deeper into economic ramifications and show that the blockchain technology can benefit to create new business models, (such as peer-to-peer energy trading), assist prosumers, and increase market liquidity, by reducing trading costs (Piotr 2024). Furthermore, the implementation of blockchain can also enhance traceability and transparency in renewable energy credits and carbon offset markets which promote further accountability and trust from investor on sustainable initiatives (Alenazi, 2023; Safari et al., 2023). Finally, the transformational potential of blockchain frames the need for further study and advancement to truly realize its disruptive potential for a stronger, fairer and environmentally sustainable global energy system (Anjimoon et al. 2024). Nevertheless, while blockchain has numerous virtues, there are significant challenges in deploying blockchain technologies into the energy markets in practice such as scalability and handling high transaction volums resulting from frequent energy generation scheduling/trading (Jiang et al., 2024). Sharding is the most notable solution to these challenges, partitioning network nodes into different clusters to execute transaction in parallel on it In addition, cross-chain message passing protocol can improve interoperability by allowing interconnection and communication between various energy specific blockchains without any friction during data and transaction exchange (Jiang et al., 2024).

CONCLUSION

Decentalized energy systems are a radical departure from the way we have traditionally produced, distributed and consumed electricity – relying upon centralized grids that depend on fossil fuels towards more sustainable, resilient and community-based models. These systems improve grid flexibility, save transmission costs and are fit for renewable power input like solar or wind, since prosumers can generate and distribute energy within the local area. Blockchain technology is going to be a key enabler of this transition, as it offers a reliable, transparent and decentralized platform for energy sharing and grid management. The blockchain technology utilizing the smart contracts and mutual system can provide reliable peer-to-peer power trading, accurate settlement, and less necessity of centralized agency. Its distributed ledger makes the equipment trustable for all participants, and meanwhile it realizes real-time transaction data sharing to optimize grid management like demand response of electricity and certificates tracing of green power production. Taken together, decentralized energy systems and blockchain provide a compelling path towards sustainable power grids that are more efficient, resilient and inclusive. While the challenges of scalability, policy alignment and energy consumed within blockchain networks persist, sustained technological development and enabling policy frameworks are gradually fixing these issues. As adoption increases, the marriage of decentralized energy and blockchain has an opportunity to expedite the worldwide shift to clean energy, give local communities more power over their own futures, and revolutionize the evolution of environmentally responsible power generation.

Funding

This work was supported by the Deanship of Scientific Research, the Vice Presidency for Graduate Studies and Scientific Research, King Faisal University, Saudi Arabia, under Grant No. KFU254857

REFERENCES

Ahl, A., Yarime, M., Tanaka, K., & Sagawa, D. (2019). Review of blockchain-based distributed energy: Implications for institutional development. *Renewable & Sustainable Energy Reviews*, *107*, 200–211. DOI: 10.1016/j.rser.2019.03.002

Akpan, N. E. (2024). *MACHINE LEARNING APPROACHES FOR FAULT DETECTION AND DIAGNOSIS IN POWER SYSTEMS*.

Alenazi, M. M. (2023). IoT and Energy. In *IntechOpen eBooks*. IntechOpen., DOI: 10.5772/intechopen.113173

Alkhaleel, B. A. (2025). Perspective Chapter: The Role of Machine Learning in Developing Sustainable and Resilient Interconnected Infrastructure Systems. In *IntechOpen eBooks*. IntechOpen. DOI: 10.5772/intechopen.1012259

Almasoudi, F. M. (2023). Enhancing Power Grid Resilience through Real-Time Fault Detection and Remediation Using Advanced Hybrid Machine Learning Models. *Sustainability (Basel)*, *15*(10), 8348. DOI: 10.3390/su15108348

Anjimoon, S., Chandrashekar, R., Singh, N., Parmar, A., Sharma, N., & Mohammad, Q. (2024). Secure and Sustainable Energy Distribution through Blockchain Technology in Smart Grids. *E3S Web of Conferences*, *505*, 2002. DOI: 10.1051/e3sconf/202450502002

Arévalo, P., & Jurado, F. (2024). Impact of Artificial Intelligence on the Planning and Operation of Distributed Energy Systems in Smart Grids. *Energies*, *17*(17), 4501. DOI: 10.3390/en17174501

Armanios, D. E., & Tucci, C. L. (2025). Building organizational and project readiness for AI: A dialectical approach. *International Journal of Project Management*, *43*(8), 102792. DOI: 10.1016/j.ijproman.2025.102792

Attah, R. U., Garba, B. M. P., Gil-Ozoudeh, I., & Iwuanyanwu, O. (2024a). Strategic frameworks for digital transformation across logistics and energy sectors: Bridging technology with business strategy. *Open Access Research Journal of Science and Technology*, *12*(2), 70. DOI: 10.53022/oarjst.2024.12.2.0142

Attah, R. U., Garba, B. M. P., Gil-Ozoudeh, I., & Iwuanyanwu, O. (2024b). Digital transformation in the energy sector: Comprehensive review of sustainability impacts and economic benefits. *International Journal of Advanced Economics*, *6*(12), 760–776. DOI: 10.51594/ijae.v6i12.1751

Ayadi, R., Forouheshfar, Y., & Moghadas, O. (2025). Enhancing system resilience to climate change through artificial intelligence: A systematic literature review. *Frontiers in Climate*, *7*, 1585331. Advance online publication. DOI: 10.3389/fclim.2025.1585331

Bâră, A., & Oprea, S. (2024). Machine Learning Algorithms for Power System Sign Classification and a Multivariate Stacked LSTM Model for Predicting the Electricity Imbalance Volume. *International Journal of Computational Intelligence Systems*, *17*(1), 80. Advance online publication. DOI: 10.1007/s44196-024-00464-1

Biswas, P., & Rashid, A. masum, abdullah al, Nasim, M. A. A., Ferdous, A. S. M. A., Gupta, K. D., & Biswas, A. (2025). An Extensive and Methodical Review of Smart Grids for Sustainable Energy Management-Addressing Challenges with AI, Renewable Energy Integration and Leading-edge Technologies. *arXiv (Cornell University)*. /arxiv.2501.14143DOI: 10.48550

Bo, X. (2025). A Review of Detection, Evolution, and Data Reconstruction Strategies for False Data Injection Attacks in Power Cyber-Physical Systems [Review of *A Review of Detection, Evolution, and Data Reconstruction Strategies for False Data Injection Attacks in Power Cyber-Physical Systems*]. *arXiv (Cornell University)*. Cornell University. /arxiv.2501.10441DOI: 10.20944/preprints202501.1098.v1

Cabrera, C., Bastidas, V., Schooling, J., & Lawrence, N. D. (2025, January 1). The Systems Engineering Approach in Times of Large Language Models. *Proceedings of the ... Annual Hawaii International Conference on System Sciences/Proceedings of the Annual Hawaii International Conference on System Sciences*. DOI: 10.24251/HICSS.2025.666

Cascio, E. L., Girardin, L., Ma, Z., & Maréchal, F. (2021). How Smart is the Grid? *Frontiers in Energy Research*, *9*, 637447. Advance online publication. DOI: 10.3389/fenrg.2021.637447

Cavus, M. (2025). Advancing Power Systems with Renewable Energy and Intelligent Technologies: A Comprehensive Review on Grid Transformation and Integration [Review of *Advancing Power Systems with Renewable Energy and Intelligent Technologies: A Comprehensive Review on Grid Transformation and Integration*]. *Electronics, 14*(6), 1159. Multidisciplinary Digital Publishing Institute. DOI: 10.3390/electronics14061159

Dasari, K., Raj, V. H., Nijhawan, G., Kalra, R., Pahwa, S., & Abdul-Zahra, D. S. (2024). Integration of AI in Distributed Energy Resource Management for Enhanced Load Balancing and Grid Stability. *E3S Web of Conferences, 505*, 3008. DOI: 10.1051/e3sconf/202450503008

Diantoro, K., Supriyanti, D., Ardi, A., Sanjaya, Y. P. A., & Watini, S. (2023). Implications of Distributed Energy Development in Blockchain-Based Institutional Environment. [ATT]. *Aptisi Transactions On Technopreneurship*, *5*(2sp), 209–220. DOI: 10.34306/att.v5i2sp.343

Dong, J., Ning, P., Zhao, H., & Song, C. (2025). Decentralized peer-to-peer energy trading: A blockchain-enabled pricing paradigm. *Journal of King Saud University. Computer and Information Sciences*, *37*(1-2), 10. Advance online publication. DOI: 10.1007/s44443-025-00025-2

Donnot, B., Guyon, I., Schoenauer, M., Panciatici, P., & Marot, A. (2017). Introducing machine learning for power system operation support. *arXiv (Cornell University)*. /arxiv.1709.09527DOI: 10.48550

Dutt, A., & Karuna, G. (2024). Machine learning approaches for fault detection in renewable microgrids. *MATEC Web of Conferences, 392*, 1192. DOI: 10.1051/matecconf/202439201192

Dutt, A., Rani, M. N. S., Bisht, M. S., Chandna, M., & Singla, A. (2024). Machine Learning Approaches for Fault Detection in Renewable Microgrids. *E3S Web of Conferences, 511*, 1030. DOI: 10.1051/e3sconf/202451101030

Faizan, M., Brenner, T., Foerster, F., Wittwer, C., & Koch, B. (2019). Decentralized bottom-up energy trading using Ethereum as a platform. *The Journal of Energy Markets*, *12*(2), 19–48. DOI: 10.21314/JEM.2019.193

Gopstein, A., Nguyen, C., O'Fallon, C. M., Hastings, N., & Wollman, D. A. (2021). *NIST framework and roadmap for smart grid interoperability standards, release 4.0*. DOI: 10.6028/NIST.SP.1108r4

Goyal, G. (2025). AI-Driven Fault Detection and Diagnosis in Smart Grids for Enhanced Power System Reliability. *Journal of Information Systems Engineering & Management*, *10*(42s), 845–859. DOI: 10.52783/jisem.v10i42s.8203

Hafez, S., Alkhedher, M., Ramadan, M., Gad, A., Alhalabi, M., Yaghi, M., Jama, M., & Ghazal, M. (2025). Advancements in grid resilience: Recent innovations in AI-driven solutions. *Results in Engineering*, *26*, 105042. DOI: 10.1016/j.rineng.2025.105042

Hamid, M., & Ganne, A. (2023). ARTIFICIAL INTELLIGENCE IN ENERGY MARKETS AND POWER SYSTEMS. *International Research Journal of Modernization in Engineering Technology and Science*. Advance online publication. DOI: 10.56726/IRJMETS35943

Honari, K., Zhou, X., Rouhani, S., Dick, S., Liang, H., Li, Y., & Miller, J. L. (2022). *A Scalable Blockchain-based Smart Contract Model for Decentralized Voltage Stability Using Sharding Technique*. 124. DOI: 10.1109/Blockchain55522.2022.00026

Jiang, T., Luo, H., Yang, K., Sun, G., Yu, H., & Huang, Q. (2024). Blockchain for Energy Market: A Comprehensive Survey. *arXiv (Cornell University)*. /arxiv.2403.20045DOI: 10.48550

Johnson, G. E., Sebastian-Cardenas, D. J., Balamurugan, S. P., Harun, N. F., Mukherjee, M., Blonsky, M., Markel, T., & Johnson, B. J. (2022). A Unified Testing Platform to Mature Blockchain Applications for Grid Emulation Environments: Preprint. *OSTI OAI (U.S. Department of Energy Office of Scientific and Technical Information)*. https://www.osti.gov/biblio/1889263

Kaur, S., Kumar, R., Singh, K., & Huang, Y. (2024). Leveraging Artificial Intelligence for Enhanced Sustainable Energy Management. *Journal of Sustainability for Energy*, *3*(1), 1–20. DOI: 10.56578/jse030101

Kuzior, A., Sira, M., & Brożek, P. (2022). Using Blockchain and Artificial Intelligence in Energy Management as a Tool to Achieve Energy Efficiency. *Virtual Economics*, *5*(3), 69–90. DOI: 10.34021/ve.2022.05.03(4)

Li, J., Fu, Q., & Ruan, P. (2024). Research on Algorithm Driven Intelligent Management and Control Technology for Future Power Grid. *EAI Endorsed Transactions on Energy Web*, *11*. Advance online publication. DOI: 10.4108/ew.5824

Li, Y., Ma, W., Li, Y., Li, S., & Chen, Z. (2023). Enhancing Cyber-Resilience in Integrated Energy System Scheduling with Demand Response Using Deep Reinforcement Learning. *arXiv (Cornell University)*. /arxiv.2311.17941DOI: 10.48550

Liu, M., Yuan, Z., & Ping, W. (2025). Artificial intelligence and green total factor energy efficiency: Evidence from non-linear models. *Applied Economics*, *1*. Advance online publication. DOI: 10.1080/00036846.2025.2480218

Lowder, T., & Xu, K. (2020). *The Evolving U.S. Distribution System: Technologies, Architectures, and Regulations for Realizing a Transactive Energy Marketplace*. DOI: 10.2172/1659880

Lv, Y. (2023). Transitioning to sustainable energy: Opportunities, challenges, and the potential of blockchain technology. *Frontiers in Energy Research*, *11*, 1258044. Advance online publication. DOI: 10.3389/fenrg.2023.1258044

Montakhabi, M., Madhusudan, A., Mustafa, M., Vanhaverbeke, W., Almirall, E., & van der Graaf, S. (2023). Leveraging blockchain for energy transition in urban contexts. *Big Data & Society*, *10*(2), 20539517231205503. Advance online publication. DOI: 10.1177/20539517231205503

Mukherjee, M., Sebastian-Cardenas, D. J., Balamurugan, S. P., Johnson, G. E., Gourisetti, S. N. G., Credle, S., Cutler, D., Bhattarai, B., & Johnson, B. J. (2022). A Blockchain Based Co-Simulation Framework for Integrating DERs to Wholesale Electricity Markets: Preprint. *OSTI OAI (U.S. Department of Energy Office of Scientific and Technical Information)*. https://www.osti.gov/biblio/1889264

Nguyen, L.-H., Nguyen, V.-L., Hwang, R., Kuo, J.-J., Chen, Y., Huang, C., & Pan, P.-I. (2024). Towards Secured Smart Grid 2.0: Exploring Security Threats, Protection Models, and Challenges. *arXiv (Cornell University)*. /arxiv.2411.04365DOI: 10.48550

Nguyen, T. C. (2025). AI as a solution for tackling the sustainable energy challenge in developing countries. *Energy Sustainable Development. Energy for Sustainable Development*, *87*, 101761. DOI: 10.1016/j.esd.2025.101761

Nolden. (2021). Blockchain Contradictions in Energy Service and Climate Markets. In *CRC Press eBooks* (p. 1). Informa. DOI: 10.1201/9781003208365-1

Omar, N. N., & Saleh, M. A.Nuhayd Nadi OmarMahsen Abdulkarim Saleh. (2024). Power to the People: Advancing Resilient and Sustainable Decentralized Energy Distribution Systems. *Proceedings of London International Conferences*, *11*(11), 125–139. DOI: 10.31039/plic.2024.11.251

Piotr. (2024). Innovative Solutions for the Future Development of the Energy Sector. *EUROPEAN RESEARCH STUDIES JOURNAL*, 297. DOI: 10.35808/ersj/3435

Porawagamage, G. D., Dharmapala, K., Chaves, J. S., Villegas, D. C., & Rajapakse, A. (2024). A review of machine learning applications in power system protection and emergency control: opportunities, challenges, and future directions [Review of *A review of machine learning applications in power system protection and emergency control: opportunities, challenges, and future directions*]. *Frontiers in Smart Grids*, *3*. Frontiers Media. DOI: 10.3389/frsgr.2024.1371153

Rana, S. (2025). AI-DRIVEN FAULT DETECTION AND PREDICTIVE MAINTENANCE IN ELECTRICAL POWER SYSTEMS: A SYSTEMATIC REVIEW OF DATA-DRIVEN APPROACHES, DIGITAL TWINS, AND SELF-HEALING GRIDS [Review of AI-DRIVEN FAULT DETECTION AND PREDICTIVE MAINTENANCE IN ELECTRICAL POWER SYSTEMS: A SYSTEMATIC REVIEW OF DATA-DRIVEN APPROACHES, DIGITAL TWINS, AND SELF-HEALING GRIDS]. *American Journal of Advanced Technology and Engineering Solutions*, *1*(1), 258–289. DOI: 10.63125/4p25x993

Rashid, A., & Biswas, P. masum, abdullah al, Nasim, M. A. A., & Gupta, K. D. (2024). Power Plays: Unleashing Machine Learning Magic in Smart Grids. *arXiv (Cornell University)*. /arxiv.2410.15423DOI: 10.48550

Rasool, S., & Jacobsen, R. H. (2024). Perspective Chapter: Leveraging Self-Sovereign Identity to Introduce ReCert – A Foundational Framework for Decentralized BCTE Solutions. In *IntechOpen eBooks*. IntechOpen. DOI: 10.5772/intechopen.1007156

Rawat, A., Arora, M., Bhadula, G., & Sharma, S. (2023). A Blockchain Based Energy Management System for Microgrids. *International Journal for Research in Applied Science and Engineering Technology*, *11*(4), 1313–1319. DOI: 10.22214/ijraset.2023.50314

Safari, A., Gharehbagh, H. K., Nazari-Heris, M., & Oshnoei, A. (2023). DeepResTrade: A peer-to-peer LSTM-decision tree-based price prediction and blockchain-enhanced trading system for renewable energy decentralized markets. *Frontiers in Energy Research*, *11*, 1275686. Advance online publication. DOI: 10.3389/fenrg.2023.1275686

Sgueglia, A., Sorbo, A. D., Visaggio, C. A., & Canfora, G. (2022). A systematic literature review of IoT time series anomaly detection solutions. *Future Generation Computer Systems*, *134*, 170–186. DOI: 10.1016/j.future.2022.04.005

Shan, S., Yang, S., Becerra, V. M., Deng, J., & Li, H. (2023). A Case Study of Existing Peer-to-Peer Energy Trading Platforms: Calling for Integrated Platform Features. *Sustainability (Basel)*, *15*(23), 16284. DOI: 10.3390/su152316284

Singh, A. R., Kumar, R. S., Bajaj, M., Khadse, C. B., & Зайцев, Є. (2024). Machine learning-based energy management and power forecasting in grid-connected microgrids with multiple distributed energy sources. *Scientific Reports*, *14*(1). Advance online publication. DOI: 10.1038/s41598-024-70336-3 PMID: 39160194

Sobati-M, S. (2025). FOGNITE: Federated Learning-Enhanced Fog-Cloud Architecture. *arXiv (Cornell University)*. /arxiv.2507.16668DOI: 10.48550

Tabassum, S., Babu, A. R. V., & Dheer, D. K. (2024). A comprehensive exploration of IoT-enabled smart grid systems: Power quality issues, solutions, and challenges. *Science and Technology for Energy Transition*, *79*, 62. DOI: 10.2516/stet/2024056

Tomar, P. S., & Grover, V. (2024). Transforming the Energy Sector: Addressing Key Challenges through Generative AI, Digital Twins, AI, Data Science and Analysis. *EAI Endorsed Transactions on Energy Web*, *10*. Advance online publication. DOI: 10.4108/ew.4825

Uzondu, N. C., & Lele, D. D.Nwankwo Charles UzonduDominic Dummene Lele. (2024). Comprehensive analysis of integrating smart grids with renewable energy sources: Technological advancements, economic impacts, and policy frameworks. *Engineering Science & Technology Journal*, *5*(7), 2334–2363. DOI: 10.51594/estj.v5i7.1347

Xu, Y., Ahokangas, P., Yrjölä, S., & Koivumäki, T. (2019). The fifth archetype of electricity market: The blockchain marketplace. *Wireless Networks*, *27*(6), 4247–4263. DOI: 10.1007/s11276-019-02065-9

Yang, Q., & Wang, H. (2021).. . *Exploring Blockchain for The Coordination of Distributed Energy Resources*, *1*, 1–6. Advance online publication. DOI: 10.1109/CISS50987.2021.9400211

Zhao, P., Li, S., Cao, Z., Hu, P. J., Gu, C., Yan, X., Huo, D., Luo, T., & Wang, Z. (2023). Socially Governed Energy Hub Trading Enabled by Blockchain-Based Transactions. *IEEE Transactions on Computational Social Systems*, *1*. Advance online publication. DOI: 10.1109/TCSS.2023.3308608

Zibaeirad, A., Koleini, F., Bi, S., Hou, T., & Wang, T. (2024). A Comprehensive Survey on the Security of Smart Grid: Challenges, Mitigations, and Future Research Opportunities. *arXiv (Cornell University)*. /arxiv.2407.07966DOI: 10.48550

Zimba, A., Phiri, K. O., Mulenga, M., & Mukupa, G. (2025). A systematic literature review of blockchain technology and energy efficiency based on consensus mechanisms, architectural innovations, and sustainable solutions. *Discover Analytics*, *3*(1), 14. Advance online publication. DOI: 10.1007/s44257-025-00041-6

Chapter 8
Greening Intelligent Machines:
Recent Innovations, Challenges, and Future Directions

Sohaib Ahmed
http://orcid.org/0000-0003-1830-7388
Taylor's University, Malaysia

Adnan Ahmad
COMSATS University Islamabad, Lahore, Pakistan

Faizan Ahmad
University of East London, UK

Hina Shakir
http://orcid.org/0000-0001-6493-8602
Bahria University, Karachi, Pakistan

ABSTRACT

The notion of greening intelligent machines becomes prominent for promoting the sustainable development of intelligent environments. The integration of innovative approaches with artificial intelligence (AI) to develop solutions that can mitigate the environmental impacts of AI systems. Recently, several innovative approaches have been adopted for supporting Green AI such as low-power hardware and edge computing, model optimization techniques, data-driven sustainability practices, and sustainable data centers. Despite these innovations, significant barriers are still in the progress of sustainable AI developments. These include the trade-off between energy efficiency and computational performance, scalability issues, data

DOI: 10.4018/979-8-3373-8490-0.ch008

and transparency limitations, as well as economic and technological concerns. Despite growing interest in Green AI, very limited literature has explored its recent advancements, barriers and future prospects. This chapter, therefore, presents recent innovations, highlights key challenges, and gives an overview of future directions of greening intelligent machines

1. INTRODUCTION

With the rapid advancement of artificial intelligence (AI), its adoption has become widespread across various industries, enabling the creation of new opportunities through capabilities that were previously unavailable (Hooli, 2025). These significant improvements open the doors for the development of AI models, specifically large language models (LLMs) in large number. However, these models consume substantial amount of energy resources for training and processing (Bolón-Canedo et al., 2024). As a result, the extensive use of such models can generate huge amounts of carbon emissions that can contribute towards environmental issues, one of the major global challenges (Hooli, 2025).

The concept of Green AI has evolved over the years, emphasizing the need to address the sustainable development of AI technologies without compromising their performance and capabilities (Hrithika, 2025). Green AI can minimize carbon footprints of AI technologies by optimizing energy resources, virtualizing servers, and managing electronic waste (Jha et al., 2022). In the literature, these Green AI practices can be divided into two main categories; Green-in AI and Green-by AI (Bolón-Canedo et al., 2024). The development of energy-efficient AI systems throughout their lifecycle is known as Green-in AI. Hardware optimization, model optimization, data center optimization and pragmatic scaling factor reductions are the techniques used in Green-in AI (Peykani et al., 2026). On the other hand, Green-by AI focuses on eco-friendly practices by leveraging AI solutions. It involves reducing carbon emissions and improving efficiency in these areas and applications including energy efficiency, smart mobility, sustainable agriculture, climate change, and environmental policies (Bolón-Canedo et al., 2024).

Currently, many innovative approaches have been discussed in the literature that promotes Green AI practices. One of the approaches is to adopt low-power hardware resources that can comparatively perform as LLMs but significantly reduce energy consumption (Hooli, 2025). Edge computing is another innovative approach that enables AI models to be processed and trained at edge layer instead of the centralized cloud infrastructure. This may utilize less network bandwidth that can contribute towards a sustainable solution (Hua et al., 2023). Model optimization techniques such as knowledge distillation (Cacciatore et al., 2025), model pruning (Scala et al., 2025), and quantization (Xiao et al., 2023) are the other key approaches that optimizes models to implement Green AI practices. Additionally, carbon-aware scheduling (Tanaka, 2026), renewable energy-powered data centers (Hooli, 2025) and orchestration architectures (Madanchian & Taherdoost, 2025) are the other approaches to develop sustainable data centers that can consume significantly less power as compared to conventional data centers. Transparency in energy consumption is another approach that has been discussed in the literature (Mitu & Mitu, 2024). This approach encourages stakeholders to disclose information transparently about carbon emissions of AI systems, enabling them to make more informed and responsible decisions (Oliveira et al., 2026). All of these approaches promote Green AI practices that contribute to environmental sustainability at large.

Despite significant advancements in implementing Green AI practices, researchers, developers and policymakers are still facing several key challenges that can hinder its progress. One of the key challenges is to maintain a right balance between AI model performance and its energy efficiency, as these models are primarily focused on computational performance rather than energy efficiency (Paul et al., 2023). The other challenge is related to the adoption of model optimization techniques such as knowledge distillation, model pruning, and quantization. These techniques demonstrate their capabilities at par; however, they are not yet validated at large scale, resulting in an ongoing scalability challenge (Peykani et al., 2026). In a similar vein, development and implementation of energy-efficient technological infrastructures such as sustainable data center at global scale remains a significant challenge (Oliveira et al., 2026). The other barriers in implementing Green AI practices are the lack of transparency in reporting of carbon emissions of AI systems, and no unified standardized mechanism for such disclosures (Kopka & Grashof, 2022). Further, achieving these advancements in Green AI, substantial financial investments and infrastructure are required to develop sustainable AI solutions globally is a serious concern (Paul et al., 2023).

The current literature highlights new emerging directions of Green AI that aim to develop sustainable AI solutions. Emerging approaches include explainable AI (X. Chen et al., 2026), neuromorphic computing (Luo et al., 2023), quantum computing (Hussaine & Aldeen , 2026), and energy-harvesting AI devices (Bolón-Canedo et al., 2024). Although these innovations are still in their early stages, they hold significant potential to drastically reduce computational costs and improve the sustainability of AI systems. In the literature, there are very limited studies that highlighted key innovations, challenges, and future pathways in the field (Bolón-Canedo et al., 2024; Paul et al., 2023; Peykani et al., 2026). This chapter, therefore, offers insights into recent advancements, while also exploring key challenges and presenting the future outlook for promoting Green AI.

This chapter is organized as follows: it begins by defining the fundamentals of Green intelligent machines and Green AI. It then explores various types of Green AI, including Green-in AI and Green-by AI. Subsequently, the chapter highlights recent innovations in Green AI technologies that support the sustainable development of AI systems. The following section discusses the challenges associated with these emerging trends. In the second-to-last section, future directions for designing next-generation Green AI technologies are outlined. Finally, the chapter concludes by summarizing the key innovations, challenges, and future prospects of Greening intelligent machines.

2. GREENING INTELLIGENT MACHINES

2.1 Foundation of Green AI

Green Computing has gained interest over the years because of the discussion on the negative environmental impact of information and communication technology (ICT) in literature. Green computing is also referred to as sustainable computing, where designing, developing and optimizing computer systems, networks and software can be achieved in such a way that can utilize energy efficiently by limiting adverse environmental impact (Paul et al., 2023). Such green practices can play a vital role for promoting environmental sustainability through various ways including optimization of energy resources, enhancing software productivity, virtualizing servers and effectiveness of electronic waste (Jha et al., 2022).

With the advancement of technologies, the use of artificial intelligence (AI) and machine learning techniques have been employed for improving accuracy and efficiency in different sectors including finance (Cao, 2023), education (Bekteshi, 2025) (Bashir et al., 2020), healthcare (Zhao et al., 2026), gamification (Fatima et al., 2024), project development (Hamid et al., 2020), computer programming (Saba et al., 2026) etc. The computational requirements of designing such AI models, specifically large language models (LLMs), produced large amounts of carbon footprints that may become one of the major concerns for the environmental impact (Hooli, 2025). In one of the studies mentioned that the training of transformer model with neural architecture search can produce 626,155 pounds of carbon emission equal to nearly five times carbon emission of an American car used in its lifetime (Patterson et al., 2021). In another study, it is highlighted that even smaller models like BERT can generate 1438 pounds of carbon emission during training which is almost equal to trans-American flight for one person (Hooli, 2025). The comparison between different activities of life contributed to carbon emissions as depicted in Figure 1. This highlighted that training of an AI model resulted in a substantial amount of carbon emissions. The carbon footprint of AI arises from the total greenhouse gases (GHGs) emissions, or carbon dioxide equivalent (CO_2e) produced during the manufacturing of computing technologies, as well as the emissions generated during AI model training and throughout the operational lifetime of these models(Mitu & Mitu, 2024) .

Figure 1. Carbon emissions benchmarks (Mitu & Mitu, 2024)

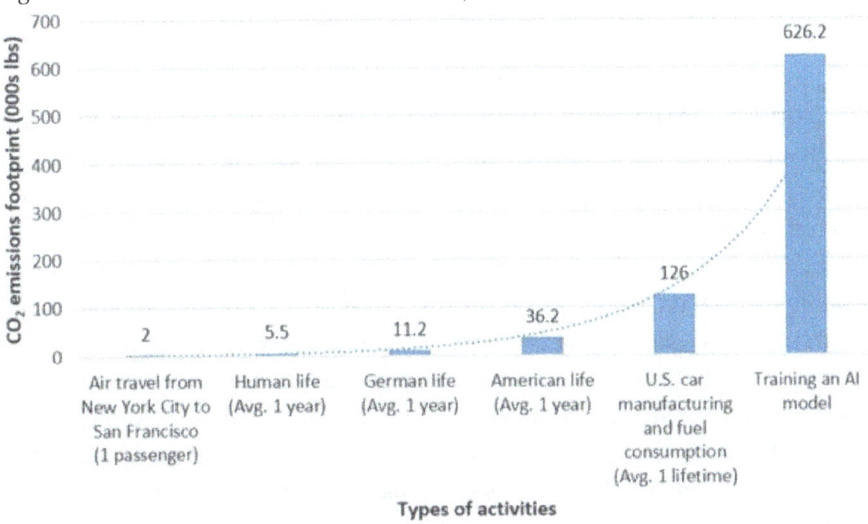

These concerns have led the researchers to advocate the new paradigm "Green AI", which mainly focuses on sustainable practices and techniques in model design, training and deployment that can reduce carbon emissions and other environmental factors (Clemm et al., 2024). Green AI can offer energy-efficient solutions, optimizing algorithms, reducing computational complexity, producing low carbon footprint, implementing data management practices that can reduce environmental impacts (Bolón-Canedo et al., 2024).

2.2 Green-by AI versus Green-in AI

The rapid expansion of artificial intelligence has significantly increased reliance on computational power, resulting in substantial energy demands and intensified environmental impacts due to increased use of complex AI models (Nikolinakos, 2023). For designing and training complex AI models, there is a need to develop energy-efficient algorithms that can achieve high precision while reducing environmental impacts (Hrithika, 2025). Such algorithms are known as green algorithms in the literature that described as "capable of maximizing the energy efficiency and reducing the environmental impact of AI models, while supporting the use of technology to respond to different environmental challenges" (Lannelongue et al., 2021). These algorithms are further categorized into two algorithms namely green-in AI and green-by AI (Bolón-Canedo et al., 2024). These terms are mentioned in the Paris agreement on the climate change, the UN sustainable development goals (https://sdgs.un.org/goals) and European green deal (https://commission.europa.eu/strategy-and-policy/priorities-2019-2024/european-green-deal_en).

Green-in AI emphasizes the development of algorithms that remain energy-efficient throughout training and usage (Bolón-Canedo et al., 2024). This suggests that AI algorithms are designed that not only demand excessive energy resources but also be optimized for efficient energy utilization, making it daunting task to balance performance with sustainability. This discussion centers on the distinction between redAI and green AI (Dhar, 2020). The notion of red AI emphasizes better output while utilizing excessive amounts of computational power. However, green AI is mainly focused on achieving sustainable AI usage by reducing computational costs from both software and hardware perspectives (Schwartz et al., 2020).

In literature, there are many strategies used in the literature that can support to reduce environmental impact in implementing green-in AI models (Bolón-Canedo et al., 2024). They are: (i) *algorithm optimization*, in which algorithms are designed to reduce the environmental footprint and enhance efficiency of AI models without compromising its performance(Hrithika, 2025). This can be achieved through many techniques including model pruning (Scala et al., 2025), quantization (Xiao et al., 2023), sparse training method (Bibikar et al., 2022) and knowledge distillation(Cacciatore et al., 2025) (ii) *hardware optimization*, in which hardware resources like efficient graphical processing units (GPUs) are chosen that can contribute power consumption (Bolón-Canedo et al., 2024). Another way is to use tensor processing units (TPUs) that can have an ability to customize AI models with optimized hardware utilization (Jouppi & Swing, 2023). Edge computing is another strategy that can be applied to perform computation at the location where data is being collected or stored instead of data transferring to cloud or data center (Wan et al., 2022). (iii) *data center optimization*, in which carbon emission of a data center mainly depends on its operational efficiency and the carbon intensity of the region where it operates (Bolón-Canedo et al., 2024). For optimizing data center resources, there are techniques used in the literature such as adjusting cooling system (Y. Li et al., 2020) and dynamic server load management(Dong et al., 2021). (iv) *pragmatic scaling factor reductions*, in which the execution of algorithms are limited number of times to minimize energy usage (Bolón-Canedo et al., 2024). Another strategy to maintain limited time spent on algorithms by optimizing hyper-parameters (L. Li et al., 2018).

On the other hand, Green-by AI denotes the development and application of AI algorithms specifically focused on addressing environmental issues (Bolón-Canedo et al., 2024). There are few approaches discussed in literature that can address environmental challenges through designing and training of AI algorithms. They are (i) *energy efficient*, in which smart grids, integrated with AI technologies that can help balance electricity between demand and supply, ultimately enhancing overall efficiency. A key strategy for reducing carbon emissions is to prioritize renewable energy sources like wind, solar, or hydroelectric power that can not only power AI infrastructure but also optimize the use of energy resources (Mustafa et al., 2022). (ii) *smart mobility*, in which AI models can be utilized to forecast traffic patterns, improve public transportation through adaptive scheduling, and optimize traffic routes (Yin et al., 2022). AI can also contribute to autonomous vehicles (AV), which can increase road safety of the passengers and reduce power consumption to promote sustainability(Mirindi et al., 2025). (iii) *sustainable agriculture*, in which AI models are developed to predict crop health and condition, accurately determine the requirements of water, pesticides and fertilizers and enable continuous monitoring of farm fields, thereby minimizing the impact of environmental challenges and enhance resource utilization (Muhie, 2022). (iv) *climate change*, in which AI models are developed to reduce electricity consumption by forecasting demand and supply, while also helping to detect and limit methane leaks in natural gas pipelines and compressor stations(Bolón-Canedo et al., 2024). (v) *environmental policies*, in which AI models can utilize to identify hidden patterns that can assist policymakers to formulate environmental policies that proactively address current and future challenges (Sánchez-Maroño et al., 2022).

3. RECENT INNOVATIONS IN GREEN AI

The adoption of artificial intelligence (AI) in various sectors has drastically transformed the modern world, as its applications can significantly contribute to efficient processing and simplifies the management of complex tasks (Saba et al., 2026). AI systems rely heavily on extensive computational power to develop and train large language models, which contribute to increased environmental impact. Green AI has emerged as a paradigm that aims to leverage its potential to reduce carbon emission from AI systems while maintaining their performance and capabilities (Hooli, 2025). In recent years, many Green AI innovations have been discussed in literature that can help to make sustainable use of AI systems. These innovations are categorized as lower power hardware and edge computing, model optimization techniques, sustainable data centers, and data-driven sustainability practices as mentioned in Table 1. These categorized are also described as under:

Table 1. Impact of Green AI innovations on energy consumption reduction adapted from (Bolón-Canedo et al., 2024; Hooli, 2025)

Green AI Innovation Category	Technology/Method	Impact/Benefit	Measurement Factor
Low Power Hardware and Edge Computing	Energy-efficient architectures	Comparable performance	Significantly less power
	Edge computing	Reduced latency and increased efficiency	Response time and throughput
Model Optimization Techniques	Knowledge distillation	Maintains high accuracy	Dramatically reduced inference energy
	Pruning	Maintains high accuracy	Weight-based metrics
	Quantization	Low power consumption	Reduces model size and computational demands
Sustainable Data Centres	Carbon-aware scheduling	Reduces emissions	No model changes required
	Orchestration architecture	Shits to renewable sources	Peak availability periods
	Renewable-powered data centres	Direct carbon reduction	Solar, wind, and hydroelectric
Data-Driven Sustainability Practice	Transparency in energy metrics	Enables data-driven decisions	Workload optimization

3.1 Low-Power Hardware and Edge Computing

3.1.1 Low-Power Hardware

Graphics Processing Units (GPUs) and Tensor Processing Units (TPUs) are important requirements of AI systems because they are specifically designed to handle the complex calculations required by machine learning algorithms (Hrithika, 2025). However, the production of these units demands substantial resources, as they rely on rare earth metals and other materials. The extraction process of these resources often destroys habitat, contaminates water, and degrades soil. Further, the manufacturing process of hardware components further contributes towards huge amounts of carbon emissions (Matuštík & Kočí, 2021). The production of such hardware components accounts for about 36% of total global emissions, posing a serious threat to environmental sustainability(Y. Chen et al., 2022). In Green AI, energy-efficient model architecture represents a key innovation, enabling the design of neural network that achieves comparable performance to large language models while requiring less computational power (Hooli, 2025).

With the rapid growth of AI systems, these GPUs, TPUs and other hardware components become obsolete within a short period. This leads to a growing environmental concern in the form of electronic waste (e-waste). When e-waste is not disposed properly then it can emit harmful chemicals that endanger the environment as well as human-being (Hrithika, 2025). For mitigating such environmental hazards, life cycle analysis (LCA) is a well-known method used to assess the carbon emissions of entire life cycle of AI hardware. The hardware life cycle has four phases including manufacturing, transport, product use, and recycling (Wu et al., 2022).

Another important hardware concern is related to the parallelization, in which the complex task is divided into smaller sub-tasks and distributed to multiple cores to minimize algorithm runtime (Bolón-Canedo et al., 2024). For tracking and analyzing carbon dioxide emissions, CarbonTracker, an energy consumption calculation tool, is developed that uses atmospheric measurements and statistical in numerous ways (Anthony et al., 2020). This study indicates that increasing the number of cores to 15 can improve the efficiency of algorithmic runtime and reduce the carbon footprint of AI hardware. However, marginal improvements in runtime can lead to large amounts of carbon emissions. This highlights that further improvements are needed to improve in reducing carbon footprint of AI hardware.

3.1.2 Edge Computing

With the advancement of sensing technologies, the demand for Internet of Things (IoT) devices and sensors increased exponentially. These context-aware devices and sensors can detect a user's location and surrounding conditions and convert the collected sensing information into usable data (Ahmed et al., 2019). Such widespread adoption of IoT applications relies on centralized cloud or fog infrastructure to process, store and distribute vast amounts of resources (Jamil et al., 2020). However, this dependence raises significant concerns, including high latency, bandwidth limitations, reduced throughput, centralized vulnerabilities and additional expenses such as data transmission and storage cost (Hua et al., 2023).

The notion of edge computing has attracted widespread attention to solving such cloud-related challenges. The core principle of edge computing is to carry out the processing close to the source where it is generated or stored, rather than transmitting it to a centralized cloud. This approach helps reduce both computation and data transmission delays in IoT devices (Bolón-Canedo et al., 2024). The architecture of edge computing is based on three layers(Hua et al., 2023); *end*, *edge* and *cloud*. This *end* layer initially observes the physical space and collects information from its surroundings. This layer also receives information from *cloud* or *edge* layers to perform the given task. The edge layer resides between *end* and *cloud* layers. This layer contains certain amount of storage and network resources to execute tasks delegated by a *cloud* layer. The *cloud* layer refers to cloud servers that can have powerful storage and computing capabilities.

The use of AI models, particularly deep learning, has been prevalent for solving complex computational problems. However, this growing adoption demands significant computational power and large storage capabilities. Traditionally, cloud computing has provided abundant processing and storage resources, making it a suitable platform for developing and deploying AI-based application (Hua et al., 2023). Nevertheless, such cloud infrastructures raise delays in data processing and data privacy. To address these concerns, the concept of edge computing has emerged. This computing focuses on the shifting of model training and processing to the edge layer rather than relying on centralized cloud infrastructure. This approach not only reduces network bandwidth but also supports efficient and sustainable development of AI systems (Wan et al., 2022).

3.2 Model Optimization Techniques

Current complex AI models demand energy-efficient algorithms that not only achieve high accuracy but also reduce the carbon emissions footprint (Hrithika, 2025). These algorithms are known as green algorithms that can fully optimize energy resources and design sustainable AI models that can be used to resolve environmental issues (Lannelongue et al., 2021). One of the significant approaches of these green algorithms is to design model optimization techniques that can minimize the carbon footprint (Bolón-Canedo et al., 2024). These techniques include knowledge distillation, model pruning and quantization. The description of these techniques is mentioned as below:

3.2.1 Knowledge Distillation

With the increasing demand for training AI models, specifically large-scale deep learning systems, significant energy is consumed due to their complexity and the vast amount of data they process (Hrithika, 2025). This highlights the need to develop energy-efficient algorithms that promote environmental sustainability by reducing overall energy consumption(Mitu & Mitu, 2024). Knowledge distillation is one of the effective optimization techniques that transfers the performance of a larger model ("teacher") to a smaller model ("student"). The smaller model can utilize fewer computational resources without compromising high accuracy, thereby reducing the environmental footprint (Hooli, 2025).

This method has been used in literature in numerous ways. In one of the studies, knowledge distillation is applied to address object detection and image classification tasks, evaluating performance based on accuracy, mean average precision and environmental cost (Rafat et al., 2023). They study estimated the carbon footprints of overall computation process using metrics such as FLOPs (in millions), energy consumption (kWh), and CO_2 emission (grams). The results indicated that smaller model ("student") can reduce carbon emissions by 13 to 18 times compared to the larger model ("teacher"). In other research work, knowledge distillation, combined with a deep supervision approach, is used to minimize the computational cost of HRNet, a convolutional neural network designed to diagnose neurological impairments in premature babies. The results show that smaller model "student" requires only 61.7% of the FLOPs compared to the larger model "teacher", leading to a significant reduction in the carbon footprint of computational resources (Cacciatore et al., 2025). These studies indicated that knowledge distillation is one of the prominent model optimization techniques that can significantly mitigate the environmental impacts of AI applications without compromising its computational performance (Bolón-Canedo et al., 2024).

3.2.2 Model Pruning

The model pruning is another model optimization technique that can be applied to machine learning models, particularly deep neural network with a larger number of parameters (Hrithika, 2025). In this approach, redundant or unnecessary parameters are removed, reducing computational requirements during both training and inference tasks. There are different types of pruning discussed in the literature (Qi et al., 2021) including (i) *unstructured pruning*, in which less important weights are identified and removed from the model (ii) *structured pruning*, in which redundant channels, filters, kernels or layers are removed for reducing computational complexity and energy efficiency (iii) *channel pruning*, in which only those channels are considered for modelling that are necessary for convolutional layers (iv) *filter/kernel pruning*, in which important filters or kernels are selected that contribute minimal impact on model's output and (v) *layer pruning*, in which entire layer can be removed for less computational power, however, this approach is less commonly used because it can degrade model performance.

In the literature related to model pruning, a recent study demonstrated that a better trade-off performance between precision and energy efficiency by combining data pruning method along with AI-like incremental training (Scala et al., 2025). This study proposes an approach called *play it straight*, in which two training phases namely *boot* and *fine-tune* are defined. In *boot* phase, Repeated Sampling of Random Subsets (RS2) *based* deep neural network (DNN) is partially trained on the entire data sets to reduce the number of model optimization steps. In the *fine-tune* phase, small subsets of instances are repeatedly selected based on their importance scores and used to incrementally update the DNN model. This process continues until the desired accuracy is achieved or the maximum energy budget specified by the user is reached. Another study emphasized that model pruning is a compression technique capable of substantially lowering computational demands by eliminating up to 90% of model parameters, while incurring less than 2% drop in accuracy (Hooli, 2025). These studies indicated that this technique has potential to reduce the carbon footprint of AI models by significantly managing computational performance.

3.2.3 Quantization

AI models need significant computational process to process floating-point numbers when aiming for high precision. Quantization is an optimization technique that burdens by converting these floating-point values into fixed-point representations. By lowering the number of bits required for mathematical operations, quantization decreases computational demands, allowing models to run faster and more efficiently even on limited hardware resources(Hrithika, 2025).

SmoothQuant (Xiao et al., 2023) is a post-training optimization technique for large language models that can balance the values in the model so that quantization becomes easier and efficiency, without compromising model's performance. This study highlights that activation is harder to quantize because of few outliers. Instead, it suggests that there is a need to shift difficulty from activation to weights that can be handled easily. In addition, this can speed up the process by 1.5 times and memory saving on PyTorch about 1.96 times (Xiao et al., 2023). Hence, this optimization approach can reduce the significant computational power. In another study, Outlier -Aware Weight Quantization (OWQ) approach is applied to enhance model performance while aiding minimal storage and computational cost (Lee et al., 2024). Furthermore, this approach employs Weak-Column Tuning (WCT) to enable task-specific adjustments while keeping storage requirements low and maintaining performance. These studies suggest that quantization is an effective approach for reducing energy consumption and promoting Green AI practices.

3.3 Sustainable Data Centers

AI models have long been widely used to accelerate task execution. However, they typically require substantial energy consumption, necessitating deployment in data center infrastructures where extensive computational resources are available to support high levels of processing power, storage capacity, and overall systems performance (Dong et al., 2021). Conventional data centers consume large amounts of energy, contributing to increased carbon emissions and negative environmental impacts (Hrithika, 2025). For tackling these challenges, sustainable data centers have been introduced in the literature using various strategies including carbon-aware scheduling, orchestration architecture, and renewable-powered data centers.

3.3.1 Carbon-aware Scheduling

Conventional electrical grids can operate through one-way flow of electricity from generators to consumers; however, they pose challenges for managing energy resources efficiently (Bolón-Canedo et al., 2024). On the other hand, smart grids can support two-way communication that can enable better management of electricity usage. By integrating with AI technology, smart grids can enhance energy efficiency by optimizing demand and supply of electricity, resulting in managing energy waste (Hrithika, 2025). AI-driven optimization can enhance the integration of renewable energy sources dynamically balancing solar and wind power generation in response to electricity demand, particularly for sustainable data centers (Bolón-Canedo et al., 2024).

Carbon-aware scheduling is one of the methods in which AI tasks are performed at that time and location where renewable energy sources are available, thus reducing carbon emissions and reducing the environmental impacts of data centers (Hooli, 2025). The study proposed a carbon-aware scheduling framework that forecasts renewable energy generation for sustainable energy management in smart grids and data centers. The forecasting is a crucial step in producing renewable energy, because its output depends heavily on environmental factors such as solar radiation, wind speed, humidity, temperature, and atmospheric pressure (Tanaka, 2026). In another study, a meta-algorithm based on the FTL (Follow the Leader) strategy that enhances carbon-aware scheduling in computing infrastructures like data centers. The results indicated that this strategy reduces around 14% of average carbon emissions (Bostandoost et al., 2024). These studies emphasize that carbon-aware scheduling is crucial for optimizing AI workloads by effectively utilizing fluctuations in renewable energy availability across different times and location. This will lead to minimizing energy consumption and potentially reducing carbon emissions for sustainable data centers.

3.3.2 Orchestration Architecture

In literature to sustainable data centers, the use of orchestration is one of the approaches where computational workloads are transferred to data centers to take advantage of renewable energy when it is abundantly available (Hooli, 2025). With the integration of AI, data and algorithms can be effectively orchestrated with existing systems to enhance operational efficiency and effectiveness, potentially making a significant positive contribution to environmental sustainability (Madanchian & Taherdoost, 2025). Such orchestration can leverage peak availability periods and shift resource usage toward renewable energy sources, thereby improving environmental impact(Hooli, 2025).

Kubernetes-based edge orchestration is an architecture that enables the automated management of large-scale container deployments, supporting application deployment, scaling, and lifecycle operations while reducing network latency and optimizing bandwidth usage (Bohm & Wirtz, 2022). Kubernetes-based edge orchestration can be classified into three categories: (i) open-source frameworks and solutions that provide orchestration capabilities for edge computing; (ii) solutions that modify and extend Kubernetes for edge orchestration; and (iii) solutions that focus exclusively on edge layers through adaptations in orchestration mechanisms (Bohm & Wirtz, 2022). For energy-efficient orchestration, a study extended Kubernetes-based edge orchestration to develop a framework called Cognitive Decentralized Edge-Cloud Orchestration (CODECO) that considers three types of Greenness-based metrics including node energy, link energy, and network energy (D. Ali & Sofia, 2025). This indicates that orchestration architecture like Kubernetes-based edge orchestration architecture can provide an effective approach to distribute and manage workloads that can reduce carbon footprints by utilizing renewable energy resources during peak hours' time (Hooli, 2025).

3.3.3 Renewable-Powered Data Centers

The rapid advancement of AI has significantly increased energy demand, particularly in the expansion of computational infrastructure such as data centers where AI models are trained and deployed. The energy consumption of AI systems varies depending on several factors, including hardware efficiency and the type of energy source powering these facilities (Mitu & Mitu, 2024). Conventional data centers mainly rely on fossil fuels and utilize energy in abundance. However, renewable-powered energy data centers can consume less energy by leveraging renewable energy sources like wind, hydro power, and solar. Consequently, reduce the carbon emissions of AI systems that contribute significantly to the cause of environmental sustainability (Hrithika, 2025).

Recently, many big tech companies were investing in renewable-powered data centers to build more sustainable AI solutions (Hooli, 2025). For instance, Google has been completely shifted towards renewable energy sources to improve electricity and cooling efficiency through AI optimization. Further, Google utilizes outside air for cooling its servers, thereby reducing energy consumption for their computational tasks (Hrithika, 2025). In contrast, some studies highlighted that certain sustainable data centers can consume power equal to energy usage by 370,000 households. Such energy demands by these data centers raise a concern that how energy providers distribute energy between consumers and data center operators (Monstadt & Saltzman, 2025). These concerns can be addressed through defining clear guidelines and regulations for utilizing renewable energy sources, resulting in the sustainable development of AI operations (Bolón-Canedo et al., 2024).

3.4 Data Driven Sustainability Practice

The use of AI in numerous fields including healthcare, education, agriculture, transportation and many more can play a significant role in providing reliable and accurate information to stakeholders. Such information can assist them in making data-driven decisions, supporting the targeted allocated resources, reducing environmental impacts and energy-efficient resource management (Bolón-Canedo et al., 2024). Green AI supports disclosure of information regarding energy usage of AI solutions. This transparency enables stakeholders to make informed decisions that facilitate environmental sustainability (Mitu & Mitu, 2024). In literature, three frameworks are described that supports transparency in energy usage. Amongst them, measurement-focused framework deals with energy metrics consumed by graphical processing units (GPUs) and central processing units (CPUs) during training phase. Life cycle assessment (LCA) is another framework that focuses on the environmental impact of resources throughout their life cycle while governance-led framework addresses institutional accountability (Oliveira et al., 2026).

In literature, these frameworks are discussed in many studies that facilitate transparency in energy usage for contributing to AI sustainability. For example, in a healthcare study, transparent reporting of carbon emissions of computational power in hospitals can reduce operational cost, better resource allocation and enhanced energy efficiency (Richie et al., 2025). The other study described experiment-impact-tracker framework that can simplify the consistent tracking and reporting of energy usage, carbon emissions and computational resources. This framework improves transparency that can help reduction in carbon footprint of AI operations (Henderson et al., 2020). In addition, one study implemented governance-led framework that can assist stakeholders to make well-informed decisions through system-level carbon emissions visibility and auditing (Raper et al., 2022).

To conclude, current innovations in Green AI have potential to reduce carbon footprint of AI systems that promote AI sustainability. These advancements are energy-efficient architecture, dependence on renewable energy sources, and model optimization techniques that not only promote environmental responsibility but also maintain computational performance of AI systems (Hooli, 2025). However, there are still some challenges that require serious consideration to tackle before promoting Green AI practices.

4. CHALLENGES IN GREENING INTELLIGENT MACHINES

Many advancements have been discussed in the literature including model optimization techniques, low power hardware and edge computing, sustainable data centers, and data driven sustainability practice that can reduce environmental impacts of AI systems without compromising their capabilities and performance (Hooli, 2025). Despite numerous innovations of Green AI initiatives, there are still some challenges including trade-offs between performance and efficiency, scalability issues, data and transparency limitations, and economic and technological barriers. They are mentioned as below:

4.1 Trade-offs Between Performance and Efficiency

The use of AI models has increased significantly for solving complex problems and improving performance; however, these models often demand substantial resources, leading to higher computational costs for training and maintenance (Bolón-Canedo et al., 2024). The substantial energy consumption associated with these AI models can lead to environmental concerns due to increased carbon emissions. The notion of Green AI supports AI usage by reducing operational costs and minimizing carbon footprints of these AI models (Mitu & Mitu, 2024). When designing AI models for computational tasks, there is an inherent trade-off between energy efficiency and performance, making it challenging to achieve an optimal balance between sustainability and high-performance (Paul et al., 2023).

A key concern for researchers and Green AI designers is the development of energy-efficient algorithms and models to reduce power consumption. However, lowering energy usage may lead to some reduction in processing speed or accuracy, potentially resulting in less effective and efficient models (Paul et al., 2023). Furthermore, devices and platforms often operate with different configurations and have limited lifespans, which creates additional challenges for designers. These constraints make it difficult to develop energy-efficient models or algorithms that can be effectively deployed across diverse platforms (Hrithika, 2025). Additionally, due to the rapid pace of technological advancements, many platforms and devices become obsolete within a short period of time, which may hinder the full implementation of Green AI initiatives (Paul et al., 2023).

Apart from these technological challenges, another significant barrier is the implementation of standardized policies that can inform rules and regulations related to sustainable development of AI solutions. Few years back, The European Union's AI Act was developed that enforces responsibility on digital industries to disclose information regarding carbon emissions of AI solutions that can promote sustainable development of AI systems (Ebers et al., 2021). In contrast, many big companies are competing with each other to develop more efficient and accurate AI systems, resulting in enhanced carbon footprint of these computational resources. Consequently, this poses a challenge and also encourages us to adopt Green AI initiative. This can be achieved through an optimal balance between model performance and energy efficiency of these AI solutions (Peykani et al., 2026).

4.2 Scalability Issues

The advancements in AI models on large-scale infrastructure have raised a growing debate on the principles of Green AI and efficient AI (Wu et al., 2022). Efficient AI mainly focuses on computational performance by optimizing cost, speed, throughput and scalability. On the other hand, Green AI addresses these aspects along with the reduction of carbon emissions of AI operations for promoting environmental sustainability (Peykani et al., 2026). One of the approaches for achieving Green AI is to apply model optimizing techniques such as model pruning, knowledge distillation, and quantization to mitigate environmental impacts (Bolón-Canedo et al., 2024). However, there is a challenge to validate these optimization techniques at large scale, still considering scalability as an unsettled issue (Peykani et al., 2026).

As the use of data has grown over the years, the demand for data centers has increased exponentially. Traditional data centers provide huge resources to manage, store, process, and distribute data effectively, transform it into meaningful information to facilitate decision making process (Almalki et al., 2023). The significant need for computational resources in data centers incurs additional financial burden and enhanced carbon footprints of AI operations, posing a serious concern for sustainable AI development (Hrithika, 2025). Further, the limited lifespan of hardware equipment in these data centers often leads to enhanced electronic waste. To tackle such challenges, sustainable data centers are discussed in the literature that aim to enhance energy efficiency through numerous ways including the use of cooling and power management systems, adoption of renewable energy sources, develop sustainable architecture, and utilize energy-efficient hardware resources (Almalki et al., 2023).

The adoption of energy-efficient algorithms and sustainable data centers demand scalable solutions that can significantly reduce carbon emissions of AI operations at a large-scale (Clemm et al., 2024). However, numerous challenges related to scalability exist such as limited renewable energy sources and the integration of these sources for large-scale infrastructure require substantial investment, which presents a significant challenge for implementing sustainable data centers at a large scale (Paul et al., 2023). The hardware efficiency and effective use of model optimization techniques are other barriers to implement at scalable infrastructures (Peykani et al.,

2026). Hence, there is a need to explore more energy-efficient approaches at scale remains an open research area.

4.3 Data and Transparency Limitations

Over the years, the need for data has increased as it is used in different stages of AI modelling such as processing, predicting, managing, analyzing, and storing. However, the development and maintenance of such massive datasets for training and deploying their AI models are crucial because they consume substantial amounts of computational power and costs, resulting in increase in carbon emissions (Bolón-Canedo et al., 2024). In contrast, Green AI supports reduction of carbon footprints of AI models and evaluation of energy-efficient hardware such as data centers or GPUs (Hooli, 2025). Even though many organizations are not willing to measure and disclose carbon emissions of AI operations. The absence of this data makes it a challenge to accurately assess the environmental impact of computational resources, leading to a lack of sustainable infrastructure to support Green AI initiatives (Henderson et al., 2020).

In recent years, various sustainability evaluation metric frameworks have been developed to assess the lifecycle of AI models. These frameworks generally fall into three main categories (Oliveira et al., 2026): measurement-focused, life cycle assessment, and governance-led approaches. Measurement-focused frameworks emphasize computational efficiency and energy consumption, while life cycle assessment frameworks examine factors such as carbon emissions. In contrast, governance-led frameworks operate at the institutional level, where sustainability budgets are allocated beyond consideration of carbon emissions and computational efficiency. In literature, many benchmark studies have been found that indicate the various advantages of using such evaluation metrics (Peykani et al., 2026). For instance, MLPerf (Reddi et al., 2021)is one of the prominent industrial frameworks that provides standards for the measurement of computational efficiency and energy consumption. Similarly, AI Benchmark(Koohestani et al., 2026) evaluates energy consumption and computational efficiency including execution speed and memory utilization. On the other hand, CarbonTracker (Anthony et al., 2020) and CodeCarbon (Liu & Yin, 2024) mainly focus on energy consumption and carbon analysis. A major challenge is that existing benchmarks focus on different dimensions of sustainability assessment. No single framework in the literature comprehensively addresses all relevant aspects of sustainability (Peykani et al., 2026). Consequently, this fragmentation limits the field of Green AI, as multiple benchmarks must be used in combination to achieve a holistic approach to environmental sustainability.

The other significant barrier in implementing Green AI is the lack of transparency in training processes and the absence of standardized procedures and guidelines for reporting carbon emissions of AI models (Peykani et al., 2026). At international level, no comprehensive international guidelines are found in the literature, although initiatives like the European Union's Act encourage the disclosure of carbon emissions consumed by AI operations for transparent implementation of Green AI initiatives (Ebers et al., 2021). Hence, there is a need to establish standardized policy guidelines for reporting carbon footprints of AI operations. In a similar vein, AI models mainly focus on the training phase in the AI lifecycle while ignoring the other critical stages like data preparation, model retrieval, and maintenance for reporting carbon emissions (Peykani et al., 2026). As a result, energy consumption of these models is often misled. On the other hand, energy measurement tools such as CodeCarbon (Liu & Yin, 2024) and CarbonTracker (Anthony et al., 2020) are used in large-scale projects for transparent information regarding carbon emissions of AI systems. However, these tools frequently face challenges due to theiroperating system dependencies, implementation constraints and other relevant computational barriers (Peykani et al., 2026). Another significant hurdle for transparency is institutional accountability, specifically related to incorrect energy usage or high levels of carbon emissions (Oliveira et al., 2026). In literature, many existing studies advocated the significance of transparent disclosure of information regarding carbon emissions of AI operations and strong organizational accountability. For instance, one study emphasizes that organizations should be responsible to regularly disclose and update information about carbon footprints and energy usage of their computational resources (Henderson et al., 2020). Such organizational accountability can play a vital role in advancing sustainable AI development.

4.4 Economic and Technological Barriers

The emergence of AI-driven infrastructures such as data centers demands significantly increased computational resources to process and manage huge amounts of data (Mitu & Mitu, 2024). For sustainable AI solutions, renewable-powered data centers came into existence because of lower energy consumption due to rely on renewable energy sources such as wind, solar, and hydropower (Bolón-Canedo et al., 2024). The big technology companies such as Microsoft and Google shifted towards renewable-powered data centers to support sustainable AI development (Hooli, 2025). For example, Google has completely shifted their data centers on renewable energy sources to increase power and cooling efficiency. In a similar vein, Microsoft has planned to completely transfer their conventional data centers into renewable-powered data centers by 2030 (Hrithika, 2025). This shows that many organizations are taken interest to promote Green AI practices. In contrast, there are growing concerns discussed in the literature about the consumption of energy by these renewable-powered data centers. For instance, a study indicated that such data centers consume energy equal to that used by approximately 370,000 households (Monstadt & Saltzman, 2025). This makes it a challenge for electricity providers to justify electricity distribution between consumers and data center owners. In addition, the transition from conventional data centers to renewable-powered data centers presents financial challenges for small and medium enterprises and even large organizations to continuously improve infrastructure to maintain sustainable energy sources (Hooli, 2025).

Another significant barrier in adopting Green AI is to consider regional factors such as industrial and technological conditions instead of relying only on computational resources. It is important to deliberate local issues such as climate changes or rising pollution levels of a particular region before adopting Green AI practices because the energy consumption of such practices may not contribute to the overall environmental sustainability of that region (Kopka & Grashof, 2022). This raises an important question for organizations and governments, as they need to take a holistic view before investing in Green AI. Further, such huge investments may fail to showcase positive environmental impact to reduce carbon emissions from the surroundings (Mitu & Mitu, 2024). Another challenge to implement Green AI technologies globally, as wealthy countries can invest in billions; however, poorer countries may face various challenges including high costs, limited resources, and different environmental conditions to adopt Green AI practices. In some cases, developed countries shifted the environmental impacts of AI solutions to less developed countries, allowing them to avoid local environmental hazards (Hooli, 2025). This situation may raise a question about the absence of global laws, regulations and policy frameworks. Consequently, Green AI practices without any proper guidelines, international standards and transparent reporting mechanisms become a significant hurdle to its implementation (Kopka & Grashof, 2022).

In order to adopt Green AI practices, it is necessary to develop energy-efficient AI models and optimize hardware resources to ensure environmental sustainability (Clemm et al., 2024). Additionally, sustainable data centers can reduce carbon footprints of AI operations through dependence on renewable energy sources and energy-efficient architectures (Almalki et al., 2023). However, implementing Green AI faces technological challenges due to the short lifespan of hardware infrastructure, such as GPUs, TPUs, data centers and system architectures, which often results in increased electronic waste (Oliveira et al., 2026). Similarly, adopting renewable energy sources for large-scale infrastructure requires significant investment, which poses a major challenge for the implementation of Green AI initiatives (Paul et al., 2023). Therefore, there is a need for researchers and policymakers in the field of Green AI to adopt forward-looking strategies and techniques to tackle these challenges for achieving environmental sustainability.

5. FUTURE DIRECTIONS

Over the years, substantial progress has been made in advancing Green AI initiatives, which continue to play a transformative role in developing environmentally sustainable intelligent machines. The existing body of research highlights several key areas that offer promising opportunities for future innovation, enabling the design of Green AI technologies that minimize environmental impact for future generations. This section presents these future directions including explainable AI, neuromorphic computing, quantum computing and energy-harvesting AI devices, drawing on current trends in the development and adoption of Green AI solutions.

5.1 Explainable AI

A growing future direction in Green AI is model explainability often referred to as explainable AI (XAI), which emphasizes transparency in the decision-making process and is becoming an essential component of technological accountability (Peykani et al., 2026). These AI systems support environmental decision-making by understanding and interpreting relevant information, applying logical reasoning, and clearly explaining to users how these explainable tools can reach a final decision. This may help users not only to understand the judgment of reaching at a solution by AI models but also deliver crucial information that led towards a sustainable environment (Bolón-Canedo et al., 2024). The use of XAI techniques can reduce computational costs by approximately 30% to 50% by eliminating redundant features, applying model optimization strategies, and making simple models (Peykani et al., 2026).

In the literature on Green AI, XAI is applied for different purposes. In this study, XAI is combined with digital corporate social responsibility (CSR) to support consumer protection, employee welfare, and community engagement, which together enhances technological responsibility and promotes environmental sustainability practices in the hotel industry (X. Chen et al., 2026). Another study was conducted to examine the use of XAI in CSR within the financial market. The results indicate that the integration of CSR with XAI can enhance financial performance of companies that can also lead towards sustainable AI practices (Lachuer & Jabeur, 2022). These studies demonstrated that there is a need to integrate Green AI with XAI to promote transparency in decision-making processes and support sustainable AI development.

5.2 Neuromorphic Computing

Another emerging area for Green AI implementation is the development of neuromorphic computers. Neuromorphic computing is an advanced technology inspired by the human brain, known for its energy efficiency when performing complex tasks compared to traditional computing systems (Bolón-Canedo et al., 2024). Neuromorphic computing is based on two key components (Luo et al., 2023): neuromorphic algorithms and neuromorphic hardware. Neuromorphic algorithms, commonly known as Spiking Neural Networks (SNNs), the third generation of artificial neural networks (ANNs), process information by representing "0" as the absence of a spike and "1" as the presence of a spike in neurons. Neuromorphic hardware uses a distributed memory architecture, combined with emerging memory technologies such as non-volatile resistive RAM (RRAM) and magneto-resistive memory (MRAM), to significantly reduce data movement (Luo et al., 2023).

Neuromorphic chips is one of the neuromorphic hardware, are designed for energy-efficient information processing by mimicking the functions of neurons in the human brain, enabling AI-driven tasks such as pattern recognition, sensing, and learning. This results in lower energy consumption of computational resources compared to traditional computer systems (Hrithika, 2025). These chips perform computations in a parallel and distributed manner, enhancing energy efficiency and improving performance by 2 to 5 times compared to traditional GPUs for real-time processing and decision-making (Bolón-Canedo et al., 2024). In another study, a neuromorphic chip called Novena was developed to process large-scale data, including text, images, and bio-signals, which typically require deep learning models, thereby contributing to environmental sustainability (Luo et al., 2023). This study also focuses on designing a neuromorphic processor based on RRAM memory, which enables scalable solutions for reducing environmental impacts. This demonstrates that research and development in neurocomputing are one of the most existing areas to explore for promoting AI sustainability.

5.3 Quantum Computing

There is a paradigm shift from classical computing to quantum computing, a new emerging direction that can support Green AI initiatives. Classical computing depends on binary bits to process data while quantum computing relies on qubits (quantum bits) that can be represented in multiple states at the same time (Hrithika, 2025). These quantum bits compute exponentially faster and reduce computational time because of its parallel nature. This efficiency can contribute to lower computational resources, enabling sustainable AI development (Hussaine & Aldeen , 2026). These qubits remain stable if they operate at near-absolute zero temperatures, thus require highly energy-efficient colling system. Although, the concept of quantum computing is still infancy; however, this computing has a strong prospect as they are much faster than supercomputers and energy-efficient in reducing carbon footprints of AI operations (Esho et al., 2026).

The development of quantum algorithms that works on the principles of quantum mechanics is another promising use of quantum computing. These algorithms can produce optimized AI models while mitigating environmental impacts during training and deployment (Hrithika, 2025). In the literature, some of the well-known quantum algorithms are discussed including Shor's integer factorization algorithm, the Quantum Approximate Optimization Algorithm (QAOA), and Grover's database search algorithm. These algorithms provide computational tasks with reduction of energy consumption in different fields such as cryptography, database search, and AI modelling (Hussaine & Aldeen, 2026). In future, renewable energy sources may be implemented in quantum computing that can further minimize carbon footprints of AI systems. Further, there is a need to establish new frameworks and guidelines that can ensure the responsible use of quantum computing for reducing carbon emissions (Esho et al., 2026). The potentials of quantum computing indicate that this computing may be further explored for developing sustainable AI solutions.

5.4 Energy-Harvesting AI Devices

Another important future avenue is the emergence of energy-harvesting AI devices that can capture and transfer energy from the surroundings to develop energy-efficient intelligent machines (Bolón-Canedo et al., 2024). Energy can be harvested from a wide range of sources, including mechanical, thermal, vibrational, solar, chemical, radio frequency, wind, and acoustic energy (A. Ali et al., 2023). Integrating these energy sources into AI systems can enable them to operate as self-sufficient devices, thereby supporting the advancement of sustainable AI technologies (Bolón-Canedo et al., 2024).

Two primary energy-harvesting technologies including piezoelectric nanogenerators (PENG) and triboelectric nanogenerators (TENG), can generate power from mechanical vibrations present in the surrounding environment (Divya et al., 2023). PENG generates electricity through the piezoelectric effect when mechanical stress is applied to certain materials, whereas TENG produces energy via the triboelectric effect, where two different materials repeatedly come into contact and then separate (A. Ali et al., 2023). Data collected from nanogenerators often comes in different formats when used in smart AI devices such as smartphones and wearables. In the future, these formats could be standardized into a unified structure that AI systems can easily process, leading to reduced energy consumption and faster operation (Divya et al., 2023).

6. CONCLUSION AND LOOKING AHEAD

In conclusion, this chapter presents recent innovations, identifies major challenges, and future directions of greening intelligent machines. The growing interest in the use of artificial intelligence (AI) models, specifically large language models (LLMs) has significantly demonstrated that they perform complex computational tasks with speed and efficiency. However, these models require extensive amounts of computational resources that lead towards significant environmental concerns. For instance, training a single AI model can produce about 626.2 thousand pounds of carbon emissions is nearly equivalent to five times lifetime emissions of an American car.

To address such concerns, the new paradigm Green AI has been adopted in various industries to reduce carbon footprints of AI systems. This can be achieved by applying different approaches such as hardware optimization, model optimization, development of renewable energy infrastructure, and sustainable data management practices. Green AI is mainly divided into two categories; Green-in AI and Green-by AI. In Green-in AI, energy-efficient algorithms are designed throughout the life cycle of AI systems. On the other hand, Green-by AI focuses on the use of AI technologies for solving environmental issues. Both these categories can play a vital role in promoting sustainable practices and providing environmentally viable solutions.

In recent times, Green AI implemented numerous innovative approaches, including low-power hardware and edge computing, model optimization techniques, sustainable data centers, and data-driven sustainability practices. In low-power hardware approach, energy-efficient model architectures are used to minimize power consumption of computational resources. In addition, edge computing utilizing the 'edge layer' of the cloud infrastructure to train and process AI models rather depending on the centralized architecture. This approach facilitates low carbon emissions with improved computational efficiency. Another recent approach is to integrate model optimization techniques such as knowledge distillation, model pruning, and quantization. These techniques have strong foundation to reduce carbon emissions of AI operations. The development of sustainable data centers that depend on renewable energy sources is one of the innovative approaches to handle large-scale computation resources with energy efficiency. This can be achieved through different strategies such as carbon-aware scheduling, use of orchestration architecture, and development of renewable-powered infrastructure. One of the other innovative approaches is to implement data-driven sustainability practices. This approach can mitigate carbon footprints f AI operations by promoting transparency in reporting mechanisms. There are three frameworks discussed in the literature that define structured reporting mechanisms including measurement-focused, life cycle assessments, and governance-led frameworks. These frameworks enable stakeholders to assist them in decision making process through transparent system-level energy usage and carbon emissions of AI operations.

Despite current advancements in Green AI, several challenges are discussed in the literature that obstacle the progress of sustainable AI development. One of the challenges is the trade-off between AI model efficiency and performance. Green AI supports the development of energy-efficient algorithms and systems to produce less carbon emissions. For achieving such efficiency, it may be possible that efficiency can affect processing speed or model accuracy in AI systems that degrade overall performance. Furthermore, configurations may differ according to the platforms used. These variations may pose a challenge for developers to design and implement AI solutions. Another major challenge in promoting Green AI is the limited timespan of hardware resources, resulting in increased electronic waste. In addition, sustainable data centers are dependent on renewable energy sources to enhance energy efficiency for large-scale data processing and management. However, it is another big concern to ensure the availability of renewable energy sources for sustainable data centers due to substantial requirements in infrastructure.

The other challenges are related to limitations in data availability and transparency. Existing literature highlights different benchmarks evaluation tools that mainly focus on one or two aspects of sustainability, such as energy efficiency, computational performance, or carbon emissions. However, no unified benchmark tool is available that can comprehensively cover all aspects of sustainability. Moreover, the lack of transparency in reporting energy metrics at the organizational level can hinder the progress of sustainable development of AI systems. Another significant challenge is the implementation of Green AI practices on the global stage. Underdeveloped countries are struggling to adopt sustainable AI practices due to limited resources, infrastructure, and investment. In contrast, developed countries have relocated their renewable energy data centers to underdeveloped countries to reduce local environmental impacts; however, this practice does not contribute towards environmental sustainability globally. In addition, lack of comprehensive international rules and regulations, policy frameworks and standardized reporting mechanisms are one of the major hurdles to adopting Green AI practices worldwide.

Current research highlights several emerging directions for advancements in green intelligent machines. One of them is the integration of explainable artificial intelligence (XAI) with Green AI that can facilitate stakeholders to make informed decisions by trusting and understanding AI-drive outcomes. Another promising direction is the application of neuromorphic computing, which is inspired by the functioning of human brain. In future, neuromorphic devices such as neuromorphic chips can improve computational performance approximately by five times as compared to traditional GPUs. These advancements can play a significant role in contributing to reducing carbon footprints of AI operations. The use of quantum computing and energy-harvesting AI devices are other future directions that can have potential to limit environmental impacts, resulting in the implementation of sustainable AI practices.

Overall, the implementation of Green AI practices through innovative strategies has great potential to reduce environmental impacts and holds a great promise for shaping sustainable AI development for upcoming generations. However, mutual efforts are required from various stakeholders such as hardware designers, researchers and policymakers for developing scalable and optimized infrastructures, ensuring awareness and transparent disclosure of carbon emissions and establishing policies and guidelines that can promote Green AI initiatives.

REFERENCES

Ahmed, S., Javaid, S., Niazi, M. F., Alam, A., Ahmad, A., Baig, M. A., Khan, H. K., & Ahmed, T. (2019). A qualitative analysis of context-aware ubiquitous learning environments using Bluetooth beacons. *Technology, Pedagogy and Education, 28*(1), 53–71. DOI: 10.1080/1475939X.2018.1557737

Ali, A., Shaukat, H., Bibi, S., Altabey, W. A., Noori, M., & Kouritem, S. A. (2023). Recent progress in energy harvesting systems for wearable technology. *Energy Strategy Reviews, 49*, 101124. DOI: 10.1016/j.esr.2023.101124

Ali, D., & Sofia, R. C. (2025). Experimenting with engergy-awareness in edge-cloud containerized application orchestration. *ArXiv, 2511*(09116), 1–9.

Almalki, F., Alsamhi, S. H., Sahal, R., Hassan, J., Hawbani, A., Rajput, N. S., Saif, A., Morgan, J., & Breslin, J. (2023). Green IoT for Eco-Friendly and Sustainable Smart Cities: Future Directions and Opportunities. *Mobile Networks and Applications, 28*(1), 178–202. DOI: 10.1007/s11036-021-01790-w

Anthony, L. F., Kanding, B., & Selvan, R. (2020). Carbontracker: Tracking and predicting the carbon footprint of training deep learning models. *ArXiv, 2007*(03051), 1–11.

Bashir, F., Ahmed, S., & Marouf, M. (2020). An Exploratory Study of Students' Learning Performance In Flipped Classroom Using Decision Tree and Regression. *2020 3rd International Conference on Computing, Mathematics and Engineering Technologies (ICoMET)*, 1–6. DOI: 10.1109/iCoMET48670.2020.9074083

Bekteshi, L. (2025). Education in the Era of AI and Immersive Technologies: A Systematic Review. *Journal of Research in Engineering and Computer Sciences, 3*(1), 1–14. DOI: 10.63002/jrecs.31.779

Bibikar, S., Vikalo, H., Wang, Z., & Chen, X. (2022). Federated dynamic sparse training: Computing less, communicating less, yet learning better. *In Proceedings of the AAAI Conference on Artificial Intelligence*, 6080–6088. DOI: 10.1609/aaai.v36i6.20555

Bohm, S., & Wirtz, G. (2022). Cloud-Edge orchestration for smart cities: A review of Kubernetes-based orchestration Architectures. *EAI Endorsed Transactions on Smart Cities, 6*(18), 1–19. DOI: 10.4108/eetsc.v6i18.1197

Bolón-Canedo, V., Morán-Fernández, L., Cancela, B., & Alonso-Betanzos, A. (2024). A review of green artificial intelligence: Towards a more sustainable future. *Neurocomputing, 599*, 128096. DOI: 10.1016/j.neucom.2024.128096

Bostandoost, R., Hanafy, W. A., Lechowicz, A., Bashir, N., Shenoy, P., & Hajiesmaili, M. (2024). Data-driven Algorithm Selection for Carbon-Aware Scheduling. *ACM SIGEnergy Energy Informatics Review*, *4*(5), 148–153. DOI: 10.1145/3727200.3727222

Cacciatore, A., Berardini, D., Scaraggi, V., Mancini, A., Moccia, S., & Migliorelli, L. (2025). Online Knowledge Distillation and Deep Supervision in HRNet: Green AI for Preterm Infants' Pose Estimation. *ACM Transactions on Computing for Healthcare*, *6*(4), 1–19. DOI: 10.1145/3757067

Cao, L. (2023). AI in Finance: Challenges, Techniques, and Opportunities. *ACM Computing Surveys*, *55*(3), 1–38. DOI: 10.1145/3502289

Chen, X., Jaffari, A. A., Muzaffar, A., & Rahman, N. (2026). Integrating explainable AI and corporate social responsibility for sustainable practices. *Technological Forecasting and Social Change*, *223*, 124444. DOI: 10.1016/j.techfore.2025.124444

Chen, Y., Yan, W., Zhang, H., Liu, Y., Jiang, Z., & Zhang, X. (2022, August 14). A Data-Driven Design Approach for Carbon Emission Prediction of Machining. *Volume 2: 42nd Computers and Information in Engineering Conference (CIE)*. DOI: 10.1115/DETC2022-90465

Clemm, C., Stobbe, L., Wimalawarne, K., & Druschke, J. (2024). Towards Green AI: Current Status and Future Research. *2024 Electronics Goes Green 2024+ (EGG)*, 1–11. DOI: 10.23919/EGG62010.2024.10631247

Dhar, P. (2020). The carbon impact of artificial intelligence. *Nature Machine Intelligence*, *2*(8), 423–425. DOI: 10.1038/s42256-020-0219-9

Divya, S., Panda, S., Hajra, S., Jeyaraj, R., Paul, A., Park, S. H., Kim, H. J., & Oh, T. H. (2023). Smart data processing for energy harvesting systems using artificial intelligence. *Nano Energy*, *106*, 108084. DOI: 10.1016/j.nanoen.2022.108084

Dong, H., Munir, A., Tout, H., & Ganjali, Y. (2021). Next-Generation Data Center Network Enabled by Machine Learning: Review, Challenges, and Opportunities. *IEEE Access : Practical Innovations, Open Solutions*, *9*, 136459–136475. DOI: 10.1109/ACCESS.2021.3117763

Ebers, M., Hoch, V. R. S., Rosenkranz, F., Ruschemeier, H., & Steinrötter, B. (2021). The European Commission's Proposal for an Artificial Intelligence Act—A Critical Assessment by Members of the Robotics and AI Law Society (RAILS). *J*, *4*(4), 589–603. DOI: 10.3390/j4040043

Esho, E. O., Akinyelu, A. A., & Dinis, M. A. P. (2026). Sustainable generative AI and quantum computing: Review assessment on the environmental impact of generative AI and quantum technologies. *Frontiers in Sustainability*, *7*, 1726832. Advance online publication. DOI: 10.3389/frsus.2026.1726832

Fatima, A., Shaheen, A., Ahmed, S., Fazal, B., Ahmad, F., Liew, T. W., & Ahmed, Z. (2024). Exploring the use of gamification in human-centered agile-based requirements engineering. *Frontiers of Computer Science*, *6*, 1442081. Advance online publication. DOI: 10.3389/fcomp.2024.1442081

Hamid, M., Zeshan, F., Ahmad, A., Munawar, S., Aimeur, E., Ahmed, S., Abu Elsoud, M., & Yousif, M. (2020). An intelligent decision support system for effective handling of IT projects. *Journal of Intelligent & Fuzzy Systems*, *38*(3), 2635–2647. DOI: 10.3233/JIFS-179550

Henderson, P., Hu, J., Romoff, J., Brunskill, E., Jurafsky, D., & Pineau, J. (2020). Towards the systematic reporting of the energy and carbon footprints of machine learning. *Journal of Machine Learning Research*, *21*, 1–43.

Hooli, M. (2025). The Carbon Footprint of AI: Can Smart Tech Be Truly Sustainable? *Sarcouncil Journal of Engineering and Computer Sciences*, *4*(8), 629–635.

Hrithika, L. (2025). SUSTAINABLE AI: STRATEGIES TO REDUCE THE ENVIRONMENTAL FOOTPRINTS OF AI. *International Journal of Advances in Business and Management Research*, *03*(SUP-02), 216–224. DOI: 10.62674/ijabmr.2025.v3isupp2.0021

Hua, H., Li, Y., Wang, T., Dong, N., Li, W., & Cao, J. (2023). Edge Computing with Artificial Intelligence: A Machine Learning Perspective. *ACM Computing Surveys*, *55*(9), 1–35. DOI: 10.1145/3555802

Hussaine, S., & Aldeen , N. (2026). Exploring the Role of Quantum Computing in Advancing Green Computing Technologies. *Journal of Sustainable Smart Systems in Education & Environment*, *1*(01), 56–68. DOI: 10.66823/jsssee.7

Jamil, S., Qureshi, R., & Ahmed, S. (2020). Analyzing Effects of Resource Distribution on Performance of Fog Infrastructure. *2020 Global Conference on Wireless and Optical Technologies (GCWOT)*, 1–5. DOI: 10.1109/GCWOT49901.2020.9391604

Jha, A., Gupta, G., Rau, S., Meshram, P., & Labhane, N. (2022). Green Computing Approach in ICT Components. *Vidyabharati International Interdisciplinary Research Journal*, 564–569.

Jouppi, N. P., & Swing, A. (2023). A Machine Learning Supercomputer with an Optically Reconfigurable Interconnect and Embeddings Support. *2023 IEEE Hot Chips 35 Symposium (HCS)*, 1–24. DOI: 10.1109/HCS59251.2023.10254691

Koohestani, R., de Bekker, P., Koç, B., & Izadi, M. (2026). Benchmarking AI Models in Software Engineering: A Review, Search Tool, and Unified Approach for Elevating Benchmark Quality. *IEEE Transactions on Software Engineering*, *52*(2), 651–674. DOI: 10.1109/TSE.2025.3644183

Kopka, A., & Grashof, N. (2022). Artificial intelligence: Catalyst or barrier on the path to sustainability? *Technological Forecasting and Social Change*, *175*, 121318. DOI: 10.1016/j.techfore.2021.121318

Lachuer, J., & Ben Jabeur, S. (2022). Explainable artificial intelligence modeling for corporate social responsibility and financial performance. *Journal of Asset Management*, *23*(7), 619–630. DOI: 10.1057/s41260-022-00291-z

Lannelongue, L., Grealey, J., & Inouye, M. (2021). Green Algorithms: Quantifying the Carbon Footprint of Computation. *Advanced Science (Weinheim, Baden-Württemberg, Germany)*, *8*(12), 2100707. Advance online publication. DOI: 10.1002/advs.202100707 PMID: 34194954

Lee, C., Jin, J., Kim, T., Kim, H., & Park, E. (2024). OWQ: Outlier-Aware Weight Quantization for Efficient Fine-Tuning and Inference of Large Language Models. *Proceedings of the AAAI Conference on Artificial Intelligence*, *38*(12), 13355–13364. DOI: 10.1609/aaai.v38i12.29237

Li, L., Jamieson, K., DeSalvo, G., Rostamizadeh, A., & Talwalkar, A. (2018). Hyperband: A novel bandit-based approach to hyperparameter optimization. *Journal of Machine Learning Research*, *18*(185), 1–52.

Li, Y., Wen, Y., Tao, D., & Guan, K. (2020). Transforming Cooling Optimization for Green Data Center via Deep Reinforcement Learning. *IEEE Transactions on Cybernetics*, *50*(5), 2002–2013. DOI: 10.1109/TCYB.2019.2927410 PMID: 31352360

Liu, V., & Yin, Y. (2024). Green AI: Exploring carbon footprints, mitigation strategies, and trade offs in large language model training. *Discover Artificial Intelligence*, *4*(1), 49. DOI: 10.1007/s44163-024-00149-w

Luo, T., Wong, W.-F., Goh, R. S. M., Do, A. T., Chen, Z., Li, H., Jiang, W., & Yau, W. (2023). Achieving Green AI with Energy-Efficient Deep Learning Using Neuromorphic Computing. *Communications of the ACM*, *66*(7), 52–57. DOI: 10.1145/3588591

Madanchian, M., & Taherdoost, H. (2025). Ethical theories, governance models, and strategic frameworks for responsible AI adoption and organizational success. *Frontiers in Artificial Intelligence*, *8*, 1619029. Advance online publication. DOI: 10.3389/frai.2025.1619029 PMID: 40741285

Matuštík, J., & Kočí, V. (2021). What is a footprint? A conceptual analysis of environmental footprint indicators. *Journal of Cleaner Production*, *285*, 124833. DOI: 10.1016/j.jclepro.2020.124833

Mirindi, D., Khang, A., & Mirindi, F. (2025). *Artificial Intelligence (AI) and Automation for Driving Green Transportation Systems: A Comprehensive Review* (pp. 1–19). DOI: 10.1007/978-3-031-72617-0_1

Mitu, N. E., & Mitu, G. T. (2024). The hidden cost of AI: Carbon footprint and mitigation strategies. *Revista de Ştiinţe Politice*, *84*, 9–16.

Monstadt, J., & Saltzman, K. (2025). HOW DATA CENTERS HAVE COME TO MATTER : Governing the Spatial and Environmental Footprint of the 'Digital Gateway to Europe.'. *International Journal of Urban and Regional Research*, *49*(4), 757–778. DOI: 10.1111/1468-2427.13316

Muhie, S. H. (2022). Novel approaches and practices to sustainable agriculture. *Journal of Agriculture and Food Research*, *10*, 100446. DOI: 10.1016/j.jafr.2022.100446

Mustafa, A., Kazmi, M., Khan, H. R., Qazi, S. A., & Lodi, S. H. (2022). Towards a Carbon Neutral and Sustainable Campus: Case Study of NED University of Engineering and Technology. *Sustainability (Basel)*, *14*(2), 794. DOI: 10.3390/su14020794

Nikolinakos, N. Th. (2023). *A European Approach to Excellence and Trust: The 2020 White Paper on Artificial Intelligence* (pp. 211–280). DOI: 10.1007/978-3-031-27953-9_5

Oliveira, A. P., Carraquico, T., & Martinez-Perez, C. (2026). Beyond Efficiency: A Systematic Review of Energy Consumption and Carbon Footprint Across the AI Lifecycle. *Sustainability (Basel)*, *18*(3), 1359. DOI: 10.3390/su18031359

Patterson, D., Gonzalez, J., Le, Q., Liang, C., Munguia, L., Rothchild, D., So, D., Texier, M., & Dean, J. (2021). Carbon emissions and large neural network training. *ArXiv*, *2104*(10350), 1–22.

Paul, S. G., Saha, A., Arefin, M. S., Bhuiyan, T., Biswas, A. A., Reza, A. W., Alotaibi, N. M., Alyami, S. A., & Moni, M. A. (2023). A Comprehensive Review of Green Computing: Past, Present, and Future Research. *IEEE Access : Practical Innovations, Open Solutions*, *11*, 87445–87494. DOI: 10.1109/ACCESS.2023.3304332

Peykani, P., Emrouznejad, A., Ghanidel, S., Javadi-Sisi, I., & Mirjalili, S. (2026). Green Artificial Intelligence: A Comprehensive Review of Metrics, Tools, Challenges, Trends, and Future Prospects. *Archives of Computational Methods in Engineering*. Advance online publication. DOI: 10.1007/s11831-026-10546-2

Qi, C., Shen, S., Li, R., Zhao, Z., Liu, Q., Liang, J., & Zhang, H. (2021). An efficient pruning scheme of deep neural networks for Internet of Things applications. *EURASIP Journal on Advances in Signal Processing*, *2021*(1), 31. DOI: 10.1186/s13634-021-00744-4

Rafat, K., Islam, S., Mahfug, A., Al, , & Hossain, M. (2023). Mitigating carbon footprint for knowledge distillation based deep learning model compression. *PLoS One*, *18*(5), e0285668. DOI: 10.1371/journal.pone.0285668 PMID: 37186614

Raper, R., Boeddinghaus, J., Coeckelbergh, M., Gross, W., Campigotto, P., & Lincoln, C. N. (2022). Sustainability Budgets: A Practical Management and Governance Method for Achieving Goal 13 of the Sustainable Development Goals for AI Development. *Sustainability (Basel)*, *14*(7), 4019. DOI: 10.3390/su14074019

Reddi, V. J., Cheng, C., Kanter, D., Mattson, P., Schmuelling, G., & Wu, C.-J. (2021). The Vision Behind MLPerf: Understanding AI Inference Performance. *IEEE Micro*, *41*(3), 10–18. DOI: 10.1109/MM.2021.3066343

Richie, C., Hinrichs-Krapels, S., Dobbe, R., French, P., Wei, J. C. J., Diehl, J. C., & Kong, R. (2025). Environmental impacts of artificial intelligence in health care: Considerations and recommendations. *Health and Technology*, *15*(6), 1087–1093. DOI: 10.1007/s12553-025-01003-4

Saba, G., Ahmed, S., Sania, H., Khan, T. A., & Mohamad Nasir, H. B. (2026). An empirical comparison of AI assisted software refactoring tools. *Scientific Reports*, *16*(1), 20971. Advance online publication. DOI: 10.1038/s41598-026-49590-0 PMID: 42032054

Sánchez-Maroño, N., Rodríguez-Arias, A., Dumitru, A., Lema-Blanco, I., Guijarro-Berdiñas, B., & Alonso-Betanzos, A. (2022). How Agent-based modeling can help to foster sustainability projects. *Procedia Computer Science*, *207*, 2546–2555. DOI: 10.1016/j.procs.2022.09.313

Scala, F., Flesca, S., & Pontieri, L. (2025). *Play it Straight: An Intelligent Data Pruning Technique for Green-AI* (pp. 69–85). DOI: 10.1007/978-3-031-78977-9_5

Schwartz, R., Dodge, J., Smith, N. A., & Etzioni, O. (2020). Green AI. *Communications of the ACM*, *63*(12), 54–63. DOI: 10.1145/3381831

Tanaka, K. (2026). Carbon-aware energy scheduling for renewable-dominant smart grid infrastructure. *Natural Journal of Renewable Energy Systems and Innovation, 2*(3), 43–50.

Wan, Y., Qu, Y., Gao, L., & Xiang, Y. (2022). Privacy-preserving blockchain-enabled federated learning for B5G-Driven edge computing. *Computer Networks, 204*, 108671. DOI: 10.1016/j.comnet.2021.108671

Wu, C., Raghavendra, R., Gupta, U., Acun, B., Ardalani, N., Maeng, K., Chang, G., Aga, F., Huang, J., Bai, C., Gschwind, M., Gupta, A., Ott, M., Melnikov, A., Candido, S., Brooks, D., Chauhan, G., Lee, B., Lee, H., & Hazelwood, K. (2022). Sustainable AI: Environmental implications, challenges and opportunities. *Proceedings of the Machine Learning Systems*, 1–19.

Xiao, G., Lin, J., Seznec, M., Wu, H., Demouth, J., & Han, S. (2023). SmoothQuant: Accurate and efficient post-training quantization for large language models. *Proceedings of Machine Learning Research, 202*, 38087–38099.

Yin, X., Wu, G., Wei, J., Shen, Y., Qi, H., & Yin, B. (2022). Deep Learning on Traffic Prediction: Methods, Analysis, and Future Directions. *IEEE Transactions on Intelligent Transportation Systems, 23*(6), 4927–4943. DOI: 10.1109/TITS.2021.3054840

Zhao, L., Liu, S., Xin, T., Tan, J., Wang, X., Li, Y., Bian, Z., Chen, Y., Kong, F., Bian, J., Qian, C., & Zhang, Z. (2026). AI agent in healthcare: Applications, evaluations, and future directions. *Npj Artificial Intelligence, 2*(1), 31. DOI: 10.1038/s44387-026-00076-4

Chapter 9
Digital Pollution:
Examining the Impact on Environmental Awareness

Fatma Alkan
http://orcid.org/0000-0003-2784-875X
Hacettepe University, Turkey

Eylül Özülüş
http://orcid.org/0009-0003-9943-4254
Hacettepe University, Turkey

Ayşem Seda Yücel
http://orcid.org/0000-0002-7654-582X
Hacettepe University, Turkey

ABSTRACT

Digital pollution refers to the energy consumption and environmental impacts caused by the excessive and unnecessary use of digital devices. Its sources include unnecessary data storage, high-resolution content, spam emails, messages, and unused applications. These practices increase energy consumption in digital devices and infrastructure. Since this energy is largely derived from fossil fuels, it leads to carbon emissions and an increased carbon footprint. Reducing the environmental impact of digital pollution is possible through awareness and conscious behavior. This study aimed to determine teacher candidates' awareness of digital pollution. Data were collected from 256 teacher candidates using a 10-item "Digital Pollution Knowledge Survey" developed by the researchers. The data were analyzed using the Descriptive Survey Model. Results showed that teacher candidates have a low level of awareness regarding digital pollution. The study contributes to the field by identifying awareness levels and offering practical suggestions to enhance them.

DOI: 10.4018/979-8-3373-8490-0.ch009

INTRODUCTION

In the current era of Industry 4.0, or the digital age, technological developments are transforming many aspects of our daily lives. The impact of digitalization can be seen in various sectors, including education, healthcare, law, government, banking, and public services. Digital technologies such as the Internet of Things, artificial intelligence, big data, cloud computing, social networks, blockchain, and virtual and augmented reality enable the processing, storage, and transmission of data in digital environments. While digitalization offers positive aspects such as easier access to information, increased efficiency and speed, and enhanced communication, it also brings negative consequences such as digital pollution, security risks, privacy issues, and information pollution (Peters, 2016; Pereira & Romero, 2017; Schwab, 2016).

Digital Pollution: Conceptual Definition and Environmental Impacts

Digital pollution, a result of digitalization, harms the environment similarly to how fossil fuel use, household, industrial, and medical waste release into nature, and chemical waste from factories contribute to environmental pollution. Digital pollution arises from excessive and unnecessary use of digital devices, leading to energy consumption and environmental impacts. The excessive consumption of data, data sharing, storage, and the widespread use of digital currencies increase energy consumption and cause environmental issues. Everyday devices such as phones, computers, and tablets contribute to digital pollution through unnecessary data storage, high-resolution content, large photo and video sharing, unnecessary emails, messages, constantly open background tabs, unnecessary notifications, unused digital apps, and files (Oo et al., 2023; Obringer et al., 2021). With the rapid advancement of technology, digital devices quickly lose popularity. The disposal of old electronic devices without recycling results in the creation of electronic waste (e-waste), which is another factor contributing to digital pollution (Das, 2024; Perkins et al., 2014).

Using digital technologies requires high amounts of electricity, and today, the majority of electricity is generated from fossil fuels such as coal, oil, and natural gas. The use of fossil fuels has numerous negative impacts on both the environment and human health. Fossil fuels used to meet the electricity demands of digital technologies result in carbon emissions. Carbon emissions, or the release of carbon dioxide (CO_2) into the atmosphere, generally occur due to the burning of fossil fuels like coal, oil, and natural gas. The released CO_2, along with other greenhouse gases such as methane (CH_4), nitrous oxide (N_2O), hydrofluorocarbons (HFCs), sulfur hexafluoride (SF_6), and perfluorocarbons (PFCs), traps heat in the atmosphere. These greenhouse gases (GHGs) create the greenhouse effect by reflecting heat back to the

Earth. Excessive emission of greenhouse gases leads to the accumulation of heat, resulting in global warming (Jeffry et al., 2021; Ritchie et al., 2020). In other words, carbon emissions resulting from the use of digital devices and technologies increase greenhouse gases, thereby contributing to global warming. To protect nature and ecosystems, the emission of greenhouse gases must be prevented, and necessary measures should be taken to reduce carbon emissions.

The digital waste and pollution generated by the use of digital devices and technologies also contribute to an increase in the carbon footprint. The carbon footprint measures the extent of harm to the environment and nature, expressed as the amount of carbon dioxide (CO_2) emitted. It helps us understand our impact on nature and serves as an indicator for tackling environmental issues such as global warming, the greenhouse effect, and climate change (Xu et al., 2015; Pandey et al., 2011).

Technologies and Digital Activities Causing Digital Pollution

Today, digital platforms such as YouTube, Facebook, Netflix, and TikTok, which are frequently used for entertainment purposes, provide high-resolution content and require large infrastructures, resulting in digital pollution (Batmunkh, 2022). Every search we make on the browser, every series and movie we watch on Netflix, and every video we watch on YouTube contributes to the emission of carbon dioxide (CO_2) into the atmosphere, thereby increasing our carbon footprint (Naeem, 2023).

In a study by Obringer et al. (2021), they calculated that 1 GB of internet usage results in the emission of between 28 to 63 grams of CO_2. Additionally, the researchers calculated that watching one hour of high-quality video streaming requires 441 grams of CO_2, and they noted that reducing the video quality from high to standard significantly reduces the amount of CO_2 emissions. According to The Shift Project's 2019 report, watching one hour of high-quality video consumes 0.519 kg CO_2/kWh and results in the emission of 280 grams of CO_2.

In his study, Batmunkh (2022) analyzed the carbon footprints of popular social media platforms like YouTube, Netflix, Facebook, and TikTok using four different calculation methods: Obringer's Method, The Shift Project Method, Andrae's Method, and Hintemann's Method. Due to the different assumptions in these methods—such as energy consumption sources, system components included (network infrastructure, servers, device usage, etc.), and data transfer rates—the results varied.

Table 1. Carbon footprint of the most popular applications by watching hours

Activity	User Engagement		Carbon Footprint (CO_2) per hour			
User number	**Watch hours per day**	**Uploaded data**	**Obringer**	**Shift Project method**	**Andrae method**	**Hintemann and Hinterholze**
YouTube	1 B hours	1.65 GB/HD/	168 g	280.26 g	72 g	135 g
Netflix	6 B hours	1.5 GB/hour/ HD/	1008 g	1681.56 g	432 g	810 g
Facebook	0.1 B hours	210 TB	16.8 g	28.026 g	7.2 g	13.5 g
Tiktok	2 B hours	500 TB	336 g	560.52 g	144 g	270 g

Source: (Batmunkh, 2022)

Based on these data, it can be concluded that Netflix, due to its high viewing times, is the platform that causes the highest carbon emissions per hour. Digital platforms such as TikTok and YouTube also contribute significantly to carbon emissions (Batmunkh, 2022; Oo et al., 2023).

Email and messaging, which we frequently use in daily life, also contribute to digital pollution, just like social media platforms. Berners-Lee (2010) calculated that a spam email produces 0.3g of carbon dioxide (CO_2), a regular email produces 4g of CO_2, and an email with attachments or photos generates 50g of CO_2.

Today, cryptocurrency mining, especially Bitcoin (BTC) mining, is also one of the factors contributing to digital pollution. Cryptocurrency mining requires significant computational power, which leads to high energy consumption. This energy is mostly derived from fossil fuels rather than renewable sources, resulting in carbon dioxide (CO_2) emissions and an increased carbon footprint. In a 2019 study by Köhler and Pizzol, the carbon emissions from Bitcoin mining were calculated to be 17.29 $MtCO_2$(e). In a 2022 study by Vries, Gallersdörfer, Klaaßen, and Stoll, the carbon emissions from Bitcoin mining were found to be 65.4 $MtCO_2$(e) (CBECI, 2022). Additionally, Bitcoin mining accelerates the obsolescence of digital devices, contributing to the creation of electronic waste (e-waste) and further digital pollution (Chamanara et al., 2023; BBC, 2021). Although various initiatives have emerged to reduce the environmental harm caused by Bitcoin mining, the rapid development of blockchain technology and the widespread use of digital currencies present challenges to implementing these initiatives (Arfelli et al., 2024; Jiang et al., 2021).

Artificial intelligence technologies, which are widely used in many fields, also contribute to digital pollution. The training of AI models requires the use of powerful computers that consume large amounts of energy. A portion of this required energy is derived from fossil fuels, and the use of these fuels leads to carbon emissions, thereby contributing to digital pollution (Barnett-Itzhaki & Tsoury, 2025; Chung et al., 2024; Strubell, Ganesh, & McCallum, 2020). In addition, not only the training phase but also the operational use of AI systems requires continuous energy consumption. In a study conducted by Tomlinson et al. (2024), the carbon emissions per query of AI systems were estimated to be approximately 2.2 g CO2 for ChatGPT and 1.6 g CO2 for BLOOM. Although these values may seem relatively low on an individual basis, the continuous and intensive use of these systems by millions of users leads to a rapid increase in total emissions (Erdem, 2024). In this context, without disregarding the benefits provided by AI technologies, developing solutions to improve energy efficiency and promoting conscious usage habits are of great importance in reducing digital pollution.

The COVID-19 pandemic, which emerged in China in late 2019, accelerated digitalization and increased the use of digital technologies. During this period, the use of digital devices became more widespread across many sectors, particularly education (Diaz-Garcia et al., 2022; Márquez-Ramos, 2021). Schools and universities had to shift to online education, and the use of Massive Open Online Courses (MOOCs) became more common. Teachers were expected to increase their digital competence. Here, they were expected to spend more time with digital tools to increase technological pedagogical content knowledge (Sağlam Kaya, 2019), they were encouraged to use web tools more in their lessons (Alkan and Mustafaoğlu, 2023; Altundağ and Koçer, 2023), and sample lessons were planned for the use of multimedia tools, including science lessons (Altundağ-Koçak and Elmas, 2024). The use of high-resolution video lessons, digital assessments, interactive content, and the storage of educational materials during the teaching process increased energy consumption. Additionally, social media use surged, digital content platforms became more popular, online shopping increased, online meetings were held in many sectors, and new applications like health-tracking apps became part of daily life. Studies show that online content consumption doubled during this period (DoubleVerify, 2020). The increased use of digital devices and technologies during this time further contributed to the rise in digital pollution.

There are several measures that can be taken to reduce digital pollution, which has many negative effects on the environment. The energy efficiency of the digital devices we frequently use in daily life can be improved, and energy consumption can be reduced by ensuring that digital devices are turned off when not in use. Instead of fossil fuels, which account for the majority of today's energy needs and have serious negative impacts on the environment and nature due to their CO_2 emissions, renewable energy sources that are continuously replenished and do not harm the environment, such as solar energy, wind energy, hydroelectric energy, geothermal energy, and biomass energy, can be used. Additionally, to reduce digital pollution, individuals using digital technologies and the internet should avoid unnecessary data storage, regularly clean their email inboxes, avoid sending unnecessary emails, refrain from always using videos in high resolution, adjust video resolution, delete unused digital applications and files, and download videos for offline viewing instead of streaming them (Das, 2024).

Digital Pollution Awareness and the Role of Teachers

It is important to raise awareness about digital pollution in society. For all these behaviors to be practiced to reduce digital pollution, individuals must be aware of the issue. Awareness is the state of an individual being conscious of their own thoughts, feelings, and the events occurring in their environment (Lopes et al., 2018; Winfrey & Jung, 2022). In a study by Nhi, Chuloy, and Glomann (2024), they explored the impact of social media platforms on digital pollution from the user's perspective. The study found that users were unaware of the environmental harm caused by social media platforms, video streaming, and digital devices and technologies.

It is crucial for individuals to realize the effects of digital pollution, as a high level of awareness can reduce the factors that harm the environment and nature. Individuals who are aware of digital pollution are more conscious of the environmental effects caused by digital technologies and the energy waste that results from excessive and unnecessary use of digital technologies. Awareness of digital pollution among individuals is one of the fundamental conditions for ensuring environmental sustainability. Being aware of this issue leads individuals to act more consciously while using digital devices and technologies, choose technologies that consume less energy, and reduce the negative environmental effects of digital technologies. Individuals who are aware of digital pollution can reduce their carbon footprint by adopting digital habits that decrease the environmental harm caused by digital technologies, using the internet and digital technologies more efficiently, and avoiding unnecessary energy consumption (Wathuge, Sedera, 2021). Furthermore, individuals' awareness of digital pollution can create a sense of social responsibility. Increased societal awareness of digital pollution can lead to the adoption of new policies and regulations to ensure the sustainability of digital technologies, which can help foster a digital culture that supports environmental sustainability.

In the digital age we live in, it is extremely important that teachers, who play a key role in shaping society, have a high level of awareness regarding digital pollution. By transferring their knowledge about digital pollution to students, teachers can make significant contributions to raising conscious generations with high levels of awareness on this issue. When considered within the context of environmental education, digital pollution is not only a technological problem but also a matter directly related to the environment and sustainability. Therefore, it is of great importance for teachers to systematically convey the environmental impacts of digital pollution, its causes, and its relationship with individual and social behaviors to students. One of the main objectives of environmental education is to develop environmental sensitivity and a sense of responsibility in individuals (Chen, Shahbaz, & Haq, 2025; Singh, 2012; Yadav et al., 2022).

At this point, awareness constitutes the first and most critical step in behavior change. Individuals with higher levels of awareness not only acquire knowledge but also tend to reflect this knowledge in their daily lives by making more conscious choices (Blakemore & Frith, 2003; Haberman, Lev, & Langley, 2003). Students who develop awareness and knowledge about digital pollution can recognize the environmental harm caused by digital technologies and may change their usage habits to reduce these impacts. They may adopt more sustainable digital lifestyles by preferring energy-efficient devices, reducing energy consumption by turning off unnecessary devices, extending the lifespan of digital tools to minimize electronic waste (e-waste), avoiding unnecessary data storage, and limiting digital media usage. In this sense, teachers' high level of awareness about digital pollution increases social responsibility and contributes to the development of environmental consciousness.

In this context, teachers can support students in developing critical thinking skills and sustainable lifestyle habits by discussing the sources of digital pollution, the behaviors that lead to it, and the necessary preventive measures. In the process of transforming awareness into behavior, activities that promote active student participation, real-life problem connections, and practical applications play a crucial role. Thus, students not only become aware of digital pollution but also learn to take responsibility as environmentally conscious individuals (Pe'er, Goldman, & Yavetz, 2007).

Based on all these considerations, this study aims to determine the levels of digital pollution awareness among teachers, particularly teacher candidates, who play a critical role in fostering awareness in individuals. In this respect, the study goes beyond merely identifying the current situation and contributes to the field by offering practical recommendations to enhance awareness of digital pollution. Focusing on teacher candidates is particularly important in terms of raising awareness at an early stage among individuals who will influence large segments of society in the future. Indeed, teachers play a key role in facilitating behavioral change at the societal level by transferring their knowledge and awareness to students. In this regard, the study is significant in that it draws attention to digital pollution as a contemporary and increasingly important environmental issue, reveals the awareness levels of teacher candidates on this topic, and provides recommendations for educational processes, thereby contributing both to the literature and to practice.

METHOD

The research model is designed as a descriptive survey model. Survey studies are conducted to describe, compare, categorize, and interpret a group or an event (Cohen, Manion & Morrison, 2007). The research follows the survey method. In the survey model, the existing situation, whether past or present, is identified as it is (Karasar, 2013). The survey method is used to define the characteristics of a group. In this context, data is collected from a sample to reveal the attitudes, views, and descriptive characteristics of a group (Creswell, 2012; Fraenkel, Wallen, & Hyun, 2012, pp. 393). The aim of this study is to determine the perceptions of university students regarding digital pollution.

STUDY GROUP

The sample of the study consists of 275 teacher candidates studying at the Faculty of Education of a state university in Turkey. During the data collection process, 19 teacher candidates were excluded from the study because they did not wish to continue with the survey. As a result, the final sample group consists of 256 teacher candidates. Since this study is a survey method study, a convenience sampling approach was preferred for determining the sampling method. This approach was chosen to reach a group that is relevant to the research problem and easily accessible (Fraenkel and Wallen, 2012). Table 2 summarizes some demographic characteristics of the sample group.

Table 2. Demographic characteristics of the sample group

Categories		F	%
Gender	Female	169	66
	Male	87	34
Age	18-20	71	27,7
	21-22	129	50,4
	23-24	45	17,6
	25 and above	11	4,3

continued on following page

Table 2. Continued

Categories		F	%
Programme	Chemistry Teaching	57	22,3
	Foreign Language Teaching	38	14,8
	Mathematics Teaching	30	11,7
	Primary School Teaching	25	9,8
	Biology Teaching	22	8,6
	Science Teaching	19	7,4
	Computer and Instructional Technologies Education	14	5,5
	Preschool Teaching	13	5,1
	Turkish Language Teaching	11	4,3
	Special Education Teaching	11	4,3
	Physics Teaching	10	3,9
	Physical Education and Sports Teaching	6	2,3
Grade	1. class	55	21,5
	2. class	72	28,1
	3. class	92	35,9
	4. class	37	14,5
Total		256	100

DATA COLLECTION TOOLS

The data for the research was collected using the "Digital Pollution Knowledge Survey," developed by the researchers. Digital platforms are environments where almost every moment of life is shared, including who you are with, where you are, past activities, future plans, and many other actions. The "Digital Pollution Knowledge Survey" was developed to determine whether teacher candidates had heard of the concept of digital pollution and whether they had opinions on whether digital environments lead to pollution. The survey consists of 10 questions. While developing the survey, the first question asked whether the concepts under study had been heard before, and the participants were asked to explain these concepts. The survey is divided into three sections. In the first section, the concept of digital pollution is examined. In the second section, social networks are investigated. The third section focuses on emails.

In this framework, in the first section of the survey, teacher candidates were asked, "Have you heard of the concept of digital pollution?" The following question about digital pollution is "What is digital pollution? Please explain."

The second section of the survey deals with social networks. Here, the pollution caused by social networks was examined. The question "Does sharing something on social media (Facebook, Instagram, Twitter, etc.) create pollution?" was asked, followed by "How is pollution created? Please explain."

The last section of the survey deals with emails. Questions related to email addresses, email checking, etc., were included. These questions are as follows: "How many email addresses do you have?", "Do you regularly check your emails?", "How many emails are in your inbox?", "Do you regularly clean your inbox?", "Is there a relationship between cleaning emails and pollution?", "Please explain your answer."

Table 3. Digital pollution knowledge survey

Digital pollution section				
"Have you heard of the concept of digital pollution?"	Yes		No	
"What is digital pollution? Please explain."	The question was answered with a written answer			
Social networks pollution section				
"Does sharing something on social media (Facebook, Instagram, Twitter, etc.) create pollution?"	Yes		No	
"How does pollution occur? Please explain."	The question was answered with a written answer			
e-mail pollution section				
How many email addresses do you have?	1	2	3	4 and above
Do you check your emails regularly?	Yes		No	
How many emails do you have in your inbox?	0-99	100-499	500-999	1000 and above
Do you regularly clean your emails?	Yes		No	
Is there a relationship between cleaning emails and pollution?	Yes		No	
Please explain your answer.	The question was answered with a written answer			

APPLICATION PROCESS

The Digital Pollution Knowledge Survey developed for the research was created using Google Forms. The form was shared with the teacher candidates who participated in the study. Teacher candidates were given 30 minutes to complete the form. It was clearly stated that no personal information would be requested, and if any was required, the candidates had the option to abandon the form.

DATA ANALYSES

The data for the research was collected through the Digital Pollution Knowledge Survey form completed by teacher candidates. The data obtained from the form was analyzed descriptively. Descriptive analysis was conducted for "Yes-No" type questions in the form. For the questions with written responses (marked as "Please explain"), the answers were first categorized as "True-False" and then analyzed.

Validity and Reliability

Content Validity

Content validity refers to the extent to which the items in a measurement instrument represent the subject area intended to be measured (Lawshe, 1975). The questionnaire used in this research consists of closed-ended questions answered with "yes/no" to elicit students' opinions on specific situations, and open-ended questions to justify these answers. Since the instrument is not a scale, statistical analyses for construct validity were not used; instead, the focus was on ensuring content validity. Accordingly, the questionnaire items were created based on the relevant literature and submitted to the opinions of field experts. In line with expert evaluations, the scope, representativeness, and comprehensibility of the items were reviewed, and the questionnaire was finalized after necessary adjustments. In addition, a pilot study was conducted with 50 prospective teachers before the data collection process. In the pilot study, the applicability of the questionnaire was tested; the participants' level of understanding of the questions was evaluated, and necessary linguistic and structural adjustments were made in line with the feedback received.

Reliability

The reliability of the questionnaire used in the study was addressed in accordance with the structure of the data collection instrument. Classical internal consistency coefficients were not calculated for closed-ended (yes/no) questions, as they directly aim to reveal a specific situation. Instead, the consistency and comprehensibility of the items were observed through expert opinion and a pilot study process, and necessary improvements were made. In the analysis of open-ended questions, the coding process was carried out independently by multiple researchers to ensure reliability, and inter-coder agreement was considered (Miles & Huberman, 1994). In line with these processes, the data obtained from the questionnaire were deemed consistent and reliable.

FINDINGS

The findings section presents the descriptive analysis of the data obtained from the Digital Pollution Knowledge Survey tool. The survey consists of three sections. When reporting the analysis results in the findings section, the results are presented in the following order: first, the results of the "Digital pollution section," followed by the results of the "Social networks pollution section," and finally, the results of the "Email pollution section."

Findings related to the Digital Pollution Section

The Digital Pollution section contains two questions. In the first question, teacher candidates were asked whether they had heard of the concept of digital pollution (Q1), and they answered "Yes" or "No." The results of this question are presented in Table 4.

Table 4. Descriptive statistics for digital pollution section question 1.

Digital pollution section		
Q1. Have you heard of the concept of digital pollution?	**F**	**%**
Yes	193	75.4
No	63	24.6
Total	256	100

When Table 4 is examined, it is observed that the vast majority of teacher candidates, 75.4% (f:193), had heard of the concept of "digital pollution," while 24.6% (f:63) had not. The fact that more than half of the sample group had previously heard of the concept suggests that they are likely aware of what digital pollution is and what can be done to prevent it.

The concept of digital pollution was analyzed using a chi-square test for question Q1, examining gender differences. The chi-square results showed that 65.8% (f:127) of women and 34.2% (f:66) of men were already aware of digital pollution. Conversely, 66.7% of women and 33.3% of men were unaware of it.

The second question of the Digital Pollution section, Q2, is "What is digital pollution? Please explain." Those who correctly explained digital pollution were categorized as "True," while those who provided an incorrect explanation or wrote nothing were categorized as "False." Descriptive statistics related to this are provided in the first section of the table.

In the Q1 question, 75.4% of teacher candidates had heard of the concept of digital pollution. To determine how many of the teacher candidates who had or had not heard of the concept provided a correct or incorrect explanation of digital pollution, a descriptive analysis was also conducted. The results of this analysis are provided in the second section of Table 5.

Table 5. Descriptive statistics for digital pollution section question 2.

Digital pollution section					
Q2. What is digital pollution? Please explain.	F		%		
True	97		37.9		
False	159		62.1		
Total	256		100		
	True	%	False	%	
Q1. Yes	76	78.4	117	73.6	193
Q1. No	21	21.6	42	26.4	63
Total	97	100	159	100	256

When Table 5 is examined, it is seen that 62.1% of teacher candidates (f:159) either did not know or incorrectly explained the concept of "digital pollution," while 37.9% (f:97) knew or explained it correctly. In Table 4, under Digital Pollution Section Q1, the majority of students had previously heard of the concept. The rates at which students, whether they had or had not heard of digital pollution before, explained what it is correctly are examined in the second part of the table.

The gender analysis of question Q2 was conducted using chi-square analysis. According to the analysis results, 58.1% (f:54) of women correctly identified digital pollution, while 41.9% (f:39) of men correctly identified it.

Upon examining the second section of Table 5, it is observed that the majority of teacher candidates (f:117) who had heard of the concept of digital pollution explained it incorrectly, while 76 teacher candidates explained it correctly. Among the teacher candidates who had not heard of the concept, 42 explained it incorrectly, while 21 explained it correctly.

Findings related to Social Networks Pollution Section

The Social Networks Pollution section consists of two questions. In the first question, teacher candidates answered "Yes-No" as to whether sharing something on social media (Facebook, Instagram, Twitter, etc.) creates pollution (Q3). The results for this question are presented in Table 6.

Table 6. Descriptive statistics for social networks pollution section question 3.

Social networks pollution section		
Q3. Does sharing something on social media (Facebook, Instagram, Twitter, etc.) create pollution?	**F**	**%**
Yes	188	73.4
No	68	26.6
Total	256	100

When Table 6 is examined, it is seen that 73.4% of teacher candidates (f:188) stated that sharing something on social media (Facebook, Instagram, Twitter, etc.) creates pollution, while 26.6% (f:68) expressed that it does not create pollution. According to the chi-square test for gender variation of question Q3, 70.6% of women (f:113) and 29.4% of men (f:47) believe that sharing things on social media leads to crime.

The second question of the Social Networks Pollution section is "How does pollution occur? Please explain." (Q4). Those who correctly explained how pollution occurs were categorized as "True," while those who gave incorrect explanations or did not write anything were categorized as "False." Descriptive statistics for this are presented in the first part of Table 7.

In question Q3, 73.4% of teacher candidates knew that sharing something on social media (Facebook, Instagram, Twitter, etc.) creates pollution. To determine how many of those who knew or did not know this also correctly or incorrectly explained how pollution occurs, a descriptive analysis was conducted. The results of the analysis are presented in the second part of Table 7.

Table 7. Descriptive statistics for social networks pollution section question 4.

Social networks pollution section					
Q4. How does pollution occur? Please explain.	F		%		
True	46		18		
False	210		82		
Total	256		100		
	True	%	False	%	
Q3. Yes	44	95.7	144	68.5	188
Q3. No	2	4.3	66	31.5	68
Total	46	100	210	100	256

When Table 7 is examined, it is observed that 82% of teacher candidates (f:210) did not know or gave an incorrect explanation about how pollution occurs, while 18% (f:46) knew or gave the correct explanation. In the Social Networks Pollution Section Q3, the majority of teacher candidates had already heard about the pollution caused by social media. The rates at which teacher candidates who were aware or unaware of the pollution caused by social media correctly explained how pollution occurs are given in the second part of the table. When the second part is examined, it is seen that the majority of teacher candidates (f:144) who had heard about the pollution caused by social media incorrectly explained how pollution occurs, while (f:44) teacher candidates gave the correct explanation. Among those who had not heard about the pollution caused by social media, (f:66) teacher candidates gave an incorrect explanation, and (f:2) gave the correct explanation.

Findings related to email pollution section

The email pollution section consists of six questions (Q5-Q10). Descriptive statistics for all the questions are presented in Table 8.

Table 8. Descriptive statistics for the email pollution section

email pollution section			
Q5.	**How many email addresses do you have?**	**F**	**%**
	1	113	44.1
	2	87	34
	3	31	12.1
	4 and above	25	9.8
	Total	256	100
Q6.	**Do you check your emails regularly?**	**F**	**%**
	Yes	172	67.2
	No	84	32.8
	Total	256	100
Q7.	**How many emails do you have in your inbox?**	**F**	**%**
	0-99	144	56.3
	100-499	56	21.9
	500-999	21	8.2
	1000 and above	35	13.7
	Total	256	100
Q8.	**Do you regularly clean your emails?**	**F**	**%**
	Yes	109	42.6
	No	147	57.4
	Total	256	100
Q9.	**Is there a relationship between cleaning emails and pollution?**	**F**	**%**
	Yes	175	68.4
	No	81	31.6
	Total	256	100
Q10.	**Please explain your answer.**	**F**	**%**
	True	66	25.8
	False	190	74.2
	Total	256	100

When Table 8 is examined, it is observed that among the teacher candidates participating in the study (Q5), 44.1% (f:113) have one email address, 34% have two email addresses, 12.1% have three email addresses, and 9.8% have four or more email addresses.

Q6 investigates whether emails are regularly checked. Here, 67.2% of teacher candidates indicated that they regularly check their emails, while 32.8% stated that they do not. While 69.2% of women check their emails regularly, only 30.8% of men do so.

Q7 asked how many emails are in the inbox. It was found that 56.3% of teacher candidates have between 0-99 emails in their inbox, 21.9% have between 100-499, 8.2% have between 500-999, and 13.7% have 1000 or more emails.

Q8 is about whether emails are regularly cleaned. Among the participants, 57.4% stated that they do not clean their emails, while 42.6% said they clean them regularly.

Q9 examines the relationship between cleaning emails and the occurrence of pollution. Here, 68.4% of teacher candidates answered yes, while 31.6% answered no.

Q10 deals with explaining the relationship between email cleaning and pollution. Those who correctly explained the relationship were categorized as "True," while those who gave incorrect answers or left it blank were categorized as "False." It was found that 74.2% of teacher candidates did not know or gave an incorrect explanation of the relationship between email cleaning and pollution, while 25.8% were able to explain the relationship correctly.

CONCLUSION AND DISCUSSION

The findings indicate that teacher candidates' overall awareness and knowledge of digital pollution remain limited. According to the Digital Pollution Section Q1 data, 75.4% of participants had previously heard of the concept; however, responses to Q2 reveal that only 37.9% were able to explain it correctly. This suggests that although the majority of teacher candidates are familiar with the term, there are significant gaps in accurately understanding and explaining it. Examining the gender distribution, female candidates correctly defined the concept of digital pollution at a higher rate (58.1%) compared to male candidates (41.9%). Data from the Social Networks Pollution Section show that while most teacher candidates (73.4%) recognize that sharing content on social media creates pollution, only 18% were able to accurately explain how this pollution occurs in response to Q4. This highlights a notable gap between awareness and conceptual understanding.

Findings related to email pollution show a similar pattern. While 67.2% of participants reported regularly checking their emails, only 42.6% indicated that they regularly clean them. Moreover, just 25.8% were able to correctly explain the relationship between email cleaning and digital pollution. These results suggest that teacher candidates' awareness in daily digital habits does not fully align with their level of knowledge. In other words, possessing knowledge alone does not guarantee behavioral change; individuals must actively develop conscious practices and take responsibility for managing digital resources.

These results demonstrate that teacher candidates require both conceptual knowledge and behavioral awareness regarding digital pollution. It is therefore crucial that educational programs are designed not only to teach the concept of digital pollution but also to equip future teachers with strategies to integrate this knowledge into daily life practices. In this way, teachers can improve their own digital behaviors and play a leading role in fostering environmental responsibility and sustainable digital living habits among their students.

With technological advancements continuing to be based on digital foundations, significant progress has been made in many different fields, ranging from science to healthcare. The digital environment encompasses a wide range of systems, from devices such as desktop computers, laptops, tablets, and smartphones, to servers that connect to websites and cloud systems. This environment also includes wireless information sharing systems. Digital waste, on the other hand, refers to a pile of unused or forgotten digital data that unnecessarily occupies storage space in these environments. These data may include emails, various documents, texts, photos, graphics, songs, videos, animations, and more. Particularly, this digital waste accumulates in personal accounts and cloud environments, consuming energy as it is stored over time.

While digitalization has brought numerous conveniences and innovations, it has also introduced some problems. One of these is the concept of digital pollution. For example, every email sent accumulates as digital waste on servers worldwide and increases over time. However, this issue has not been fully calculated and remains an unnoticed problem for most people. The increased use of electronic devices and other technological tools, driven by the impact of digitalization, has a negative effect on global warming and climate change. The use of these devices leads to greenhouse gas emissions, contributing to atmospheric pollution, thus introducing a new concept: "carbon footprint." In particular, carbon emissions from electronic devices used for heating and communication lead to environmental damage (Gökçek, Bozdağ, & Demirbağ, 2019; Sharma, Dash, 2023).

In recent years, environmental scientists have focused on the concept of "footprint," which is considered a tool to measure how much natural resources humans use and conserve. This measurement is generally examined in three main dimensions: carbon footprint, ecological footprint, and water footprint. In connection with digital pollution, the carbon footprint has become an important concept in terms of sustainability. The carbon footprint relates to how much nature is used and preserved, and in this context, digital pollution is also brought to the forefront. The digital waste that arises as a result of digitalization is a concrete example of digital pollution. Digital pollution is both a problem and a unique concept that contains many unknowns, with efforts to find solutions to this issue. Therefore, the investigation of digital pollution, the creation of awareness, and the implementation of serious measures is a matter that needs to be addressed.

Digital pollution is a concept that highlights the negative impacts of the digital process, alongside the opportunities it brings. It addresses problems such as information pollution resulting from the unnecessary and excessive production and consumption of digital content, environmental impacts, and mental load. With the rapid development of digital technologies, digital pollution has shifted from being an individual issue to a global problem (Hilty & Aebischer, 2015).

To understand the environmental effects of digital technologies, it is important to remember that internet searches are only the beginning of the process. The internet, while seemingly an abstract concept, actually relies on concrete components such as cables, servers, data centers, and routers. The production and operation of these devices require large amounts of electricity. Activities such as watching videos, sending emails, cloud storage, using apps, or browsing social media have a significant negative impact on the environment, as most of the electricity used globally still comes from fossil fuels (Oo et al., 2023). The use of digital technologies has important implications for environmental sustainability. The large increase in the number of information and communication technology devices, as well as the energy required to operate infrastructure such as data centers, servers, and power and cooling equipment, leads to substantial energy consumption (Berkhout & Hertin, 2004; Kunkel & Matthess, 2020). The development of digital systems also means more electrification and, consequently, increased energy demand. When this high energy demand is not met by renewable energy sources, it is impossible to talk about the sustainability of energy resources (Bonilla et al., 2018; Hilty & Aebischer, 2015). The use of renewable energy has become an important focus for many companies in order to reduce environmental damage. For example, large technology companies such as Microsoft and Apple prioritize the use of renewable energy. While Apple uses renewable energy in all of its data centers, Microsoft plans to transition to 100% renewable energy in its data centers by the end of 2025 (Capgemini, 2021). High energy consumption also leads to significant carbon emissions (Wen et al., 2021).

Furthermore, even small actions we perform every day on our phones and computers contribute to CO_2 emissions. For instance, online video streaming alone generates 30 million tons of CO_2 emissions annually. The digital tools we use every day—such as the internet, smartphones, tablets, video streaming, and emails—pollute the environment in unseen ways. As devices shrink and more internal components are added, more waste is produced during manufacturing and in the environment. Modern smartphones contain an average of 54 components, compared to just 10 components in mobile phones from the 1960s. The methods of separating components and the need for more components pose significant challenges. Additionally, large-scale internet infrastructure often goes unnoticed. Data centers, servers, underwater fiber-optic cables, relay antennas, Wi-Fi boxes, and many other components work together to carry out every online transaction (Oo et al., 2023).

In conclusion, every email sent can spread this information worldwide. Moreover, the ever-growing "digital divide" between technologically advanced leading countries and less developed economies perpetuates the use of outdated, environmentally damaging, and resource-consuming technological infrastructure. Thus, all emerging technologies contribute to this digital pollution. This problem is as serious and real as urban air pollution. In fact, in 2021, people were invited to perform a "digital cleanup," and March 20 was declared as the World Digital Cleanup Day (Yanarateş, 2022).

There are two main sources of digital pollution: the operation of internet infrastructure and the production of computer hardware. The internet consumes large amounts of electricity. Although the internet has become entirely digital, it still requires infrastructure to function. This infrastructure—including cables, antennas, and data centers—powers the internet and cloud systems, while also physically polluting our planet. If the internet were a country, it would rank as the sixth-largest polluter in the world. Additionally, some elements used in the production of digital technologies are rare in nature and very difficult to recycle (e.g., lithium, rhenium, and terbium used in hardware production) (Kunkel & Matthess, 2020; Bonilla et al., 2018). As digitalization increases, the demand for these elements also rises. From an environmental perspective, ensuring the sustainability of these elements appears to be a significant challenge (Bonilla et al., 2018).

The production of information and communication technology hardware leads to environmental harm, such as greenhouse gas emissions and waste generation. For instance, semiconductors used in hardware release acid fumes, doping gases, and volatile organic compounds, which contribute to greenhouse gas emissions. Solvents, silicones, and metals used in hardware production also lead to water pollution and the creation of dirty waste (Berkhout & Hertin, 2004).

Recognizing the scale of this issue is the first step to addressing it.

Digital pollution can be grouped into four main dimensions:

1. **Electronic Waste (E-Waste)**: The rapid obsolescence of technological devices and the constant release of new products into the market have led to a rapid increase in e-waste. The rare minerals used in the production of computers, phones, televisions, and other electronic devices can cause significant environmental harm when these devices reach the end of their life. When not properly recycled, e-waste causes the release of chemicals and heavy metals into the environment.
2. **Energy Consumption of Data Centers**: Today, digital data is stored in cloud systems and data centers, which consume vast amounts of electricity. As the

energy needs of data centers grow, so do their environmental impacts. Much of this energy is supplied by electricity generated from fossil fuels, which contributes to greenhouse gas emissions.

3. **The Proliferation and Degradation of Digital Content**: As the internet grows rapidly, digital content also increases at a fast pace. Every post, video, and photo on social media platforms leaves a digital footprint, resulting in massive data piles. However, the excessive production of unnecessary content leads to significant congestion in data storage systems and increases the environmental burden. The concept of "digital trash" draws attention to the fact that not only waste but also meaningless or outdated content contributes to the growing pollution in the digital world.
4. **Digital Communication and Excessive Emails**: More people are communicating via digital platforms every day, with email being particularly overused and occupying storage spaces. This indirectly leads to energy consumption and higher carbon emissions. Additionally, digital advertisements, the growing volume of data traffic, and the monitoring of online activities are all factors that contribute to the increase in digital pollution.

When summarizing the dimensions of digital pollution, it becomes clear that there are various impacts associated with it.

Effects of Digital Pollution

1. **Environmental Impacts**: The environmental effects of digital pollution are significant. Improper processing of electronic waste leads to soil and water pollution. Furthermore, the energy consumption of data centers, relying on fossil fuels, may contribute to climate change.
2. **Health Impacts**: Constant interaction with digital devices can have negative effects on human health. The radiation emitted by electronic devices can disrupt sleep patterns and harm eye health. Additionally, excessive use of digital devices can lead to mental health issues such as stress and anxiety.
3. **Economic Costs**: Digital pollution also presents an economic burden for companies and governments. Managing and recycling e-waste requires considerable resources. Furthermore, the costs associated with unnecessary data storage and management can put organizations in difficult situations.

Measures to Prevent Digital Pollution

To prevent digital pollution, the following steps should be taken:

1. **Digital Cleanup**: The first step in reducing digital pollution is performing digital cleanup at both personal and organizational levels. Deleting unnecessary emails, archiving or deleting old files, and reducing digital clutter will help reduce the burden on the digital world. Deleting unnecessary content from social media accounts also helps clean up digital footprints (Das, 2024).
2. **Sustainable Management of Data Centers**: Equipping data centers with energy-efficient devices, using renewable energy sources, and implementing efficient cooling systems are essential steps to prevent digital pollution. Companies and organizations should ensure that their digital infrastructure is environmentally friendly (Hilty & Aebischer, 2015).
3. **Recycling of Electronic Waste**: Proper recycling of e-waste helps reduce the negative environmental impacts. Governments should offer incentives for e-waste recycling and raise public awareness on the issue (Lepawsky, 2018).
4. **Education and Awareness**: Raising general awareness about digital pollution is crucial. Schools, workplaces, and social media platforms should offer education on digital pollution, encouraging individuals to adopt more sustainable digital habits (Apaydın, 2024). Given all these impacts, it is crucial for individuals to start developing awareness of digital pollution. It is well known that the invisibility of digital pollution means that people generally have very little knowledge about it. Therefore, it has become necessary to instill this awareness in students at educational institutions. In a study conducted to reveal the awareness of teacher candidates regarding this issue, the results of a 10-question survey consisting of three sections showed that the level of awareness regarding digital pollution was very low. Even those teacher candidates who thought they were knowledgeable about digital pollution actually had insufficient knowledge. Similarly, those who tried to make logical explanations about digital pollution without any prior knowledge were still able to make some inferences about the concept. As a result, it was determined that teacher candidates have a low level of awareness. According to a study reported by Yanarates (2022), Biçer (2022) conducted research in Turkey regarding digital environment and digital waste with teacher candidates and found that their perceptions of these concepts were insufficient. Nearly half of the teacher candidates participating in the study stated that teachers, who are responsible for raising conscious individuals in society, did not have sufficient knowledge about digital concepts. The study also highlighted that there were conceptual gaps regarding the digital environment and digital waste in the literature review. Therefore, it is suggested that digital environment and digital waste-related courses should be added to educational programs to increase individuals' awareness on the matter (Roczen, Kaiser, Bogner, & Wilson, 2014).

Digital pollution has become one of the significant environmental and social problems of the modern age. With the rapid development of technology, the content, data, and devices created in the digital realm are having visible effects on the environment. In this context, it is extremely important for teacher candidates to gain awareness about digital pollution, as it will help them regulate their own digital behaviors and instill a sense of digital responsibility in their students. A study conducted to understand the concept of digital pollution, along with pollution related to social networks and emails, discussed how teacher candidates should approach this responsibility.

1. The Concept of Digital Pollution and the Role of Teacher Candidates:

Digital pollution can be defined as the environmental, energy, and health impacts resulting from the excessive, outdated, or unnecessary production and sharing of data in the digital realm. It is crucial to act consciously in the digital world to reduce or eliminate this pollution. Understanding this concept will help teacher candidates become more responsible digital citizens and impart this awareness to their students.

Developing awareness of digital pollution allows teacher candidates to review their own digital habits. Teacher candidates, who frequently use digital tools and platforms in teaching processes, must find more sustainable ways to create an eco-friendly digital world. Being aware of digital pollution not only protects the environment but also improves students' digital literacy and helps raise responsible digital citizens (Mustafaoğlu & Alkan, 2025).

2. Social Networks and Digital Pollution:

Social media is one of the most significant areas where digital pollution is most intense. The constant production of content, sharing, unnecessary interactions, and data storage increase the negative environmental effects of social networks. Teacher candidates must adopt an approach that reduces digital pollution both personally and professionally while using social networks.

What Should Be Done?

- **Prevent Information Pollution:** Teacher candidates should ensure that the information they share on social networks is accurate and reliable. Misinformation, unnecessary posts, or fake news can increase digital pollution.

- **Reduce Unnecessary Content Sharing:** Teacher candidates can reduce the load of unnecessary information and content in the digital world by sharing only important and meaningful content on their social media accounts.
- **Conduct Social Media Clean-up:** Teacher candidates should also encourage their students to clean up unnecessary content from their social media accounts, delete old posts, and store only the necessary information.

These behaviors help prevent the excessive storage of data and the production of unnecessary content in the digital world, thus reducing digital pollution.

3. Email and Digital Pollution:

Email is one of the most commonly used communication tools in the digital world. However, the sending of unnecessary emails, spam messages, and the need for massive data storage are significant factors that contribute to digital pollution. Teacher candidates need to develop awareness about email usage and play a critical role in reducing digital pollution.

What Should Be Done?

- **Organize Email Inboxes:** Teacher candidates should delete unnecessary emails from their inboxes and only keep important and necessary ones. This reduces the unnecessary energy consumption of email servers and ensures efficient use of storage space.
- **Responsibility in Email Sending:** Teacher candidates should communicate only when necessary and with the correct recipients. Sending unnecessary mass emails can increase digital pollution.
- **Be Cautious in Sharing Large Files:** Teacher candidates should only share large files when truly necessary and prefer efficient data-sharing tools like cloud systems.

Organizing email inboxes and avoiding unnecessary email traffic significantly reduces digital pollution.

4. Strategies to Increase Teacher Candidates' Awareness of Digital Pollution:

Instilling awareness of digital pollution in teacher candidates not only helps them improve their own digital habits but also enables them to set an example for their students. The following strategies are suggested for teacher candidates:

- **Organize Workshops and Seminars:** Conduct workshops or seminars explaining digital pollution, its effects, and possible solutions. This can be effective in raising awareness.
- **Develop Projects on Digital Pollution:** Teacher candidates can develop projects on digital pollution to inform their students about the environmental impacts of the digital world.
- **Provide Training on Digital Responsibilities:** While providing digital literacy education, teacher candidates should also incorporate digital responsibilities into their curricula and teach students behaviors that reduce digital pollution.

CONCLUSION

The findings of this study reveal that teacher candidates' awareness and knowledge levels regarding digital pollution are still limited; particularly, although many had previously heard of the concept, they exhibited deficiencies in accurately explaining it and applying it to their daily digital habits. For instance, only 37.9% of the participants could correctly define digital pollution, while the proportion of those who could accurately explain the pollution caused by social media sharing and the environmental impacts of email management were as low as 18% and 25.8%, respectively. This indicates a significant gap between conceptual understanding and behavioral awareness among teacher candidates. In this context, the primary contribution of the study lies in quantitatively revealing the current level of digital pollution awareness and offering concrete suggestions to enhance this awareness in terms of both individual digital habits and educational practices. By developing a sense of responsibility regarding digital pollution, teacher candidates can lead both themselves and their students toward more sustainable, environmentally friendly, and conscious digital behaviors. In this regard, the study provides a significant contribution to both the literature and educational practice by highlighting the need to increase digital pollution awareness and transform it into responsible digital behavior.

REFERENCES

Alkan, F., & Mustafaoğlu, F. M. (2023). Using web tools in lecture: Example of micro teaching lesson. In *Fostering pedagogy through micro and adaptive learning in higher education: Trends, tools, and applications* (pp. 261–286). IGI Global Scientific Publishing. DOI: 10.4018/978-1-6684-8656-6.ch012

Altundağ, C. K., & Elmas, R. (2024). Multimedia material for science and technology teaching: Concept maps enriched with digital technologies. In *Preconceptions of policies, strategies, and challenges in education 5.0* (pp. 100–117). IGI Global Scientific Publishing. DOI: 10.4018/979-8-3693-3041-8.ch007

Altundağ, C. K., & Koçer, M. Y. (2023). Web 2.0 tools supported innovative applications in science education based on the context-based learning approach. In *Designing context-rich learning by extending reality* (pp. 179–204). IGI Global Scientific Publishing. DOI: 10.4018/978-1-6684-7644-4.ch010

Andrae, A. S. (2019). Prediction studies of electricity use of global computing in 2030. *International Journal of Scientific & Engineering Investigation*, *8*(86), 27–33.

Andrae, A. S., & Edler, T. (2015). On global electricity usage of communication technology: Trends to 2030. *Challenges*, *6*(1), 117–157. DOI: 10.3390/challe6010117

Apaydın, N. (2024). Raising awareness on digital pollution: Individual and organizational productivity. *Studies in Social Science Insights*, *4*(3), 88–107. DOI: 10.53035/SOSSCI.99

Arfelli, F., Coralli, I., Cespi, D., Ciacci, L., Fabbri, D., Passarini, F., & Spada, L. (2024). The environmental stake of Bitcoin mining: Present and future challenges. *Applied Sciences (Basel, Switzerland)*, *14*(20), 9597. Advance online publication. DOI: 10.3390/app14209597

Barnett-Itzhaki, Z., & Tsoury, A. (2025). From awareness to action: A UK-based study on public perceptions of digital pollution. *Sustainability (Basel)*, *17*(17), 7839. Advance online publication. DOI: 10.3390/su17177839

Batmunkh, A. (2022). Carbon footprint of the most popular social media platforms. *Sustainability (Basel)*, *14*(4), 2195. Advance online publication. DOI: 10.3390/su14042195

BBC News. (2021). US leads Bitcoin mining as China ban takes effect. https://www.bbc.com/news/technology-58896545

Berkhout, F., & Hertin, J. (2004). De-materialising and re-materialising: Digital technologies and the environment. *Futures, 36*(8), 903–920. DOI: 10.1016/j.futures.2004.01.003

Berners-Lee, M. (2010). *How bad are bananas? The carbon footprint of everything*. Greystone Books.

Biçer, M. (2022). *Öğretmen adaylarının dijital çevre ve dijital atık kavramlarıyla ilgili algılarının incelenmesi* [Doctoral dissertation, Kastamonu Üniversitesi].

Blakemore, S. J., & Frith, C. (2003). Self-awareness and action. *Current Opinion in Neurobiology, 13*(2), 219–224. DOI: 10.1016/S0959-4388(03)00043-6 PMID: 12744977

Bonilla, S. H., Silva, H. R. O., Terra da Silva, M., Franco Gonçalves, R., & Sacomano, J. B. (2018). Industry 4.0 and sustainability implications: A scenario-based analysis of the impacts and challenges. *Sustainability (Basel), 10*(10), 3740. Advance online publication. DOI: 10.3390/su10103740

Capgemini. (2021). *Sustainable IT*. https://www.capgemini.com/de-de/wp-content/uploads/sites/5/2021/05/Sustainable_IT_Report_2021_Capgemini.pdf

Chamanara, S., Ghaffarizadeh, S. A., & Madani, K. (2023). The environmental footprint of Bitcoin mining across the globe: Call for urgent action. *Earth's Future, 11*(10), Article e2023EF003871. https://doi.org/DOI: 10.1029/2023EF003871

Chen, C., Shahbaz, P., & Haq, S. U. (2025). Transforming students' green behavior through environmental education: The impact of institutional practices and policies. *Frontiers in Psychology, 15*, 1499781. Advance online publication. DOI: 10.3389/fpsyg.2024.1499781 PMID: 39881698

Chung, J.-W., Gu, Y., Jang, I., Meng, L., Bansal, N., & Chowdhury, M. (2024). Reducing energy bloat in large model training. In *Proceedings of the ACM SIGOPS 30th Symposium on Operating Systems Principles* (pp. 144–159). https://doi.org/DOI: 10.1145/3694715.3695970

Cohen, L., Manion, L., & Morrison, K. (2007). *Research methods in education* (6th ed.). Routledge. DOI: 10.4324/9780203029053

Creswell, J. W. (2012). *Educational research: Planning, conducting, and evaluating quantitative and qualitative research* (4th ed.). Pearson.

Das, B. (2024). Digital pollution: A warning to our planet. *Asian Journal of Management, 15*(2), 211–216. DOI: 10.52711/2321-5763.2024.00033

De Vries, A., Gallersdörfer, U., Klaaßen, L., & Stoll, C. (2022). Revisiting Bitcoin's carbon footprint. *Joule*, *6*(3), 498–502. DOI: 10.1016/j.joule.2022.02.005

Díaz-García, V., Montero-Navarro, A., Rodríguez-Sánchez, J.-L., & Gallego-Losada, R. (2022). Digitalization and digital transformation in higher education: A bibliometric analysis. *Frontiers in Psychology*, *13*, 1081595. Advance online publication. DOI: 10.3389/fpsyg.2022.1081595 PMID: 36533011

DoubleVerify. (2020). Global online content consumption doubles in 2020, research shows. https://doubleverify.com/newsroom/global-online-contentconsumption-doubles-in-2020-research-shows/

Erdem, D. (2024). *Synchribble: Designing to increase digital carbon footprint awareness* [Unpublished manuscript].

Ferreboeuf, H., Berthoud, F., Bihouix, P., Fabre, P., Kaplan, D., & Lefèvre, L. (2019). *Lean ICT: Towards digital sobriety*. The Shift Project. https://theshiftproject.org/en/article/lean-ict-our-new-report/

Fraenkel, J. R., Wallen, N. E., & Hyun, H. H. (2012). *How to design and evaluate research in education* (8th ed.). McGraw-Hill.

Gökçek, M., Bozdağ, E., & Demirbag, H. (2019). Carbon footprint and digitalization: Environmental impacts. *Journal of Environmental Engineering*, *3*(1), 78–92.

Haberman, B., Lev, E., & Langley, D. (2003). Action research as a tool for promoting teacher awareness of students' conceptual understanding. *SIGCSE Bulletin*, *35*(3), 144–148. DOI: 10.1145/961290.961552

Hilty, L. M., & Aebischer, B. (Eds.). (2015). *ICT innovations for sustainability* (Vol. 310). Springer., DOI: 10.1007/978-3-319-09228-7_1

Hintemann, R., & Hinterholzer, S. (2019). Energy consumption of data centers worldwide. In *Proceedings of the 6th International Conference on ICT for Sustainability (ICT4S)*.

Jeffry, L., Ong, M. Y., Nomanbhay, S., Mofijur, M., Mubashir, M., & Show, P. L. (2021). Greenhouse gases utilization: A review. *Fuel*, *301*, 121017. Advance online publication. DOI: 10.1016/j.fuel.2021.121017

Jiang, S., Li, Y., Lu, Q., Hong, Y., Guan, D., Xiong, Y., & Wang, S. (2021). Policy assessments for the carbon emission flows and sustainability of Bitcoin blockchain operation in China. *Nature Communications*, *12*(1), 1938. Advance online publication. DOI: 10.1038/s41467-021-22256-3 PMID: 33824331

Karasar, N. (2013). *Bilimsel araştırma yöntemi* (25th ed.). Nobel Yayın Dağıtım.

Köhler, S., & Pizzol, M. (2019). Life cycle assessment of Bitcoin mining. *Environmental Science & Technology, 53*(23), 13598–13606. DOI: 10.1021/acs.est.9b05687 PMID: 31746188

Kunkel, S., & Matthess, M. (2020). Digital transformation and environmental sustainability in industry: Putting expectations in Asian and African policies into perspective. *Environmental Science & Policy, 112*, 318–329. DOI: 10.1016/j.envsci.2020.06.022

Lawshe, C. H. (1975). A quantitative approach to content validity. *Personnel Psychology, 28*(4), 563–575. DOI: 10.1111/j.1744-6570.1975.tb01393.x

Lepawsky, J. (2018). *Reassembling rubbish: Worlding electronic waste*. MIT Press. DOI: 10.7551/mitpress/11111.001.0001

Lopes, R. P., Mesquita, C., de la Cruz del Río-Rama, M., & Álvarez-García, J. (2018). Developing sustainability awareness in higher education. In *Strategies and best practices in social innovation* (pp. 131–152). Springer. DOI: 10.1007/978-3-319-89857-5_9

Marks, L. U., Makonin, S., Rodriguez-Silva, A., & Przedpełski, R. (2021). *Tackling the carbon footprint of streaming media* [SSHRC Knowledge Synthesis Grant Final Report].

Márquez-Ramos, L. (2021). Does digitalization in higher education help to bridge the gap between academia and industry? An application to COVID-19. *Industry and Higher Education, 35*(6), 630–637. DOI: 10.1177/0950422221989190

Miles, M. B., & Huberman, A. M. (1994). *Qualitative data analysis: An expanded sourcebook* (2nd ed.). Sage.

Mustafaoğlu, F. M., & Alkan, F. (2025). Prediction of middle school students' recycling behaviors with machine learning algorithms. *Science Education International, 36*(2), 209–218. DOI: 10.33828/sei.v36.i2.8

Naeem, R., Bajwa, A. M., Sattar, H., & Naeem, B. (2023). Effect of social media usage on digital carbon footprint: Analyzing awareness level of social media users of Punjab. https://doi.org/DOI: 10.21203/rs.3.rs-3225346/v1

Nhi, D. T. T., Chuloy, M., & Glomann, L. (2024). Environmental impact of video streaming from users' perspectives. *Human Side of Services Engineering, 143*.

Obringer, R., Rachunok, B., Maia-Silva, D., Arbabzadeh, M., Nateghi, R., & Madani, K. (2021). The overlooked environmental footprint of increasing Internet use. *Resources, Conservation and Recycling*, *167*, 105389. Advance online publication. DOI: 10.1016/j.resconrec.2020.105389

Oo, K. T., Jonah, K. A. Z. O. R. A., & Thin, M. M. Z. (2023). A systematic review of the pros and cons of digital pollution and its impact on the environment. *Journal of Sustainability and Environmental Management*, *2*(1), 61–73. DOI: 10.3126/josem.v2i1.53118

Pandey, D., Agrawal, M., & Pandey, J. S. (2011). Carbon footprint: Current methods of estimation. *Environmental Monitoring and Assessment*, *178*(1–4), 135–160. DOI: 10.1007/s10661-010-1678-y PMID: 20848311

Pe'er, S., Goldman, D., & Yavetz, B. (2007). Environmental literacy in teacher training: Attitudes, knowledge, and environmental behavior of beginning students. *The Journal of Environmental Education*, *39*(1), 45–59. DOI: 10.3200/JOEE.39.1.45-59

Pereira, A. C., & Romero, F. (2017). A review of the meanings and the implications of the Industry 4.0 concept. *Procedia Manufacturing*, *13*, 1206–1214. DOI: 10.1016/j.promfg.2017.09.032

Perkins, D. N., Drisse, M. N. B., Nxele, T., & Sly, P. D. (2014). E-waste: A global hazard. *Annals of Global Health*, *80*(4), 286–295. DOI: 10.1016/j.aogh.2014.10.001 PMID: 25459330

Peters, M. A. (2017). Technological unemployment: Educating for the fourth industrial revolution. *Educational Philosophy and Theory*, *49*(1), 1–6. DOI: 10.1080/00131857.2016.1177412

Ritchie, H., Rosado, P., & Roser, M. (2020). Greenhouse gas emissions. *Our World in Data*. https://ourworldindata.org/greenhouse-gas-emissions

Roczen, N., Kaiser, F. G., Bogner, F. X., & Wilson, M. (2014). A competence model for environmental education. *Environment and Behavior*, *46*(8), 972–992. DOI: 10.1177/0013916513492416

Sağlam-kaya, Y. (2019). Öğretmen adaylarının teknopedagojik eğitim yeterliklerinin çeşitli değişkenler ve öğretmen öz yeterlikleri bağlamında incelenmesi. *Journal of Theoretical Educational Science*, *12*(1), 185–204. DOI: 10.30831/akukeg.420909

Schwab, K. (2016). *The fourth industrial revolution*. World Economic Forum.

Sharma, P., & Dash, B. (2023). *The digital carbon footprint: Threat to an environmentally sustainable future*. SSRN. https://doi.org/ (replace with actual if available) DOI: 10.2139/ssrn.XXXX

Singh, A. (2012). Environmental education for sustainable development. *Jyotirmay, 18*.

Strubell, E., Ganesh, A., & McCallum, A. (2020). Energy and policy considerations for modern deep learning research. *Proceedings of the AAAI Conference on Artificial Intelligence*, *34*(09), 13693–13696. DOI: 10.1609/aaai.v34i09.7123

The Shift Project. (2019). *Climate crisis: The unsustainable use of online video—The practical case study of online video*. https://theshiftproject.org/en/article/unsustainable-use-online-video/

Tomlinson, B., Black, R. W., Patterson, D. J., & Torrance, A. W. (2024). The carbon emissions of writing and illustrating are lower for AI than for humans. *Scientific Reports*, *14*(1), 54271. Advance online publication. DOI: 10.1038/s41598-024-54271-x PMID: 38355820

University of Cambridge. (2022). *Cambridge Blockchain Network Sustainability Index (CBECI)*. https://ccaf.io/cbnsi/cbeci

Wathuge, A., & Sedera, D. (2021). An awareness-based model to minimize the environmental damage of internet usage: A longitudinal study. arXiv. https://arxiv.org/abs/2111.04453

Wen, H., Lee, C.-C., & Song, Z. (2021). Digitalization and environment: How does ICT affect enterprise environmental performance? *Environmental Science and Pollution Research International*, *28*(39), 54826–54841. DOI: 10.1007/s11356-021-14474-5 PMID: 34014481

Winfrey, O., & Jung, C. G. (2022). Self-awareness and skill development. In *Leadership: Leaders, followers, environments* (p. 47).

Xu, Z., Sun, D.-W., Zeng, X.-A., Liu, D., & Pu, H. (2015). Research developments in methods to reduce the carbon footprint of the food system: A review. *Critical Reviews in Food Science and Nutrition*, *55*(9), 1270–1286. DOI: 10.1080/10408398.2013.821593 PMID: 24689789

Yadav, S. K., Banerjee, A., Jhariya, M. K., Meena, R. S., Raj, A., Khan, N., & Sheoran, S.. (2022). Environmental education for sustainable development. In *Natural resources conservation and advances for sustainability* (pp. 415–431). Elsevier. DOI: 10.1016/B978-0-12-822976-7.00010-7

Yanarateş, E. (2022). Dijital çevre ve dijital atık. In *Dijitalleşme* (p. 159).

ADDITIONAL READING

Alkan, F., & Cangüven, M. Y. (2025). Examining the relationship between ecological literacy and ecological footprint awareness. *Pedagogical Perspective*, *4*(2), 412–423. DOI: 10.29329/pedper.2025.131

Alkan, F., & Cantürk, E. (2024). The Validity and Reliability of a Tool for Measuring Ecological Literacy. In *Cases on Collaborative Experiential Ecological Literacy for Education* (pp. 215–240). IGI Global. DOI: 10.4018/979-8-3693-1710-5.ch008

Bilbay, Ö. F. (2024). İklim krizi ve dijitalleşme. *Yönetim Bilimleri Dergisi, 22*(Özel Sayı: Endüstri 4.0 ve Dijitalleşmenin Sosyal Bilimlerde Yansımaları), 1544-1568.

Martínez-Acosta, M., Vázquez-Villegas, P., Caratozzolo, P., Lara-Prieto, V., García-García, R., & Membrillo-Hernández, J. (2022, September). Education for sustainability: calculation of the digital carbon footprint. In *International Conference on Interactive Collaborative Learning* (pp. 345-353). Cham: Springer International Publishing.

Özülüş, E., Alkan, F., & Yücel, A. S. (2024). *Digital reading: Examining the characteristics of prospective teachers.* International Liberty Interdisciplinary Studies Conference: Promoting Interdisciplinary Collaboration, New York, ABD.

Özülüş, E., Alkan, F., & Yücel, A. S. (2024, December). Artificial intelligence technologies: How ready are prospective teachers for artificial intelligence? *ITED 2024: International Teacher Education and Development Conference*, Nasarawa State University, Keffi, Nigeria.

Özülüş, E., Alkan, F., & Yücel, A. S. (2025, March). *Determining the self-perception and self-evaluations of pre-service teachers regarding technological pedagogical content knowledge (TPACK).* Teacher Digital Competency 2025: Shaping the Future of Education International Conference, Ukraine.

Pan, M., Zhao, X., Rosak-Szyrocka, J., Mentel, G., & Truskolaski, T. (2023). Internet development and carbon emission-reduction in the era of digitalization: Where will resource-based cities go? *Resources Policy*, *81*, 103345. DOI: 10.1016/j.resourpol.2023.103345

Weidema, B. P., Thrane, M., Christensen, P., Schmidt, J., & Løkke, S. (2008). Carbon Footprint. *Journal of Industrial Ecology*, *12*(1), 3–6. DOI: 10.1111/j.1530-9290.2008.00005.x

KEY TERMS AND DEFINITIONS

Carbon Emissions: The release of carbon dioxide (CO_2) and carbon-based greenhouse gases into the atmosphere as a result of energy use and human activities.

Carbon Footprint: It is the amount of CO_2 per unit that indicates the environmental impact of greenhouse gases released during the use of digital and online services.

Digital Pollution: The energy consumption and environmental impacts arising from the excessive and unnecessary use of digital devices.

Electronic Waste (E-waste): Waste generated from electronic devices that have lost their function, become obsolete, or are no longer in use.

Greenhouse effect: The process by which certain gases in the atmosphere trap heat and cause the Earth to warm.

Greenhouse Gases: Gases that trap heat in the atmosphere and lead to climate change.

Sustainability: It is an approach that aims to reduce the environmental impacts of digital technologies and to ensure that natural resources are used with consideration for future generations.

Chapter 10

Evaluation of Smart City Policies on Effectiveness, Efficiency, and Socio-Economic Impacts:

Smart City Analysis in Malang City

Febby Ananta Lestari
http://orcid.org/0009-0009-8180-9154
Independent Researcher, Indonesia

Putri Wahidatul Azizah
Independent Researcher, Indonesia

Tri Sulistya Ningsih
Independent Researcher, Indonesia

Dwi Putri Vidiastuti
Independent Researcher, Indonesia

Mohammad Jafar Loilatu
http://orcid.org/0000-0001-6921-6879
Independent Researcher, Indonesia

ABSTRACT

This study aims to evaluate the implementation of the Smart City Masterplan in Malang City as a transformation effort towards a sustainable smart city. Using qualitative descriptive methods, the evaluation focused on six main pillars: Smart Governance, Smart Economy, Smart Branding, Smart Living, Smart Society, and

DOI: 10.4018/979-8-3373-8490-0.ch010

Smart Environment. The results show that Malang City has achieved a significant level of success with program realization reaching 96.22%. In terms of digital infrastructure readiness (E-Readiness), Malang City obtained a score of 58.3%, which has exceeded the minimum readiness standard of 50%. This success is driven by the Hexahelix collaboration model that involves synergy between government, academia, communities, the private sector, media, and banking. Despite showing positive progress, this study identified major challenges in the form of obstacles to interoperability of data integration between agencies, limited human resource capacity, and the need to strengthen regulations supporting innovation.

INTRODUCTION

Smart City is a city development concept that utilizes digital technology, data, and innovation to improve the quality of life for citizens, accelerate public services, and ensure more efficient and transparent governance. The development of the Smart City initiative in Malang City emerged in response to the increasing complexity of urban issues, including congestion, urban flooding, and public services that are still not optimally integrated. Various regional vulnerability reports, such as the Climate Risk Assessment and periodic publications from the Meteorology, Climatology, and Geophysics Agency (BMKG), indicate that Malang City faces environmental and social challenges that require a data-driven governance system to accelerate risk mitigation. The implementation of the Smart City policy is then positioned as a strategic effort to improve service effectiveness, resource efficiency, and the ability of local governments to respond to community needs in real time. This policy remains questionable due to the gap between digital planning and the institutional capacity to implement transparent, participatory, and accountable governance. This aligns with the principles of good governance, where the quality of public policy is largely determined by the government's ability to ensure the integration of technology, community involvement, and the sustainable use of data in the decision-making process (Cendekiawan& Wardhani, 2025).

Problems in Malang City's Smart City policy are evident in its governance pattern, which still resembles an authority-based model, where decisions tend to be centralized, reactive, and not based on sustainable data integration. This is evident in the handling of urban issues such as flooding, congestion, and public services, which are often carried out in a particular, partial, and short-term project-based manner without systemic solutions. This gap indicates that the principles of good governance, which demand transparency, accountability, cross-sector coordination, and public participation, have not been fully implemented in Smart City implementation. As a result, digital policies that should increase effectiveness and efficiency are hampered by the weak integration of technology with governance processes. In theory, good governance serves as an analytical tool that assesses whether governance has moved towards a more collaborative and responsive model, rather than merely authoritative and administrative. The situation in Malang City demonstrates that the success of a Smart City depends on the government's ability to shift from a reactive approach to a proactive, inclusive, and data-driven management system. Thus, good governance theory provides a framework for understanding why the effectiveness, efficiency, and socio-economic impact of Smart City policies have not been optimal (Cendekiawan& Wardhani, 2025).

The issues in evaluating Malang's Smart City policy are further validated by examining Indonesia's historically strong tendency toward authority-based governance, a centralized, hierarchical decision-making model with minimal public participation. This model has its roots in the centralized government during the New Order era, where effectiveness was considered synonymous with dominant government control, resulting in a top-down development process that is not adaptive to the complexity of community needs.(Damayanti & Putra, 2023). In theory, good governance, which began to strengthen at the end of the 20th century, supported by international institutions such as the UNDP, emerged as a critique of authoritarian models and closed bureaucracies that proved incapable of promoting sustainable development. This theory emphasizes the importance of collaboration, transparency, and accountability as corrective mechanisms for governance that relies too heavily on a single authority. This theory works by assessing how government, private sector, and civil society actors interact in the policy process, and whether these mechanisms produce inclusive, effective, and efficient decisions. In the case of Malang City, the lack of integration of digital systems, weak cross-sector coordination, and the lack of space for public participation indicate that the principles of good governance have not been fully institutionalized, so that authority-based patterns remain dominant. Therefore, good governance theory is an important analytical tool for understanding why Smart City policies have not achieved optimal results and how improved governance can promote greater effectiveness, efficiency, and socio-economic impact (Cendekiawan& Wardhani, 2025).

The urgency of this research becomes increasingly apparent when observing that various performance indicators in Malang City demonstrate a mismatch between Smart City objectives and achieved results. For example, the persistently high number of public service complaints and digital inequality was recorded in the regional statistical report by the Malang City Statistics Agency (BPS). Furthermore, several digitalization programs, such as smart transportation services, disaster information systems, and administrative data integration, have not shown significant improvements in service effectiveness or efficiency, as evidenced by the stagnant public satisfaction index, which remains in the "good" category without significant improvement. This situation reinforces the argument that authority-based governance patterns still dominate, leading to technological innovation not being accompanied by improvements in coordination, accountability, and public participation mechanisms. Therefore, a thorough understanding of the characteristics of this governance is crucial for identifying structural weaknesses in decision-making, particularly those that hinder the achievement of Smart City system integration. The theory of good governance then becomes an urgent conceptual framework to use, as it is able to dissect the extent to which local governments manage collaboration between actors and utilize technology effectively. This research is urgent to assess whether Smart City implementation truly produces measurable socio-economic impacts or whether it remains merely administrative and symbolic. Thus, the evaluation of Malang City's Smart City policy is not only important for improving regional performance, but also as a basis for improving the city governance model in Indonesia.(Maharani et al., 2024).

Several previous studies have examined the implementation of Smart City in Malang from various perspectives. For example, RAISA explained that the dynamics of political decisions significantly influence the resilience of urban areas in facing development challenges. Syava (2021) found that rent-seeking practices in city governance limit the latitude of policymakers, especially in responding to strategic issues that require innovation and coordination. Meanwhile, Nauvalla (2022) revealed that local leadership capacity is still inadequate to manage technocratic issues that require cross-sector collaboration. However, the majority of these studies still focus on political and institutional aspects without deeply examining how authority-based governance patterns influence the effectiveness, efficiency, and socio-economic impact of Smart City policies. This research gap is significant because authority-based governance has the potential to be a structural factor determining the success or failure of city digitalization programs. Furthermore, several studies have integrated good governance analysis to measure the extent to which the dominance of authoritative patterns hinders the optimization of technology in public services. Thus, this study is presented to fill this gap by offering a new perspective on the relationship between authority structure, governance quality, and the performance of Malang City's Smart City policy.

Based on the limitations of previous research, this study aims to analyze the Smart City Policy Evaluation of Malang City from an authority-based governance perspective, focusing on how centralized decision-making patterns affect the effectiveness, efficiency, and socio-economic impact of the policy. This study also seeks to identify the extent to which good governance principles have or have not been integrated into Smart City implementation, thereby explaining the various structural barriers that arise. In addition, this study aims to map the relationship between institutional capacity, inter-actor coordination, and the success of digital technology in improving the quality of public services. The expected contribution is to provide new insights into the influence of urban governance approaches on the performance of technology-based policies, particularly in the context of cities still influenced by the dominance of authority structures. Another contribution offered by this study is to provide an analytical basis for local governments in designing digital governance improvements in a more collaborative, participatory, and responsive manner. Practically, this research is expected to serve as a reference in formulating strategies for strengthening Smart Cities that are more sustainable and oriented towards improving community welfare. Thus, this study not only enriches the academic literature but also supports more adaptive and evidence-based policymaking.

LITERATURE REVIEW/RELATED WORKS

Urbanization in developing cities has triggered increasing complexity of governance, particularly regarding the integration of technology into public services. The accelerated digitalization of urban services has been shown to increase the potential for bureaucratic efficiency, but still faces challenges in coordination and transparency of government institutions. Previous research shows that Smart City implementation often fails to achieve effectiveness goals due to a lack of accountability mechanisms and public participation in decision-making (Mahendra et al., 2022). In some cases, the dominance of hierarchical authority structures results in digital policies becoming merely technology projects, rather than sustainable governance systems. This aligns with findings that the success of digital transformation depends heavily on the quality of governance, not just technical capabilities or the magnitude of infrastructure investment (Noor et al., 2023). Based on these arguments, understanding the effectiveness and efficiency of digital policies such as Smart Cities can be theoretically explained through a good governance framework that emphasizes transparency, accountability, effectiveness, and participation as prerequisites for successful public policy (Winata & Friantini, 2019). Thus, the concept of good governance becomes an entry point for analyzing why Malang City's Smart City policy faces various structural and operational obstacles.

The theory of good governance began to be popularized by international institutions such as the World Bank in the early 1990s, when the failure of hierarchical bureaucracies in many developing countries demonstrated the need for a new, more transparent and accountable governance model (World Bank, 1992). This concept was later reinforced by the UNDP, which defined good governance as a process of managing power that emphasizes effectiveness, participation, responsiveness, and the rule of law (UNDP, 1997). Subsequent research has shown that good governance operates through three main pillars: responsive government, engaged civil society, and the private sector playing a role in policy collaboration (Kurniawan et al., 2023). The workings of this theory require integration between actors, transparent information flow, and cross-institutional coordination as the primary mechanisms for creating effective public policy (Anggraini, 2021). In the context of digital transformation, good governance serves as an evaluative framework for assessing whether the use of technology truly strengthens public services or simply expands government administrative power (Sari, 2020). Thus, this theory is consistently used in modern policy research focusing on local government performance and digital-based innovation. Based on this framework, the analysis of Malang's Smart City is relevant because the success of digital policies cannot be separated from the underlying governance capacity.

Various studies have shown that Smart City implementation in Indonesian cities, including Malang, still faces challenges with cross-sector coordination, low public engagement, and weak data integration (Sukmana, 2022). This situation provides a rationale for the need for good governance theory to explain why urban digital innovation often functions solely as an administrative project without generating significant improvements in service effectiveness or efficiency. Malang City, which is experiencing population growth, infrastructure pressure, and an increasing need for digital services, shows indications of a dominant authority-based governance pattern that has the potential to hinder Smart City optimization (Rahadiansyah, 2023). Scientific speculation suggests that the biggest obstacle to Smart City development lies not in technological facilities, but rather in decision-making structures that do not fully comply with the principles of participation, transparency, and accountability (Andani & Yusuf, 2020). Thus, good governance theory can be directed to uncover how authority-based governance affects the performance of digital policies in Malang City. This also reinforces the importance of an in-depth analysis of the role of local governments in managing collaboration and data integration as key elements of Smart City development (Suwanto, 2021). Therefore, the use of this theory is an appropriate conceptual basis for investigating the effectiveness, efficiency, and socio-economic impacts of Malang City's Smart City policy.

METHOD

This study uses a mixed methods approach to comprehensively understand the effectiveness, efficiency, and socio-economic impacts of the Smart City policy implementation in Malang City. This approach was chosen because it combines the depth of qualitative analysis with the precision of quantitative measurements, thus providing a more objective empirical picture of authority-based governance. The data sources used include secondary data that serves as a basis for comparison and verification of empirical findings, such as the Malang City Spatial Plan (RTRW), Regional Spatial Plan (RDTR), Regional Medium-Term Development Plan (RPJMD), Smart City Masterplan, and performance reports from relevant regional government agencies. Furthermore, online literature searches from reputable sources such as SpringerLink, ScienceDirect, Taylor & Francis, and journal articles indexed in SINTA 1–3 were used to strengthen the theoretical foundation and support the policy analysis. These secondary data provide a macro context regarding the direction of city development and ensure that data interpretation is measurable and accountable. With this data composition, the study is able to capture policy dynamics while testing the accuracy of its implementation at the operational level.

Data analysis was conducted through qualitative thematic analysis aligned with quantitative components to identify patterns, the effectiveness of policy interventions, and potential discrepancies between Smart City planning and implementation in Malang City. This analysis process aimed to uncover specific information regarding how the policy was implemented, its efficiency, and how the socio-economic impacts were perceived by the community. To maintain data validity, researchers applied source triangulation, including comparisons between policy documents, regional statistical data, and findings from scientific literature. Furthermore, cross-synthesis was used to ensure that the analysis findings were not solely dependent on a single type of data but were consistently verified across multiple sources. This effort enabled the research to produce reliable and relevant conclusions for the development of city governance. Thus, data analysis can provide a more accurate picture of the structural weaknesses and achievements of Smart City policies.

Data processing was conducted using NVivo (QDAS) version 14, which allows researchers to systematically manage textual data through a coding process. The software's coding feature was utilized to identify key emerging themes, such as the effectiveness of digital services, public response, government operational efficiency, and the economic impacts of Smart City implementation. All codes and cases were organized based on a theoretical understanding of how authority-based urban governance influences the direction and quality of Smart City implementation at the local level. After establishing coding benchmarks, the data classification process was carried out by identifying specific information relevant to indicators of effectiveness, efficiency, and socioeconomic impact. Data corresponding to the codes and cases were then linked to obtain a stable and adequate thematic structure. Through this stage, the analysis can reveal how Smart City policies are actually implemented and where implementation gaps lie.

RESULTS AND DISCUSSION

Smart City Analysis in Malang City

Based on the latest evaluation for the 2024–2025 period, Malang City has demonstrated significant progress in implementing Smart City principles nationally. One of the key strategic outcomes is Malang's achievement of an Indonesian Digital Society Index (IMDI) score of 62.67 in 2025, the highest IMDI score recorded among regions in Indonesia. This result indicates an increase in the community's digital literacy level, a high level of readiness in information technology infrastructure, and the effective utilization of digital services to support social, economic, and government activities. The following is a summary table of Malang City's Smart City evaluation based on the six main pillars and their achievement status

The implementation of the Smart City concept in Malang City has shown striking and structured progress, as evidenced by the equitable distribution of digital maturity across all key aspects of smart city development. This success is not a short-lived achievement, but rather the result of consistently designed policies, strengthened institutional capacity, and the local government's determination to make digital technology the primary foundation of urban development. In the Smart Governance aspect, Malang City achieved a program realization rate of 96.22%, indicating high effectiveness in implementing digital policies. This figure reflects the successful process of transforming governance from a traditional bureaucratic model to a digital-based government system that is more responsive, transparent, and adaptive to community needs. This transformation was realized through the launch of the Malang Mbois integrated service portal, which serves as the main gateway for citizens to access various public services. This portal integrates administrative services, licensing, complaints, and general information in a single digital platform, thereby reducing service fragmentation and increasing bureaucratic accountability.

Achievements in the Smart Governance sector have also contributed to the expansion of digital infrastructure for citizens. The Malang City Government has prepared more than 550 free Wi-Fi hotspots at the neighborhood (RW) level, which serves as a crucial means of reducing the gap in technology access between regions and social strata. This infrastructure supports the Smart Society dimension, which focuses on improving citizens' digital skills and literacy. This is proven by the 2025 Indonesian Digital Society Index (IMDI) score of 62.67, the highest nationally and exceeding the national average. This score indicates not only good technological affordability but also the community's ability to utilize technology productively and wisely. This digital literacy adaptation supports technology-based government services, the utilization of digital economic opportunities, and active participation in technology-based public interactions. In the economic sector, the synergy between

Smart Economy and Smart Branding is a key strategy in driving sustainable local economic growth. The government has made the creative economy and digital innovation the core of regional economic development in facing global challenges.

The Malang Creative Center (MCC) plays a crucial role as an incubator for 17 creative economy sub-sectors, from culinary and fashion to design and digital application development. The MCC not only provides physical facilities but also provides guidance, training, and access to marketing networks to enable local MSMEs and startups to thrive and compete in the digital age. This Smart Economy ecosystem is further strengthened through Smart Branding, which elevates cultural identity and local riches as added economic value. A concrete example of this is the digitalization of the Kayutangan heritage area, which successfully combines historical heritage with modern technology without altering the area's original character. The implementation of cashless transactions, a digital-based tourism information system, and promotional strategies through social media have made Kayutangan not only a tourist center but also a catalyst for the creative economy for local businesses, while strengthening Malang's image as a creative and historic city. Within the Smart Living dimension, Malang City maintains stable performance in ensuring the safety and comfort of its residents. The integrated emergency service, Ngalam 112, enables a rapid response to emergency situations with an average response time of around 11 minutes, considered effective for an urban area. The use of information technology improves inter-agency coordination and reduces risks to citizen safety, thus improving the quality of life and public trust in government services.

Meanwhile, the Smart Environment dimension is realized through data-based environmental management innovations, one of which is the Tree ID Card (KTP Tree) program. This program makes it easier for the government to monitor, record, and manage green spaces accurately and sustainably, supporting climate change mitigation efforts while preserving the city's environment. Overall, the Smart City achievements in Malang City demonstrate that information technology is not merely an administrative tool, but plays a key role in improving the quality of life for residents, strengthening the local economy, and sustainable urban development. This cross-dimensional synergy positions Malang as one of the best examples of Smart City implementation in Indonesia.

The image contains an evaluation matrix for the implementation of the Smart City program in Malang City, which is based on the Malang Smart City Master Plan (MKC) 2020–2025. This matrix is used as a monitoring tool and annual performance measurement for Smart City implementation, which is systematically evaluated by the Ministry of Communication and Informatics (Kemenkominfo). The evaluation process is carried out in two stages: Stage I to assess the readiness and suitability of the work plan, and Stage II to measure the extent to which the program has been implemented and produced tangible results in the field. Broadly speaking, the matrix displays the level of realization of the main program and the implementation of Smart City policies in six main dimensions: Smart Governance, Smart Living, Smart Society, Smart Environment, Smart Branding, and Smart Economy. These six dimensions constitute a national standard framework that combines aspects of governance, quality of life for citizens, environmental sustainability, economic strengthening, and city identity.

In addition to the main dimensions, the matrix also includes key supporting components such as policies/regulations, information technology infrastructure readiness, and a hexahelix collaboration model involving government, academics, business actors, society, media, and communities. This shows that the evaluation does not only measure program outputs, but also assesses the institutional foundations and collaborative ecosystem that support the sustainability of Smart City. The results of the Phase II evaluation showed an overall realization rate of 96.2%, which is categorized as very high. This figure indicates that most programs in the MKC Master Plan have been implemented consistently and have had a real impact, reflecting the continuity between strategic planning and technical implementation. The color code in the matrix is used as a visual marker of the achievement of each dimension and supporting component. Green indicates high realization, the program has been fully implemented and is operating optimally, Yellow indicates moderate or partial achievement. Gray/Purple indicates that it is still in the planning stage or requires further strengthening.

Green dominates Smart Governance, with 100% implementation thanks to the implementation of the Electronic-Based Government System (SPBE), which integrates digital technology into all government processes, from public services and data management to coordination between regional agencies. In Smart Living, green achievements are achieved through the E-JKN digital health service, which facilitates citizens' access to national health services, while simultaneously improving the comfort, security, and well-being of the community. The Smart Society dimension also shows high achievement figures, one of which is the integrated emergency service Ngalam 112, which strengthens the government's response to emergencies and proves that smart community development also includes aspects of safety and citizen engagement. Smart Environment displays green achievements through the Sipaldi system, which supports data-based environmental and infrastructure management, demonstrating Malang's commitment to sustainable environmental management.

Meanwhile, Smart Economy and Smart Branding are in the yellow category, with scores ranging from 86–89%. Both dimensions have developed quite well but still require improvement, for example through optimizing the digital economy, strengthening technology-based MSMEs, and more integrated heritage tourism management. The yellow color indicates the need to increase synergy between digital innovation and the development of a city identity with significant potential, which can strengthen Malang's competitiveness. Elements in gray or purple in the matrix indicate components that require special attention, particularly strengthening supporting infrastructure and cross-stakeholder collaboration. This emphasizes that the success of a Smart City depends on the quality of collaboration and policy consistency, not just on technological mastery.(Rachmawati et al., 2024).

The matrix grid structure is clearly designed, with the top row showing key programs and overall achievements, while the columns display Smart City dimensions and supporting factors. This layout makes it easier for policymakers and readers to identify key strengths and areas requiring intervention. Overall, this evaluation shows that Malang has four dimensions, with dominant green achievements and two yellow dimensions as areas of focus for improvement. This finding aligns with various achievements, such as ranking third in the 2018 Indonesian Smart City Index (IKCI) and entering the top 6 of the 2024 IMDI, which strengthens Malang's position as a leading city in digital transformation and improving the quality of life for its citizens.

Based on the research results, the evaluation shows that the implementation of the Smart City concept in Malang City has achieved significant success, with program realization reaching 96.22%. This indicates that the majority of activities and projects listed in the strategic plan have been implemented well and efficiently, and can be transformed into concrete programs that have a direct impact on government administration and public services. In terms of infrastructure and support system readiness, Malang City achieved an E-Readiness score of 58.3%, which means it has exceeded the minimum national readiness level of 50%. This indicates that the existing information and communication technology infrastructure, organizational capacity, and digital systems are at a sufficient level to support sustainable digital transformation. With this readiness, Malang City is not only ready to implement technology but also able to integrate it into various service sectors and city development.

The successful implementation of the Smart City concept is closely related to the use of the Hexahelix collaboration model, which emphasizes cooperation between six main parties: government, academia, communities, business actors, media, and the banking sector. This intersectoral collaboration is crucial for expanding resources, improving policy perspectives, and encouraging innovation that is more responsive to community needs. With this approach, Smart City in Malang is not only built from the top down, but also through a co-creation process involving various stakeholders. Analysis of Smart City implementation involves six important pillars, namely Smart Governance, Smart Economy, Smart Branding, Smart Living, Smart Society, and Smart Environment. These six pillars are designed to complement each other and work together to improve the efficiency of public services, strengthen local economic competitiveness, ensure environmental sustainability, and improve the quality of life of the community. The dominant achievements in the Smart Governance and Smart

Society pillars indicate that bureaucratic reform and increasing public digital literacy are the main foundations in the smart city transformation in Malang. On the other hand, the Smart Economy and Smart Branding pillars contribute to encouraging the growth of the creative economy and strengthening the city's identity through the use of digital technology. Meanwhile, the Smart Living and Smart Environment pillars play a role in creating a safe, comfortable, and sustainable urban environment through technology-based public services and more adaptive and data-based environmental management.(Rabbani et al., 2025).

While implementation results generally show very positive progress, the evaluation document also highlights several strategic challenges that require attention going forward. One key challenge is the suboptimal integration and interoperability of data across agencies, which can disrupt the effectiveness of integrated services and data-driven decision-making. Fragmentation in systems and applications persists, hindering the creation of fully integrated digital governance. Furthermore, there is a continued need to improve human resource capacity, both among civil servants and the general public. Without adequate digital competency development, the sustainability of Smart City innovation risks being hampered during operations and ongoing development. Another challenge is the need to strengthen flexible and long-term regulations to ensure the continuity of innovation, protect data security, and ensure the consistency of Smart City policies amidst dynamic government changes. Overall, Smart City development in Malang can be considered to be on the right track and demonstrates a high level of maturity. However, to ensure sustainability and improve the quality of future implementation, ongoing efforts focused on system integration, human resource capacity building, and regulatory framework improvements are needed so that Smart City truly functions as a strategic tool for inclusive, adaptive, and sustainable urban development.

The image displayed is a visual representation of a word cloud containing the main themes in the implementation of the Smart City program in Malang City. At the center of the visualization, the word “Malang” appears as the most prominent element, emphasizing Malang City's role as the core subject in the overall narrative of smart city development. This situation indicates that all Smart City strategies, concepts, and policies are strongly oriented towards local conditions and the specific needs of the region. Surrounding the main words are a number of key terms such as “smart city,” “smart city,” “city,” “implementation,” “program,” “government,” “developing,” and “towards.” The large font size of these words indicates their high frequency of occurrence in the policy documents, reports, and text data analyzed. This reflects the government's focus on the implementation process, the strategic role of local government, and the direction of Malang's transformation into a sustainable smart city.

In addition to these terms, the word cloud displays conceptual and technical terms such as "hexahelix," "infrastructure," and "governance." The presence of these words demonstrates that the discussion of Smart City focuses not only on technology but also on governance and cross-sectoral collaboration. The hexahelix concept visualizes the involvement of government, academics, the business world, society, media, and communities as the foundation for establishing an inclusive Smart City ecosystem. Technically, the font size in the word cloud represents the frequency or weight of the words in the analyzed data set.(Rabbani et al., 2025). The largest-sized terms, such as "smart," "city," and "Malang," indicate high emphasis in planning and evaluation documents, underscoring the position of the transformation towards a smart city as the central narrative in the Malang Smart City Master Plan 2020–2025.

This visualization also demonstrates the close relationship between the Smart City concept and its six key dimensions, from Smart Governance to Smart Economy. The dominance of words related to governance, programs, and implementation reflects Malang's ability to translate the smart city concept into concrete policies and programs, not just slogans or discourse. In terms of results, this word cloud supports the evaluation findings, which recorded a Smart City realization rate of 96.2%. The alignment between the text narrative and quantitative figures demonstrates the close relationship between strategic planning, program implementation, and the resulting output. Words related to government and governance also demonstrate that digitalization has become part of bureaucratic reform and the improvement of technology-based public services. Furthermore, the emergence of the terms "hexahelix" and "infrastructure" emphasizes the strong multi-stakeholder collaboration as the key to the success of Smart City Malang.

From a community development perspective, the dominance of terms related to urban development and development indicates that Smart City is oriented towards improving citizen well-being, creating a more efficient, inclusive, and responsive city to its citizens' needs. At the same time, the relatively smaller font size of terms related to the digital economy and tourism branding signals that these two aspects still require strengthening and optimization. Thus, this word cloud serves not only as a descriptive visual aid but also as a policy analysis tool, helping to identify core strengths, namely digital governance and collaboration, and highlighting priority areas for improvement, particularly in the development of the creative economy and city image. Overall, this visualization confirms that Malang's Smart City implementation path is on target. The emerging narrative is consistent with the city's achievements in various national indices such as the IKCI and IMDI, while also supporting the long-term vision of realizing a smart, competitive city oriented towards improving the quality of life for its citizens.

Based on the visual analysis of TreeMap data, it is clear that Malang City's Smart City policy focuses on implementing priority programs that directly impact the community. The dominance of the "program" and "public service" categories in the visualization confirms that the government positions citizen services as the core of digital transformation, not merely a complement to technological modernization. This reflects the understanding that Smart City is a means to address the real needs of urban communities. The focus on public services also demonstrates a strong commitment to bureaucratic reform to create effective and accountable governance (good governance). Through the integration of services in the Public Service Mall, the government simplifies procedures, accelerates service processes, and increases transparency. The bureaucratic paradigm has shifted from a process-oriented to a results-oriented and citizen-satisfying paradigm.(Irhab et al., 2024).

The success of these reforms is supported by the Malang City Government's active role in overseeing policies and ensuring consistent implementation on the ground. The TreeMap visualization illustrates the dominant role of the government as the primary driver, from policy formulation and infrastructure provision to coordination between regional agencies. Furthermore, the Malang City Government emphasizes the importance of multi-stakeholder collaboration, in accordance with the hexahelix approach, involving academics, business actors, communities, the media, and the public at every stage of development. This collaboration strengthens the legitimacy of policies while increasing the effectiveness of Smart City implementation. Furthermore, the TreeMap data raises concerns about strengthening the economic ecosystem and measurable growth. Smart City is not viewed solely as a technology project, but rather as a strategic framework for developing the creative economy, MSMEs, and local innovation to enable them to adapt to digital transformation. This approach demonstrates a balance between technological advancement and preserving the city's historical and cultural values. The integration of digital innovation with heritage areas and local identity ensures that the Smart City transformation strengthens Malang's character and is grounded in local wisdom, thus maintaining a distinct approach to other cities.

This strategy has an impact on improving the quality of life for citizens, with technology that opens up easier access to services, wider economic opportunities, and more inclusive and sustainable urban spaces. Smart City exists not only as an application and system, but also as a tangible part of people's daily experiences. Overall, TreeMap data positions Smart City as a strategic evaluation tool as well as an adaptive policy instrument. With a data-driven and visualized approach, the government can set priorities, monitor achievements, and adjust policy direction responsively to the city's challenges. Based on this foundation, Smart City in Malang is projected to become a pillar of future development, encouraging an increasingly smart, collaborative, competitive, and sustainable city, with citizens as key actors in digital transformation.

Based on this document, the assessment results indicate that the implementation of Smart City in Malang City has achieved an extraordinary level of success, as evidenced by the program realization of 96.22%. This achievement confirms that most of the plans and programs stipulated in the Malang Smart City Master Plan have been implemented consistently and in line with the city's development vision based on technology and innovation. This high level of realization also reflects the ability of local government institutions to manage digital transformation in a planned and sustainable manner. In terms of technical readiness, Malang City's digital infrastructure readiness (E-Readiness) score was recorded at 58.3%, indicating that this figure has exceeded the national minimum readiness limit of 50%. This indicates that ICT infrastructure, information systems, as well as organizational and human resource support in general are sufficient to support the provision of digital-based public services and the integration of e-government systems.

The successful implementation of Smart City is inseparable from the application of the Hexahelix collaboration model, which emphasizes multi-stakeholder synergy between the government, academics, communities, business actors, media, and the banking sector. This collaborative model enables the exchange of knowledge, strengthening of resources, and increasing innovation, so that Smart City is not only a government agenda, but also a collective movement across sectors oriented towards the public interest. This evaluation involves the six core pillars of Smart City, namely Smart Governance, Smart Economy, Smart Branding, Smart Life, Smart Society, and Smart Environment. These six elements are designed to support each other and form an overall smart city ecosystem, with the main goal of improving the quality of public services, encouraging a comprehensive economy, strengthening city identity, and creating a safe, comfortable, and sustainable urban atmosphere for residents.

While implementation results show very positive trends, evaluation data also indicates several strategic challenges that require further attention. One of the most significant challenges is the system's ability to interact and integrate data between various agencies, which has not yet been fully functional, potentially hindering the effectiveness of integrated services and the optimal use of data for evidence-based decision-making.

Furthermore, the evaluation highlighted the need for strengthened regulations and a more adaptive and sustainable policy framework to ensure the sustainability of Smart City innovation in the future. Without strong and consistent regulations, there is a risk of policy fragmentation, overlapping programs, and uncertainty in the long-term management of digital systems. Considering these achievements and challenges, a more in-depth analysis of implementation barriers and obstacles is highly relevant for formulating future improvement strategies. This in-depth analysis can include identifying root causes, mapping actors and authorities, and formulating more operational and applicable policy recommendations. As a follow-up, an in-depth analysis of implementation challenges and obstacles can provide an important basis for the Malang City Government to refine Smart City policies to make them more inclusive, integrated, and sustainable. Thus, Smart City becomes not only a quantitative achievement but also a strategic instrument for strengthening the quality of governance and the long-term well-being of the community.

Considering the achievements and various obstacles, a more detailed analysis of the obstacles and barriers to implementation is crucial for developing future improvement strategies. This in-depth analysis could include identifying the source of the problem, mapping actors and authorities, and developing more practical and implementable policy recommendations. As a next step, an in-depth analysis of the challenges and obstacles to implementation could provide a valuable foundation for the Malang City Government in refining its Smart City policy to make it more inclusive, integrated, and sustainable. In this way, Smart City will not only be a measurable achievement but also a strategic tool for strengthening the quality of governance and the long-term well-being of the community.

Based on analysis showing a 96.22% Smart City implementation achievement rate in Malang City and an infrastructure readiness (e-readiness) that exceeds national standards, policy recommendations are now directed away from the initial stages of adoption and toward strengthening the quality, integration, and sustainability of Smart City systems. Future policy should focus on addressing deeper impacts and consolidating digital management across sectors. First, local governments are required to improve interoperability and data integration between institutions through the development of a standardized data governance framework. Strengthening a single data platform and implementing an integrated SPBE architecture will reduce application fragmentation, prevent data duplication, and increase efficiency in data-driven decision-making. This policy also requires support from technical regulations governing data exchange, information security, and personal data protection.(Rabbani et al., 2025).

Second, human resource (HR) capacity development must be a sustainable strategic priority. The Malang City Government is encouraged to design a systematic program to improve the digital competency of civil servants (ASN), including training in digital leadership, data analysis, and public innovation management. From a societal perspective, expanding community-based digital literacy is vital to increasing public participation in more inclusive and sustainable Smart City services, especially in suburban areas. Third, strengthening the Smart City regulatory framework is an urgent need to ensure the continuity of innovation across different government periods. The development of regional regulations or regional head regulations specifically governing Smart City can serve as a legal basis for budgeting, application management, and collaboration between stakeholders. Flexible regulations will prevent dependence on individuals or short-term policies and ensure program sustainability.

Fourth, strengthening the Smart Economy and Smart Branding dimensions must be done by increasing integration between the creative economy, MSMEs, and digital platforms in the city. It is recommended that the local government expand the function of the Malang Creative Center (MCC) as an innovation hub directly connected to digital marketing systems, inclusive financing, and data-driven city promotion. This approach will strengthen the economic impact of Smart City in a tangible and measurable way. Fifth, the Hexahelix collaboration, which has been a crucial pillar for Malang City, needs to be formalized into a more formal institution. The establishment of a permanent collaborative forum involving government, academics, the business world, communities, the media, and the financial sector will strengthen the collaborative policymaking process and increase accountability in Smart City implementation. This strategy can also encourage the emergence of innovations that are more suitable to local needs. Overall, these policy recommendations place Malang Smart City in a phase of consolidation and strengthening the quality of implementation, rather than simply program expansion. By strengthening data integration, human resource capacity, regulations, and cross-sector collaboration, Malang City has the potential to become a model for a sustainable Smart City that is not only technically superior but also has a significant impact on community welfare and regional competitiveness in the future.(Zahrah & Putra, 2025).

Evaluation of Smart City Success

The successful implementation of Smart City in Malang City is illustrated through a mind map that positions "Smart City Malang 2020–2025" as the core. This mind map is based on the Master Plan of Mayoral Regulation Number 43 of 2020 and shows a realization rate of 96.2% based on the 2023 evaluation. From this center, main branches were formed that cover the six dimensions of success, supporting infrastructure, achievements, and positive impacts. All of these elements reflect the digital transformation of SPBE, which contributes to improving the efficiency of public services and the quality of life of the community.The following is a mind mapping of the results of the Smart City evaluation in Malang City.

The success of Malang's Smart City implementation can be measured through the alignment between strategic planning, performance achievements, and the tangible impact felt by the community. Based on the annual evaluation of the Ministry of Communication and Informatics, the program realization rate reached 96.22%, indicating that most of the programs listed in the Malang Smart City Master Plan 2020–2025 have been implemented effectively and consistently. This achievement places Malang City in the category of regions with very high Smart City maturity. In the Smart Governance dimension, the main achievement is reflected in the successful transformation of governance based on the Electronic-Based Government System (SPBE). The integration of public services through a digital portal, accompanied by increased transparency and bureaucratic efficiency, indicates that technology has played a role as a driver of bureaucratic reform, not just an administrative tool. The realization approaching 100% reflects institutional readiness and strong political commitment from the local government.

The Smart Society dimension is characterized by the achievement of the Indonesian Digital Society Index (IMDI) 2025 of 62.67, the highest at the national level. This demonstrates the successful development of digital infrastructure, including the provision of more than 550 free Wi-Fi points, which encourages literacy and increases citizens' digital participation. Malang City residents not only have access to technology but also the ability to utilize it to support social, economic and public service life. In the Smart Living dimension, stable performance is demonstrated through the integrated emergency service Ngalam 112, with an average response time of around 11 minutes. This success demonstrates that the use of information technology can increase public sense of security, comfort, and trust in dealing with emergency situations. The Smart Environment dimension is considered successful thanks to data-driven innovations such as Sipaldi and KIPOR (Tree ID Card). These programs combine technology with environmental sustainability principles, demonstrating that Smart City in Malang City focuses not only on efficiency but also on sustainable environmental management.

Meanwhile, the Smart Economy and Smart Branding dimensions showed positive achievements but remained at a partial level, at around 86–89%. The presence of the Malang Creative Center (MCC) and the implementation of digitalization in the Kayutangan heritage area have strengthened the creative economic ecosystem and the city's identity. However, both dimensions require further strengthening, particularly in terms of the scale of economic impact, digital market expansion, and the integration of data-driven tourism promotions. Overall, the evaluation results indicate that the success of Malang's Smart City is measured not only by the availability of applications and infrastructure, but also by the ability to transform digital innovation into concrete socio-economic impacts. With a strong foundation in governance and digital society, the challenge ahead is to optimize the potential of the digital economy and city branding so that the benefits of the Smart City can be felt in an inclusive and sustainable manner.

Smart City Implementation Challenges

Although Malang City has achieved an impressive level of Smart City implementation, with program implementation reaching 96.2% in the 2023 evaluation, various structural and operational dilemmas remain significant obstacles to achieving a fully integrated and sustainable Smart City. These challenges demonstrate that successful implementation depends not only on the level of program achievement but also on the quality of management, resource readiness, and infrastructure and regulatory support. One of the main challenges lies in the limited technical regulations and integration of digital applications. Although the Malang City Government has created numerous applications for public services, not all of them operate within a uniform regulatory framework and are optimally connected. This situation results in a silos system, resulting in a less efficient and integrated user experience.(Salauddin & Laurens, 2024).

Furthermore, shifts in employee transfers and rotations within local governments significantly impact the sustainability of Smart City implementation. The transfer of civil servants with technical expertise to other units often results in a loss of institutional knowledge, necessitating a process of adapting and managing digital systems from scratch. As a result, innovation and maintenance of existing applications are hampered. From a financing perspective, the limited budget for Smart City outside of cross-sectoral schemes, which amounts to IDR 570 billion, is also a major challenge. Dependence on sectoral budgets often means Smart City development must compete with other development priorities, preventing all digital innovations from being developed optimally and sustainably.

a. Infrastructure and Technology Challenges

In terms of infrastructure and technology, a key issue relates to unequal access to digital infrastructure across regions. Although Malang City has provided 551 free Wi-Fi hotspots at the neighborhood (RW) level, the distribution of high-speed internet networks remains far from uniform, particularly in suburban areas. This imbalance can exacerbate the digital divide and hinder comprehensive public participation in the Smart City ecosystem. Furthermore, the integration of 42 public service applications remains hampered by various technical issues. Several applications reportedly experience disruptions, feature limitations, or are not regularly updated. This situation undermines user trust and reduces the applications' effectiveness as digital public service tools. Limited interoperability between applications also results in data duplication and increased staff workloads, contradicting the principle of efficiency, a key goal of Smart Cities.

b. Human Resource Challenges

In the human resources sector, the most significant challenge is the low digital literacy rate, both among the general public and among civil servants. Limited ability to operate and utilize various digital services has resulted in suboptimal public participation in services such as Ngalam 112 and E-JKN, despite the existing infrastructure. Low digital literacy levels among civil servants also impact service quality and system maintenance. Without sufficient skills, digital innovation risks stalling in the early stages of implementation without further, sustained development. This problem is exacerbated by the lack of public outreach and application maintenance. Many Smart City applications are still unknown or inconsistently used by the public due to a lack of public education and support. As a result, application adoption rates remain far from their potential.

c. Governance and Collaboration Challenges

In the governance dimension, the main challenge lies in the suboptimal coordination between Regional Apparatus Organizations (OPDs). The cross-sectoral implementation of Smart City requires policy and data integration across OPDs, but in practice, sectoral egos and differing priorities between agencies are still evident. Hexahelix-based collaboration involving government, academia, the business world, the public, the media, and communities has also not been fully implemented. The involvement of non-governmental actors is still partial and has not been strongly institutionalized in the planning and evaluation process of Smart City policies. Bureaucratic resistance to digital reform, particularly in the implementation of SPBE, also poses a barrier. Changing work patterns from manual to digital systems often faces organizational cultural challenges, which require time and an adaptive managerial approach.

The impact of these management issues is clearly evident in the limited results for the Smart Economy (86%) and Smart Branding (89%). Shortcomings in digital economic development, MSME integration into digital platforms, and the utilization of data-driven tourism branding indicate that the city's economic and identity aspects require greater attention and a more in-depth improvement strategy. Overall, these challenges demonstrate that the success of Malang's Smart City is determined not only by the quantitative results of program implementation, but also by the quality of its implementation. In the future, strengthening regulations, equitable distribution of digital infrastructure, improving human resource capabilities, and collaboration and consolidation of management across sectors are crucial to ensuring the inclusive, flexible, and sustainable development of Malang's Smart City.

Challenge Mitigation Strategy

Here is a SWOT Analysis and Mitigation Strategy based on the findings in the document to strengthen the discussion section of your journal:

A SWOT analysis of the Smart City implementation in Malang City shows that the main strength lies in the high commitment and effectiveness in implementing the program, as evidenced by the realization of 96.22%. This achievement is supported by digital infrastructure readiness (E-Readiness) which reached 58.3%, exceeding the minimum national set standards. This situation indicates that Malang City has a sufficient technological, institutional, and management foundation to support the transition to a smart city. Furthermore, the implementation of the Hexahelix collaboration model is a strategic element that creates significant opportunities for innovation, especially in the creative and education sectors, and increases the community's digital literacy by actively involving academics, communities, business actors, the media, and the banking sector. However, the analysis also identified several weaknesses that could hinder the optimization of Smart City implementation. The main weakness lies in the suboptimal interoperability and integration of data between institutions or OPDs, which results in a fragmented system and reduces the effectiveness of digital-based public services. Furthermore, the capabilities of human resources, both from the state civil service and the community, are not yet fully prepared to manage, utilize, and develop digital systems sustainably. In terms of threats, the sustainability of the Smart City program is at risk if it is not supported by strong and adaptive regulations, including the challenges of cybersecurity and public data protection which are increasingly becoming a concern, along with the increasing dependence on digital systems.

Based on the analysis, the mitigation strategy aims to strengthen governance, technology, human resources, and regulations. One important step that must be prioritized is the development of the "One Data Malang City" architecture as the basis for integrating information systems between regional government agencies (OPDs). This architecture aims to ensure standardized, accurate, and real-time data exchange, support data-driven decision-making, and improve the quality of public services. Implementation of this strategy needs to be balanced with improving the digital competency of civil servants through ongoing technical training, along with broader outreach to the public to increase the adoption rate of digital services. From an institutional and legal perspective, the development of a Regional Regulation (Perda) specifically for Smart City is a crucial step to provide regulatory certainty, maintain policy continuity across various leadership periods, and strengthen the security and protection of public data. Furthermore, the mitigation strategy also includes regular maintenance, evaluation, and audits of technology infrastructure to ensure system reliability, minimize the risk of technical disruptions, and maintain and improve technology readiness scores for the future. With this integrated mitigation approach, the hope for Smart City implementation in Malang City is not only quantitatively successful, but also sustainable and adaptable to technological changes and community needs.

CONCLUSION

An evaluation of the Smart City implementation in Malang City shows that the city has achieved significant success and is relatively well-developed in the smart city development aspect. The program implementation percentage of 96.22% indicates that most of the plans listed in the Malang Smart City Master Plan have been implemented consistently and measurably. This result reflects a strong synergy between strategic planning, operational policies, and program implementation in the field. In terms of technical readiness, Malang City has met the requirements in the digital readiness aspect with a score of 58.3%, which is higher than the national minimum limit of 50%. This indicates that the existing technological infrastructure, information systems, and institutional capacity are sufficient to support the digital transformation process in government and public services in a sustainable manner.

The successful implementation of Smart City is inseparable from the implementation of the Hexahelix collaboration model, which emphasizes synergy between various actors as the primary foundation of development. The active participation of local governments, educational institutions, communities, businesses, the media, and the banking sector has created a collaborative ecosystem that supports innovation, expands resources, and strengthens the legitimacy of Smart City policies. From a policy perspective, the primary focus of implementation is allocated to strengthening the Smart Governance dimension as a pillar of bureaucratic reform. The use of digital technology in government management has driven increased efficiency, transparency, and accountability in public services. The integration of service systems is a crucial tool in building public trust in local governments.

In terms of the Smart Economy, the Smart City policy serves as a driver for the growth of the creative and digital economy. Various incubation efforts, support for MSMEs, and the use of digital media have encouraged the development of economic activities based on local innovation. This aspect has significantly contributed to strengthening Malang City's economic competitiveness at the regional and national levels. Furthermore, strengthening the Smart Society aspect aims to increase citizens' digital literacy and inclusion. Providing broader access to digital infrastructure, coupled with educational programs and community engagement, has increased the community's ability to utilize government digital services and actively participate in the Smart City ecosystem.

The Smart Branding dimension serves to strengthen Malang City's identity through the use of digital technology to promote the city, its tourism sector, and the creative economy. The integration of digital innovation with local historical and cultural values strengthens Malang City's image as a creative, adaptable city, yet still grounded in its local identity. Meanwhile, the Smart Living and Smart Environment dimensions contribute to the sustainable improvement of the community's quality of life. The application of technology in the health, security, emergency response, and environmental management sectors demonstrates that Smart City in Malang City focuses not only on administrative efficiency but also on long-term well-being and sustainability.

While implementation is considered very high, the study also identified several issues that require serious attention in the future. Key issues include the lack of interoperability and data integration between various agencies, which is not yet fully effective, which can hamper the efficiency of integrated services and data-driven decisions. Furthermore, there is a continued need to improve human resource capacity, both among civil servants and the general public. Without equitable digital competency development, the sustainability of Smart City innovations could potentially be hampered during the operational and subsequent development phases.

Another crucial aspect is strengthening flexible regulations and policies to adapt to technological advances. Firm and sustainable regulations are needed to ensure innovation, data protection, and Smart City implementation can continue amidst policy changes and increasingly complex urban challenges. Overall, the results of this evaluation indicate that Malang City is on the right track in implementing the Smart City concept and can serve as a national best practice. However, continued success depends heavily on the local government's ability to improve the quality of system integration, strengthen human resource capacity, and maintain consistent intersectoral collaboration to address future urban development challenges.

REFERENCES

Al Zahrah, R., & Putra, T. F. (2025). The Smart Economy Concept in the Creative Economy in Balikpapan City. Performance. *Journal of Economics and Management, 22*(2), 244–250. DOI: 10.30872/jkin.v22i2.15453

Cendekiawan, K. A., & Wardhani, F. A. (2025). The Role of Information Technology and Product Innovation on the Competitive Advantage of Companies at Jember East Java. *Dirya: Journal of Economic Management, 2*(1), 40–47.

Damayanti, D., Suyeno, S., & Putra, L. R. (2023). Evaluation of Malang City Government Policy on Traffic Engineering in Overcoming Congestion (Study at the Malang City Transportation Agency). *Public Response, 17*(6), 17–27.

Irhab, MD, & Nugroho, A. (2024). Application of decision tree regressor in business intelligence development for predicting tourist movement in east java. JSR: Robotic Information System Network, 8(2), 126–135.

Kurniawan, T., Sihombing, A. M., & Berliane, A. (2023). Konstruksi Politik Hukum Pidana Terhadap Delik Perzinaan Dalam Rancangan Kitab Undang-Undang Hukum Pidana. *Binamulia Hukum, 12*(1), 11–24. DOI: 10.37893/jbh.v12i1.445

Laurens, S., & Salauddin, J. (2024). Implementation of the Smart City Concept in Improving the Quality of Urban Public Services. JIIP - Scientific Journal of Educational Sciences, 7(11), 13130-13137. doi:10.54371/jiip.v7i11.6291

Maharani, ES, Fatima, S., & Pratiwi, A. (2024). Malang City Government Strategy in Improving the Quality of Public Services: Critical Analysis and Prospects for Improvement. Idarotuna: Journal of Administrative Science, 5(1), 71-84.

Mahendra, G. S., Wali, M., Idwan, H., Listartha, I. M. E., Yuliastuti, G. E., Sasongko, D., & Jude, G. A. (2022). *Keamanan komputer*. Galiono Digdaya.

Noor, Q., Ilyas, A., Javaid, Z., & Khan, H. (2023). Framing a knowledge domain visualization on green human resource management: A bibliometric analysis from 2008-2022. *Pakistan Journal of Humanities and Social Sciences, 11*(4), 4200–4212. DOI: 10.52131/pjhss.2023.1104.0684

Rabbani, H., & Yuliani, D. (2025). HEXAHELIX COLLABORATION STRATEGY BASED ON EMPOWERMENT OF THE SERIKANDI BIRU COMMUNITY IN UTILIZING GARUT REGENCY RESOURCES. *DEVELOPMENT: Journal of Community Engagement, 4*(3), 422–439. DOI: 10.46773/djce.v4i3.2564

Rachmawati, R., Susilowati, A. P. E., Radika, P., Sofie, T., & Aditya, M. Y. (2024). Sustainable Regional Development Through the Development of Smart Village and Smart Branding: A Case Study of Bangsri Village, Blora Regency. *Intelligent Systems Journal*, *7*(2), 199–214. DOI: 10.37396/jsc.v7i2.441

Rahadiansyah, M. (2023). *PENGARUH KUALITAS PRODUK, HARGA, DAN LOKASI TERHADAP KEPUTUSAN PEMBELIAN PADA RESTORAN POPOLAMAMA DI TANGERANG* (Doctoral dissertation, Universitas Satya Negara Indonesia).

Sukmana, A. H. (2022). Pengaruh strategi pemasaran pasca pandemi covid-19 terhadap minat nasabah menabung di Bank BSI Mataram. *Jurnal Perbankan Syariah*, *1*(2), 26–35. DOI: 10.20414/jps.v1i2.6324

Widodo, N. (2016). Development of e-Government in Regional Government in the Context of Realizing a Smart City (Study in the Regional Government of Malang City). *Scientific Journal of Public Administration*, *2*(4), 227–235. DOI: 10.21776/ub.jiap.2016.002.04.15

Winata, R., & Friantini, R. N. (2019). Pengaruh motivasi belajar terhadap prestasi belajar matematika siswa kelas VIII SMPN 1 Kuala Behe. [Jurnal Ilmiah Pendidikan Matematika]. *JIPM*, *7*(2), 85–92. DOI: 10.25273/jipm.v7i2.3663

Chapter 11

When Innovation Outpaces Sustainability: The Environmental Risks of Rapid AI Expansion

M. Poovayar Priya
http://orcid.org/0000-0002-3200-4634
Vel Tech Rangarajan Dr. Sagunthala R&D Institute of Science and Technology

ABSTRACT

Artificial Intelligence (AI) is widely celebrated as a transformative force capable of solving some of humanity's most pressing challenges, from climate modelling to medical diagnostics. Yet this narrative of technological promise often obscures a less visible reality: AI systems are deeply material, dependent on vast quantities of energy, water, minerals, and hardware that carry significant ecological and social costs. This chapter critically examines the environmental footprint of AI across its full lifecycle, addressing five interconnected dimensions: escalating energy consumption, water usage in cooling systems, mineral extraction for hardware, e-waste generation, and carbon emissions. It argues that the structural forces driving AI's rapid expansion—competitive scaling races, market incentives, corporate opacity, and regulatory gaps—are actively outpacing the development of sustainability safeguards.

1. INTRODUCTION

AI's Paradox of Sustainability. Global technological change has taken place from almost revolution to revolution without the hindrance made to it by its reliance on artificial intelligence, which plays no small role in industry and public

DOI: 10.4018/979-8-3373-8490-0.ch011

life, commerce, and politics. Few developments in recent history have so quickly and comprehensively gripped public imagination and investment of capital as AI technologies like that. AI is often presented as a means to solve humanity's great challenges (the optimization of energy grids, modelling of climate systems, rapid drug discovery and enhancement of agricultural outputs under increasingly unpredictable weather conditions). This framing portrays AI as compatible not just with but as an instrumental enabler of sustainability goals. But this is a paradox tale. The systems marketed as solutions for environmental issues are, in fact, among the main causes of environmental destruction themselves. Training and running big data AI models demands vast quantities of power, clean water, rare minerals and specialized hardware — and the extraction, consumption and disposal of the systems exact devastating ecological tolls. As AI's outputs are sometimes referred to as "virtual" or "cloud-based," with a sense of dissolving weightlessness, the very infrastructure underpinning all model queries, training runs, and inference tasks is very physical.

Server farms the size of small cities consume power equivalent to mid-sized nations; cooling towers drain local water tables; mining operations deplete ecosystems in order to extract those precious metals and lithium found in every chip. Recent scholarship refers to this tension as the fundamental sustainability paradox of AI: the systems themselves claimed to be an engine for ecologically solving problems simultaneously create new environmental pressures and their costs are largely unseen to practitioners (Toderas, 2025). AI's environmental footprint has been largely invisible — and hardly discussed, given its scale. In many cases that footprint is similar to or greater than those from traditional heavy industries, and yet it receives much less public scrutiny, in part because its physical realities, whether data centers, mining sites, or e-waste dumps, are long distances away from the end users that receive most of the conveniences of an AI system. The spatial and conceptual distance that exists between the benefits of AI and its costs, in itself, is a kind of feature worth interrogating because it allows unsustainable practices to move relatively unchecked.

1.1 Literature Review: Gaps, Contribution And Methodology

An increasing body of scholarship is mapping the environmental impact of AI through complementing lenses. Frimpong (2025) provides a typology that distinguishes among efficiency-driven, sustainability-driven and critical-ecological strands of "Green AI" discourse and Strader, Huang, & Tu (2024) identifies six major research themes that emerge across energy, policy and organisational practice; Bhardwaj et al. (2025) uses technical, economic, ecological and social lenses to explore the structural limits of further scaling of AI. Much of this literature, however, focuses on AI's environmental costs more as a technical/corporate accountability issue, with comparatively little to no contribution to the two dimensions this chapter seeks to address. First, existing accounts seldom make explicit connections among AI's environmental footprint and the education sector; despite universities being among the fastest-growing institutional adopters and actors best positioned to model environmentally responsible use, both are relatively underrepresented in existing literature.

Second, although AI is being described as having a direct resource cost in terms of energy, water, minerals, e-waste, and carbon, such accounts have not always been able to tie this directly to a structural account — specifically how the gains from efficiency have not manifested into actual reductions, an aspect for which this chapter investigates the dynamic of the Jevons paradox (Luccioni, Strubell, & Crawford, 2025). This chapter is a critical review of the synthesis work, utilizing peer-reviewed and industry literature spanning environmental science, computer science, and policy studies mostly from 2018 to 2025. It follows five thematic dimensions for AI's direct impact on the environment (energy, water, minerals, e-waste, and carbon), identified in a step-by-step reading of the literature, which was followed by a structural analysis of competitive and regulatory forces constraining unsustainable growth, a conversation on the education sector's part in this footprint, and a synthesis of technical, regulatory, institutional and behavioral pathways forward. As a synthesis instead of a new empirical literature review, its methodological weaknesses stem from using data coming from secondary sources with different methodological quality—and based on numbers often quoted rather than being independently verified, the transparency issue as the chapter itself observes, in Section 7, this structural issue.

The distinct contributions that make up this chapter are thus threefold: an interdisciplinary synthesis of the lifecycle environmental costs for AI; a critical account, in Section 7, of the structural and competitive forces that reproduce innovation outpacing sustainability safeguards; and, in Section 9, a more explicit treatment of what this footprint means for universities as adopters, educators, and future leaders in sustainable AI practice. What follows is divided into ten interconnected sections. Sections 2-6 analyse AI's energy, water, mineral, e-waste, and carbon footprints in turn. Structural and economic forces driving this trend are examined in Section 7. Technical, regulatory, institutional, and behavioural pathways towards sustainability are presented at Section 8. Section 9 moves from there to what this means for higher education, while Section 10 ends with a synthesis and long-range vision.

2. ENERGY AT SCALE

The computational requirements of modern AI systems have grown at a pace with few historical parallels. Large-scale generative AI and foundation models rely on training processes involving billions or trillions of parameters, optimized over weeks or months across thousands of specialized processors running continuously. The energy needed for a single training run of a state-of-the-art model can be roughly equivalent to the annual electricity consumption of several hundred households, a figure that is rising and continues to push developers to compete for larger and more capable systems. This part explores where the energy goes, where it comes from, and what its longer-term trajectory implies for grid planning and climate commitments.

2.1 Training Versus Inference

A critical distinction in evaluating AI's energy footprint lies between the energy consumed during model training and that consumed during inference — the process of using a trained model to generate outputs in response to user queries. Training is a one-time, intensive process, but inference occurs continuously at scale, across millions or billions of queries per day once a model is deployed. As AI tools become embedded in search engines, productivity software, and consumer applications, the cumulative energy cost of inference can eventually exceed that of training, even though each individual query consumes comparatively little energy (Strader, Huang, & Tu, 2024). Figure 1 illustrates this dynamic conceptually: training produces an early, concentrated spike in energy demand, while inference accumulates more gradually but, at sufficient adoption scale, can overtake cumulative training costs.

Figure 1. Conceptual illustration of training versus inference energy over time

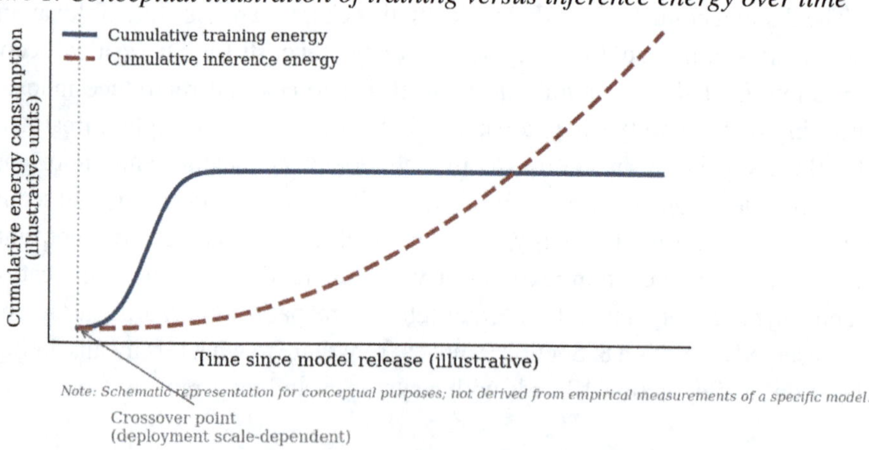

That change in where energy demand is concentrated carries implications for how energy efficiency should be assessed and where intervention can be aimed: optimizing a single training run isn't nearly as consequential, on a macroscale, as optimizing the per-query cost of inference once a model is deployed on a large scale.

2.2 Geographic Concentration of Data Centers

The deployment of AI infrastructure is not evenly spread through data center locations throughout the world. They tend to cluster in areas that provide the kind of favorable conditions we've been describing: cheap land, tax breaks, nearness to fiber-optic infrastructure, and — critically — access to cheap electricity. This concentration introduces localized pressure on the national and regional power grids, especially in regions where expansion of data centers has outstripped grid capacity planning. In some areas, demand for data center electricity has surged so fast that utilities have had to defer the retirement of fossil fuel plants or hasten new fossil-based generation just to satisfy near-term demand, undercutting longer-term decarbonization trajectories. From 2024 through 2030, global data center electricity demand is expected to more than double, with natural gas and coal providing a significant portion of the additional near-term demand, even as renewable and later nuclear generation grows (International Energy Agency [IEA], 2025).

2.3 Renewable Constraints and Long-Term Sustainability Concerns

Even though the public face of a growing number of AI firms will emphasize that it will procure renewable energy, grid mechanics make that claim less likely. Electricity grids are interconnected systems in which the marginal unit of power consumed by a new data center is not commonly the same renewable electricity that a company has purchased on paper through power purchase agreements or renewable energy certificates; where renewable generation is intermittent or inadequate to keep up with new demand, the marginal source that fills the gap is often natural gas or coal. The disconnect between these corporate sustainability claims, and grid-level electricity sourcing realities is a continuing challenge for transparency, which is made worse by the fact that over a longer period of time, the electricity consumption in a data center could double or triple within a decade if future scaling trends remain unchallenged (Goldman Sachs, 2024). It also places AI development in stark contrast to national and international climate priorities, as demand growth this magnitude is not immediately sustainable with renewable buildout alone in the same time frame. Fixing this tension will take not just the design of AI architectures better for energy optimization, but structural alterations to how the demand for energy is forecasted, regulated and priced for big computationally intensive users — a question that is resurrected in Section 8 in this chapter.

3. WATER IN THE MACHINE

But AI infrastructure does not only consume energy. Furthermore, data centers consume large quantities of water, mainly cooling the servers that provide immense heat during operation. A single large data center can use millions of liters of water per day; that is the same as that used by a small city. This water usage is not as much discussed by the general public in the media as energy consumption — despite potentially significant impact on environmental sustainability and resource equity.

3.1 Cooling Technologies and Their Water Footprints

Different cooling technologies are used throughout the data center industry, with some having different water and energy benefits and disadvantages. Evaporative cooling systems are commonly used due to their energy efficiency by evaporating water during cooling rather than circulating it. Air and liquid immersion-based cooling systems are considered solutions to minimize direct water consumption, but often move the burden toward higher overall electricity consumption or require specialized infrastructure that is not widely in use. The choice of cooling technology is thus also an assumed trade-off between water and energy footprints, and it is something that companies frequently don't expose easily to the public or regulators or both.

3.2 Regional Water Stress Impacts

Data center water usage is also of greater environmental importance in drought-prone or water-stressed regions. Some of the same favorable conditions that help drive the investment of data centers — low land prices, available space and supportive local tax policy — can sometimes be found in arid or semi-arid regions where it is already difficult to meet agricultural or residential water demands. In circumstances like these, building a massive data center can pit operators against local farmers and residents as the competing application of a scarce and dwindling resource in turn brings up critical issues of environmental fairness and distributive justice.

3.3 Transparency Issues Around Water Reporting

Water consumption by a data center is also not disclosed publicly, unlike carbon emissions, where the reporting frameworks have at the very least some standardization in place. Facility-level water consumption for companies is often not reported, and voluntary sustainability disclosures may vary widely in both coverage and depth. In the absence of transparency, regulators, researchers, and affected communities have difficulty judging the extent of AI's water footprint on a scale sufficient to be measured correctly, and no one can hold operators accountable to stewardship of their resource.

3.4 Climate–Water–AI Interdependencies

AI, water and climate change go hand-in-hand. Climate change is exacerbating drought in many of the same regions where data centers are built, while intensifying local water stress and increasing atmospheric temperatures, which in turn raises cooling requirements. This sets up a feedback loop in which climate change amplifies the resource stressors exacerbating the AI infrastructure growth, thus multiplying these risks for ecosystems and communities already more susceptible to water shortages. Water is just one of many material inputs whose extraction comes at a large environmental cost; the next section investigates the accompanying mineral supply chains that form AI hardware itself.

4. MINING INTELLIGENCE

At a deep level of abstraction under cloud computing and machine learning models is the extractive industry. AI hardware — the GPUs, TPUs and specialized chips that execute the computations used to make every model work — relies on cobalt, lithium, rare earth elements, copper and other minerals requiring extraction from the earth, often through processes that are environmentally destructive and socially exploitative.

4.1 Toxic Waste and Environmental Pollution

Mining to supply semiconductor and battery industries often generates toxic byproducts, such as heavy metals and chemical runoff that can seep into surrounding soil and waterways. In less-enforced environmental areas, this contamination can persist for decades, and can alter farming soil, drinking water supplies or local biodiversity, long after a mine is shut down. Such contamination rarely bears the environmental cleanup costs of technology companies that profit from the extracted materials as an economic benefit, leading to a structural gap between beneficiaries and ecological cost.

4.2 Deforestation and Disrupted Habitat

Mineral extraction usually involves clearing forested regions for mines, access roads and processing facilities. This clearance leads to habitat loss, reduces biodiversity and releases carbon that had been stored in forest ecosystems. In many of the world's most biodiverse areas, the growth of mining associated with global technology supply chains remains one of humankind's largest — and least discussed — contributors to ecological deterioration.

4.3 Labor Exploitation and Community Displacement

Mineral extraction comes at the economic and social costs that are as significant as the environmental ones. Mining operations historically have typically led to the dispossession of local and indigenous populations whose options for legal recourse against multinational mining interests or the governments that provide them with concessions have been narrow. Many mining regions, especially in informal or artisanal mining sectors, have had labor conditions documented to include exploitative wages, hazardous working conditions and, in some cases, child labor. These social harms also combine resource conflicts, when the control over mineral-rich territory is violence-inducing and results in political instability.

4.4 International Supply Chain Inequalities

The geography of AI's mineral supply chain shows vast inequalities. Its social and environmental costs of extraction are concentrated in the Global South, while the economic and technological wealth of AI is disproportionately accrued by firms and consumers in wealthier countries. This asymmetry is reminiscent of historical cycles of extractive resource extraction and exploitation which have characterized industrial history, and implies that AI is, notwithstanding its future-oriented framing, doing more (or less) what historical extractive economic structures have done before. But once embedded in finished hardware, those material inputs finally hit the end of their usable life — a passage that creates its own unique environmental crisis, reviewed next.

5. E-WASTE AND HARDWARE CRISIS

The same competitive pressures that make rapid AI advancement so rapid are accelerating hardware obsolescence. As faster, bigger chips are released, the useful lifespan of current GPUs, servers and other supports is reduced at an ever-more rapid pace. This has developed into an accumulating flow of electronic waste which the existing electronics recycling infrastructure is unable to accommodate.

5.1 The Acceleration of the Disposal of Hardware

Data centers on the frontiers of AI capability, for example, regularly replace hardware on cycles sometimes measured in just a few years — or even sooner — as new chip architectures provide significant performance improvements that render ongoing use of older hardware commercially untenable. This turnover is not comparable to the decades-long lives of well-maintained computing hardware found in less performance-sensitive applications, highlighting the manner in which competitive dynamics in AI development serve to shape material throughput and create waste.

5.2 Toxic Items in AI Waste Pollution Elements in AI Related Waste

AI hardware that is discarded includes a mix of materials, some valuable and recoverable, others hazardous for the environment. Printed circuit boards, batteries and certain electronics components, including circuit board materials, batteries as well as PCBs and some chips are rich in heavy metals such as cadmium, lead, mercury and flame retardants, as well as other chemicals, posing hazards to human health and the environment if not handled properly during disposal or recycling.

5.3 Flows of Global E-Waste to Developing World

A significant amount of electronic waste produced in richer countries is shipped, in accordance with local laws, to less developed nations where the cost of labor and processing is not as high and environmental requirements often less strict. In these spots, e-waste is commonly handled through an informal, manual process — including open burning to recover valuable metals — that emits poisonous fumes and that pollutes local soils, places workers and all neighbouring communities at risk of serious health crises.

5.4 Limitations of the Existing Recycling Systems

Where formal recycling programs are in place, the systems are unable to provide enough recycling to a wide range of AI hardware products. Many components are hard to disassemble, contain materials in forms that are uneconomical to recover or simply are not covered by existing extended producer responsibility regulations. What you get is a recycling gap in which a portion of the AI hardware that gets discarded gets recycled responsibly while the rest goes to accumulating quantities of ungovernable electronic waste. The carbon cost of the AI hardware lifecycle goes far beyond waste itself — from mining to manufacturing, to disposal — a fact that is often ignored when the focus is solely on operational electricity consumption, which we will examine in the next section, in greater detail.

6. THE CARBON COST OF INTELLIGENCE

The carbon footprint of AI carbon impact goes well beyond just the electricity power used to power the big data center. A full accounting of the climate impact of AI must take emissions from its entire lifecycle, including chip and server manufacturing, as well as transportation and deployment and at the disposal stage, into account.

6.1 Lifecycle Assessment (LCA)

It is a methodological approach that assesses the environmental impact of a product or system across the entire life cycle. In AI, the application of LCA shows that a vast share of total emissions originate in the manufacturing process and not from the operational energy input required in the process of manufacturing semiconductors requiring highly controlled surroundings, specialized chemicals and energy consuming processes. A full LCA approach is mandatory to not make the frequent mistake of defining the environmental impact of AI in terms of the electricity bill alone.

6.2 Embedded Versus Operational Carbon

Embedded carbon — the amount created by manufacturing hardware before it is ever turned on — can be a large fraction of a data center's overall lifecycle emissions, especially now that operational energy in many areas is shifting toward alternatives that are more renewable. This difference is important for policy and corporate accountability: a concentration on operational carbon alone risks downplaying the actual climate cost of quick hardware turnovers and aggressive scaling strategies that require constant production of new chips and servers.

6.3 AI Companies' Carbon Reporting Challenges

Corporate carbon disclosures for the AI sector are mostly incomplete, inconsistent, or not possible to verify independently. Companies may report emissions related to their own emissions without including upstream emissions from chip manufacturing or downstream emissions arising from customer use of their AI products and services. Lack of standardized compulsory and independently audited carbon reporting tailored to AI workloads makes it impossible to compare firms, monitor trajectory over time, and monitor the industry's compliance with climate pledges (Balsalobre-Lorente, Abbas, He, Pilar, and Shah, 2023).

6.4 Misinterpretations of AI as "Virtual"

A stubborn belief remains that AI's environmental footprint — since it works via software and cloud infrastructure — is lighter than what physical industries need to do. This framing masks the deeply material reality beneath every AI system: physical chips, physical data centers, physical cooling systems, physical mines. Correcting this misconception is a prerequisite that must be taken before getting real environmental accounting and meaningful policy responses. With a review of AI's absolute resource costs across five dimensions, the chapter goes on to ask why AI consumes rather than what: the structural and competitive pressures that continue driving consumption at a pace faster than the realisation of the costs.

7. WHEN INNOVATION WINS SUSTAINABILITY

Now that we have analyzed the particular environmental impacts of AI's energy, water, materials, and carbon footprints, this section returns to the structural and systemic forces behind our path of unsustainable growth. Understanding these forces is central, because it requires more than just incremental efficiency improvements to deal with AI's carbon footprint: economic and competitive logic that encourages ever-larger variants regardless of what they mean for resource costs needs to be confronted. Bhardwaj, Alexander, & Becker (2025) maintain that technical, economic, ecological and social pressures are intertwined closely in AI scaling, in which attempts to address limits on a single dimension often displace the pressure, rather than resolving it altogether. Figure 2 presents the four structural drivers studied below and the cumulated influence of these forces on the trajectory.

Figure 2. Structural drivers of AI's unsustainable growth trajectory

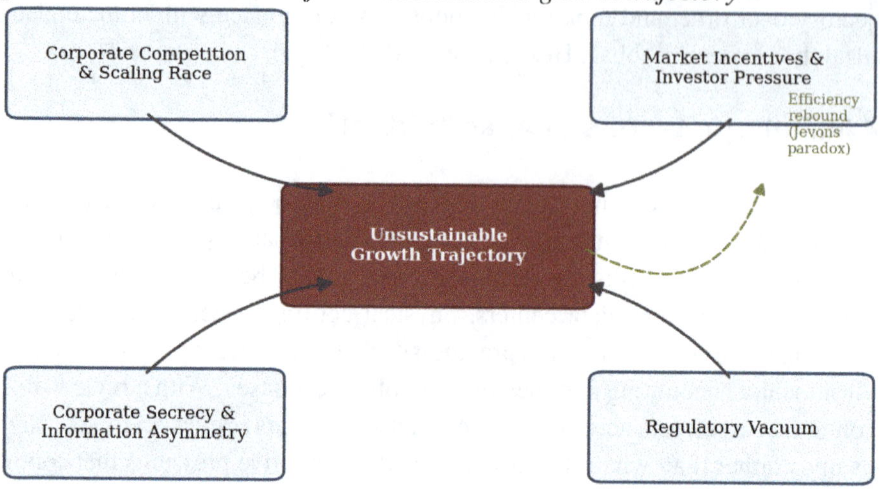

7.1 The challenge of growing AI competition in the large company space

The domain of AI has intense competition because there are just a handful of powerful companies in it trying to develop and deploy the AI models they could dominate their competition. In fact, its model scaling capabilities are often viewed as more desirable than resource usage and investment efficiency by firms that believe that a slowdown in development will be a loss in production or in the sales of AI products. The result is an epic race to the top end of the scale in a world race where environmentalism is often seen as a second-rate commodity or side item to being driven by the technology leadership requirement.

7.2 Market Incentives and Investor Pressure

The financial markets tend to like AI companies that are capable of scaling at a fast pace. This will go-up to AI industries that outperform the others as they scale rapidly, often paying much more on scale and performance indicators than sustainability credentials: they get their dividends from financial markets when they take big data for granted to have high value, the financial markets have rewarded them far outweighing those who show to have a higher level of performance. And it perpetuates a discrepancy between what investors want — and what environmental responsibility demands — as companies are pressured to focus on growth that's available to the public rather than environmental costs that are relatively hidden to investors.

7.3 Information Asymmetry and Corporate Secrecy

Many leading AI companies consider detailed data from businesses on their energy use and water and supply chain practices as either private material and or commercially sensitive; while others have decided that it is not necessary. Researchers, regulators, and consumers are unable to assess what environmental costs there are to AI in terms of the public's ability to measure their contribution to environmental degradation, through this secrecy is that researchers, regulators, and public can't properly assess the long-term environmental costs. This lack of transparency can result into an information deficit that allows existing unsustainable technologies to continue by protecting them from external scrutiny and the threat of accountability.

7.4 The Regulatory Vacuum

Laws on AI's environmental impact and regulation on its environmental impact are only beginning to emerge as it percolates and development of the industry is increasing at a rapid rate. Current government environmental rules had been designed before the advent of AI on a large scale and are likely to be insufficient to deal specifically with the resource demands – particularly water and embedded carbon – unique to this technology. This absence of regulation permits competitive dynamics and market forces to operate in near-unrestricted fashion, bolstering the other structural drivers referred to previous in the current study.

7.5 Finally Dynamic

The Re-emergence of Efficiency. One last structural force cuts across the four drivers described above instead of being separate. And where companies do make individual models or chips more efficient, even though they succeed, these improvements don't go on to necessarily result in lower aggregate resource consumption. It's precisely the AI-specific manifestation of the Jevons paradox: if compute gets cheaper and more powerful per unit in this era (or more efficient per unit), it is easier to reach, hence broadening the AI market and leading to an increase in the total amount of energy demand, while reducing the per-query energy consumed. This process can also increase overall demand for energy (Luccioni, Strubell, & Crawford, 2025). The efficiency-enhancing technologies like DeepSeek were achieved in early 2025, to do just this: instead of lowering the compute demand throughout the industry, lower inference costs led to the commitment of vast infrastructure investments from major technology companies, as lower cost per query opened up a new set of use cases and deployment to wider networks. Any solutions architecture, which will be discussed following, must therefore treat that as useful but not sufficient.

8. GREENING THE MACHINE

Despite the major issues discussed earlier in this chapter, such findings in this chapter do not imply that the environmental costs of AI are incommensurable. Instead, it posits that there are real pathways for AI innovation to be paired with environmental sustainability, given industry, policymakers, and researchers make commitments to the structural rather than cosmetic changes. To be cohesive with our approach, considering the diverse actors that implement them, the pathways below fall into four classes: technical, regulatory, institutional and behavioral interventions.

8.1 Technical Interventions

There are some substantial opportunities for architectural innovation to drive down the energy footprint of AI. Some strategies like model compression, sparse models, knowledge distillation and more efficient training algorithms have a significant effect on reducing the computing resources necessary for comparable performance. But hardware manufacturers' work might be complemented by design choices that further promote durability of equipment, reuse of components, and mitigations against reliance on extraction methods most hostile to the environment — modular hardware designs that enable component-level upgrades rather than complete system renewal could significantly decrease the pace of e-waste creation typical of very rapid AI scaling. Most importantly, though, the rebound dynamic we discuss in Section 7.5 means that the technical efficiency gains should be coupled with demand-side and regulatory measures rather than viewed as sufficient per se; efficiency without a complementary cap or pricing mechanism risks being simply a facilitating force for further scaling (Luccioni, Strubell, & Crawford, 2025).

8.2 Regulatory Interventions

Implementing standardized, compulsory, and independently verifiable environmental reporting requirements for AI companies on energy use, water consumption, embedded and operational carbon, and supply chain activities would close the gaps for transparency that have been pointed out in this chapter. These standards would allow for meaningful comparisons across companies, underpin evidence-based policymaking, and establish structures for accountability currently missing from voluntary types of corporate disclosure practices. Regulation would also contribute in narrowing the regulatory vacuum described in Section 7.4, especially on facility-level water disclosure and extended producer responsibility over AI hardware.

8.3 Institutional Challenges and Interventions

True progress on data center sustainability will need no paper renewable energy credits, but rather tangible co-location of facilities with new, additional renewable generation capacity and localizing for the data centre with new water-efficient cooling capacities using new cooling technology and an understanding of siting that considers local water stress and grid capacity, besides on the ground. A prominent institutional trend (as of 2025) has been the recent pivot of several large technology firms to long-term nuclear power agreements, from plant restarts and commitment to small modular reactors, as a means of supplying firm energy for low-carbon baseload power generation for data centers (IEA, 2025). Although this change may reduce the near-term dependence on fossil generation, it also implies long lead times and high capex, and does not itself solve the water and mineral pressure mentioned in Sec. 3 and 4. Applying principles of the circular economy to AI hardware— designing for disassembly, creating strong take-back and refurbishment programs, investing in advanced recycling technology capable of recovering critical minerals from equipment going through use—could go a long way toward alleviating both extraction pressures in mining regions and the e-waste toll on disposal sites, although this must be done with multi-component, integrated investment from chip designers to data centers to formal recycling businesses.

8.4 Behavioral and Institutional Demand Interventions

With the rebound dynamic that emerges in Section 7.5 indicating efficiency gains are predominantly being absorbed in expanded demand, sustainability strategies will also need to consider how institutional buyers adopt and deploy AI, not only how it is constructed. Large institutional buyers of AI services, from corporations to government and universities, can influence market incentives through procurement decisions that weigh environmental disclosure and efficiency alongside capability of sourcing providers and implementing tools. Section 9 explores in detail this institutional-demand lever from the higher education sector, where the tension between pedagogical innovation and environmental stewardship is especially pronounced. None of these four sets of solutions is good enough in isolation, but together they provide a unified blueprint for how responsible AI development could resemble what this article suggests: technical efficiency alongside regulatory protection from rebound effects, institutional deployment of truly low-carbon infrastructure, and demand-side responsibility for those organizations that buy and deploy AI at scale.

9. AI IN THE ACADEMY

Environmental Implications for Higher Education. The dimensions of AI's environmental footprint explored in Sections 2 through 6, and the structural drivers investigated in Section 7, are not limited to the technology field. Colleges and universities are two of the fastest-growing institutional adopters of generative AI, embedding such tools as large language model chatbots into teaching, research support, administrative workflows, and student services. This section analyzes what that adoption might look like through the view of the remainder of this chapter, and what universities, consumers and teachers alike, are in a position to do about it.

9.1 Power Costs to the Educational AI Adoption

The hidden cost that is also hidden. The classification of AI training and inference introduced in Section 2.1 is of special interest to educational organizations, which have access to AI predominantly on the inference side: a university does not train its own foundation models, but generates huge cumulative inference burden as thousands of students and staff are responding to queries daily across writing assistance, code creation, tutoring and administrative work. This demand is largely invisible in institutional sustainability accounting, which typically tracks direct campus energy use but not off-site data center load generated by licensed AI tools. Widespread adoption of generative AI in higher education raises exactly the sort of scaled, dispersed inference demand that Section 2.1 identifies as able to surpass over time the power cost of training, although no single campus query in itself seems environmentally significant. The rate of that adoption, and the factors shaping it, have also come under active investigation: reviews of generative AI's usage of higher education note a very rapid spread that occurs in pedagogy and assessment practice, mostly without associated environmental damage assessments (Bittle & El-Gayar, 2025).

9.2 Illustrative Case

Institutional AI Procurement Trade-offs Between Market Forces. The next composite situation, based upon general patterns reported in the literature of higher education technology adoption as opposed to for any particular institution (namely, any institution), demonstrates some of the trade-offs universities experience. Student support and staff productivity drive a public research university to negotiate an institution-wide license for a generative AI writing and tutoring assistant. The procurement process balances licensing cost, data privacy terms, and pedagogical features, but there is no requirement for vendors to disclose if the supplyable inference infrastructure is powered by renewable energy sources, is using water, or is burning carbon (the request for proposals does not include these variables). After deployment, use grows rapidly among thousands of students and the institution has no system for measuring its own part in the vendor's overall environmental footprint, much less how well that footprint stacks up against any alternative vendor. This situation is reflective of a wider issue: although environmental criteria are missing from most institutional AI procurement processes, the universities currently have little pressure — and little access to information — needed to incentivize more sustainable vendors, further entrenching the information asymmetry described at this institutional level in Section 7.3.

9.3 Curriculum, Awareness and Sustainability Competencies

Universities are not only the consumers of AI; they are the key institutions charged with training the next generation of AI researchers, engineers and regulators. Computer science and engineering programs covering model architecture, training and deployment without taking into consideration environmental cost would result in professionals who consider sustainability a marginal after-action rather than a design consideration, and could reinforce the innovation-outpaces-sustainability dynamic explored in Section 7 in generations to come. There is yet one practical way forward for embedding sustainability competencies in AI-related education — lifecycle assessment, energy-aware model selection, and the Green AI typology which includes efficiency-oriented, sustainability-oriented, and critical-ecological approaches (Frimpong, 2025) — instead of seeing it as an isolated ethics course as it should in technical education alone.

9.4 Systematic Strategies to Sustainable AI Adoption

A number of specific actions are derived from the analysis above. Universities might first embed environmental disclosure mandates in the procurement practices of AI, asking vendors for details on energy sourcing, water usage, and carbon intensity, as well as standard privacy standards and educational standards. Second, institutions can monitor and report on their own AI-related inference footprint in the context of broader campus sustainability disclosures, narrowing the accounting gap highlighted in Section 9.1. Third, faculty governance bodies can mandate that when new AI tool adoption goes public, these considerations are made overtly, as opposed to leaving procurement decisions solely to individual departments or vendors. Finally, individual universities may fall short of the collective visibility or purchasing power to take on the standardized environmental reporting requested in Section 8.2 — through consortia or shared procurement frameworks — which they may possess, given the relative inability of individual universities to own that market leverage. These strategies, when taken collectively, place universities not only as passive consumers of AI's environmental footprint, but also as institutions that can influence how that footprint is displayed, quantified, and ultimately mitigated.

10. CONCLUSION

This chapter has explored the environmental footprint of artificial intelligence in the context of five interrelated domains — energy, water, minerals, e-waste, and carbon — and mapped these costs to structural forces, such as corporate competition, market incentives, information asymmetry, and regulatory gaps that prioritize scale above sustainability. They have also demonstrated that efficiency gains alone cannot close this tension, particularly given the rebound effects described by the Jevons paradox, and that this footprint directly cascades into higher education — where institutions are major inference-side consumers of AI with an underutilized ability to influence vendor performance through purchasing, curricula, and collective action. The technical, regulatory, institutional, and behavioral roads laid out in Section 8, alongside education-specific roads in Section 9, on their own offer no guarantees of a sustainable AI ecosystem; each one is not enough in a vacuum, and at an institutional level they require actors — firms, policymakers, agencies, institutions — to embrace environmental accountability as a primary design imperative, not an afterthought. What this chapter asks us to call instead for a movement away from a narrative that assumes AI is intrinsically "green" through digital framing and towards a more truthful and rigorous exploration of what it means to develop and deploy AI responsibly and in a planetary context — within industry, in policy, and, as we have pointed out in this chapter, in the universities that train the next generation of builders and at scale adopt its tools.

Significance and Original Contributions to Current Research

This chapter offers a plurality of meaningful contributions to current debates over sustainable AI:

- **Cross-discipline Synthesis**: It synthesizes research emerging from environmental studies, computer science, data center engineering, ethics, and public policy into a singular analytical frame.
- **Lifecycle Perspective**: It describes AI's impact on the ecosystem from mineral extraction to disposal, showcasing a 360-degree lens seldom brought to life in a single chapter.
- **Environmental Justice Lens**: It highlights the way in which these environmental damages of AI disproportionately impact communities already marginalized, as it particularly affects mining places and the regions containing data centers.
- **An Overview of Innovation Culture**: How competitive pressures to build larger models directly harm the environment, and how efficiency gains are absorbed via rebound effects rather than absolute reductions.
- **Academic Relevance**: It directly links AI's environmental impact to higher education as a whole, and positions universities as inference-side purchasers, as educators, as the providers of the curriculum and as potential leaders in sustainable AI procurement and use.
- **Responses that Are Practical**: It provides practical, actionable routes to make AI design, operation, and governance sustainable — divided into technical, regulatory, institutional, and behavioral perspectives for sustainability — that appeal to scholars of artificial intelligence, policymakers, business leaders, industry representatives, and academia.

REFERENCES

Balsalobre-Lorente, D., Abbas, J., He, C., Pilar, L., & Shah, S. A. R. (2023). Tourism, urbanization and natural resources rents matter for environmental sustainability: The leading role of AI and ICT on sustainable development goals in the digital era. *Resources Policy*, *82*, 103445. DOI: 10.1016/j.resourpol.2023.103445

Bhardwaj, E., Alexander, R., & Becker, C. (2025). Limits to AI growth: The ecological and social consequences of scaling. arXiv preprint arXiv:2501.17980.

Bittle, K., & El-Gayar, O. (2025). Generative AI and academic integrity in higher education: A systematic review and research agenda. *Information (Basel)*, *16*(4), 296. DOI: 10.3390/info16040296

Frimpong, V. (2025). *What counts as Green AI? Mapping efficiency, sustainability, and critical-ecological strands in a fragmented discourse*. Qeios. DOI: 10.32388/8FB2P4.2

Goldman Sachs. (2024). *Generational growth: AI, data centers and the coming US power demand surge*. Goldman Sachs Research.

International Energy Agency. (2025). Energy and AI. IEA. https://www.iea.org/reports/energy-and-ai

Luccioni, A. S., Strubell, E., & Crawford, K. (2025). From efficiency gains to rebound effects: The problem of Jevons' paradox in AI's polarized environmental debate. In *Proceedings of the 2025 ACM Conference on Fairness, Accountability, and Transparency (FAccT '25)*. DOI: 10.1145/3715275.3732007

Strader, T., Huang, Y.-H., & Tu, Y.-J. (2024). Artificial intelligence and environmental sustainability: Review and research directions. *Journal of International Technology and Information Management*, *32*(4).

Toderas, M. (2025). Artificial intelligence for sustainability: A systematic review and critical analysis. *Sustainability (Basel)*, *17*(17), 8049. DOI: 10.3390/su17178049

KEY TERMS AND DEFINITIONS

Embedded (Embodied) Carbon: greenhouse gas emissions produced in the manufacturing of hardware prior to the real-life turn-on stage from chip fabrication, materials processing, and transportation.

Inference: as opposed to training, the process of running a trained AI model on a new input to obtain an output as opposed to training to adjust the model parameters appropriately with data.

Lifecycle Assessment (LCA): a methodology for measuring environmental impacts occurring throughout the entire product or a system lifecycle, from the extraction of raw material to the making, production and disposal of the product or system.

Power Usage Effectiveness (PUE): the ratio of the total energy consumption of the data center as compared with the energy used in order to support its computing equipment on a single panel; where PUE closer to 1.0 shows that the cooling and auxiliary systems incur lower overhead.

Compilation of References

Abbadi, A., & Mucsi, G. (2024). A review on complex utilization of mine tailings: Recovery of rare earth elements and residue valorization. *Journal of Environmental Chemical Engineering, 12*(3), 113118. DOI: 10.1016/j.jece.2024.113118

Acres (Land Values). (2025, July 15). *CoreWeave locks down 1.3 GW across 10 U.S. sites in $9B deal.* https://landvalues.acres.com/all

Adler, D., & Schneider, J. (2020, December 1). Amazon workers are fighting for their rights. This holiday season, think of them. *The Guardian.* https://www.theguardian.com/commentisfree/2020/dec/01/amazon-workers-fighting-for-their-rights

African Union. (2009). *Africa Mining Vision.* African Union Commission. https://au.int/sites/default/files/documents/30995-doc-africa_mining_vision_english_1.pdf

African Union. (2025). *African Green Minerals Strategy: Beneficiation-led development for critical minerals (Building on the 2009 African Mining Vision).* African Union Commission.

Agarwal, S., & Chakraborti, M. (2025). The hidden AI race: Tracking environmental costs of innovation. /arXiv.2511.22781DOI: 10.48550

Ahl, A., Yarime, M., Tanaka, K., & Sagawa, D. (2019). Review of blockchain-based distributed energy: Implications for institutional development. *Renewable & Sustainable Energy Reviews, 107*, 200–211. DOI: 10.1016/j.rser.2019.03.002

Ahmed, S., Javaid, S., Niazi, M. F., Alam, A., Ahmad, A., Baig, M. A., Khan, H. K., & Ahmed, T. (2019). A qualitative analysis of context-aware ubiquitous learning environments using Bluetooth beacons. *Technology, Pedagogy and Education, 28*(1), 53–71. DOI: 10.1080/1475939X.2018.1557737

Airaudo, F., de Soyres, F., Fisgin, E., Gaillard, A., Santacreu, A. M., Richards, K., & Young, H. (2025, December 12). Understanding trade fragmentation. FEDS Notes. Board of Governors of the Federal Reserve System. DOI: 10.17016/2380-7172.3960

Akpan, N. E. (2024). *MACHINE LEARNING APPROACHES FOR FAULT DETECTION AND DIAGNOSIS IN POWER SYSTEMS.*

Al Zahrah, R., & Putra, T. F. (2025). The Smart Economy Concept in the Creative Economy in Balikpapan City. Performance. *Journal of Economics and Management, 22*(2), 244–250. DOI: 10.30872/jkin.v22i2.15453

Alenazi, M. M. (2023). IoT and Energy. In *IntechOpen eBooks*. IntechOpen., DOI: 10.5772/intechopen.113173

Ali, A., Shaukat, H., Bibi, S., Altabey, W. A., Noori, M., & Kouritem, S. A. (2023). Recent progress in energy harvesting systems for wearable technology. *Energy Strategy Reviews, 49*, 101124. DOI: 10.1016/j.esr.2023.101124

Ali, D., & Sofia, R. C. (2025). Experimenting with engergy-awareness in edge-cloud containerized application orchestration. *ArXiv, 2511*(09116), 1–9.

Alkan, F., & Mustafaoğlu, F. M. (2023). Using web tools in lecture: Example of micro teaching lesson. In *Fostering pedagogy through micro and adaptive learning in higher education: Trends, tools, and applications* (pp. 261–286). IGI Global Scientific Publishing. DOI: 10.4018/978-1-6684-8656-6.ch012

Alkhaleel, B. A. (2025). Perspective Chapter: The Role of Machine Learning in Developing Sustainable and Resilient Interconnected Infrastructure Systems. In *IntechOpen eBooks*. IntechOpen. DOI: 10.5772/intechopen.1012259

Almalki, F., Alsamhi, S. H., Sahal, R., Hassan, J., Hawbani, A., Rajput, N. S., Saif, A., Morgan, J., & Breslin, J. (2023). Green IoT for Eco-Friendly and Sustainable Smart Cities: Future Directions and Opportunities. *Mobile Networks and Applications, 28*(1), 178–202. DOI: 10.1007/s11036-021-01790-w

Almasoudi, F. M. (2023). Enhancing Power Grid Resilience through Real-Time Fault Detection and Remediation Using Advanced Hybrid Machine Learning Models. *Sustainability (Basel), 15*(10), 8348. DOI: 10.3390/su15108348

Altundağ, C. K., & Elmas, R. (2024). Multimedia material for science and technology teaching: Concept maps enriched with digital technologies. In *Preconceptions of policies, strategies, and challenges in education 5.0* (pp. 100–117). IGI Global Scientific Publishing. DOI: 10.4018/979-8-3693-3041-8.ch007

Altundağ, C. K., & Koçer, M. Y. (2023). Web 2.0 tools supported innovative applications in science education based on the context-based learning approach. In *Designing context-rich learning by extending reality* (pp. 179–204). IGI Global Scientific Publishing. DOI: 10.4018/978-1-6684-7644-4.ch010

Analytics, F. P. (2025, July 18). Artificial intelligence and the critical minerals crunch: Strategies for resilient supply chains. Foreign Policy. https://fpanalytics.foreignpolicy.com/2025/07/18/artificial-intelligence-critical-minerals-supply-chains/

Andrae, A. S. (2019). Prediction studies of electricity use of global computing in 2030. *International Journal of Scientific & Engineering Investigation*, *8*(86), 27–33.

Andrae, A. S., & Edler, T. (2015). On global electricity usage of communication technology: Trends to 2030. *Challenges*, *6*(1), 117–157. DOI: 10.3390/challe6010117

Andrae, S. (2025). Energy-saving alternatives in Blockchain technology. In Gupta, B. B., Pramod, D., & Moslehpour, M. (Eds.), *Sustainable Information Security in the Age of AI and Green Computing* (pp. 229–246). IGI Global Publishing., DOI: 10.4018/979-8-3693-8034-5.ch012

Andrae, S. (2026a). Justice and climate change: An interdisciplinary perspective on global environmental limits. In Meçik, O. (Ed.), *Economic Growth Amidst Social and Environmental Uncertainty* (pp. 1–44). IGI Global Publishing., DOI: 10.4018/979-8-3373-1529-4.ch001

Andrae, S. (2026b). The role of local platforms and Non-Western perspectives on Platformization. In Çelik, N., Ateşoğlu, N., & Güner, S. S. (Eds.), *Sociological Dimensions of Platformization: Power, Culture, and Identity in the Digital Age* (pp. 31–70). IGI Global Publishing., DOI: 10.4018/979-8-3373-4267-2.ch002

Anjimoon, S., Chandrashekar, R., Singh, N., Parmar, A., Sharma, N., & Mohammad, Q. (2024). Secure and Sustainable Energy Distribution through Blockchain Technology in Smart Grids. *E3S Web of Conferences, 505*, 2002. DOI: 10.1051/e3sconf/202450502002

Anthony, L. F., Kanding, B., & Selvan, R. (2020). Carbontracker: Tracking and predicting the carbon footprint of training deep learning models. *ArXiv*, *2007*(03051), 1–11.

Anthropic. (2025, November 12). *Anthropic invests $50 billion in American AI infrastructure*. https://www.anthropic.com/news/anthropic-invests-50-billion-in-american-ai-infrastructure

Anute, N., Patole, J., Pawar, P., Kalshetti, P., & Sharma, M. (2026). Digital marketing and semiotics. In Boero, M., & Piccolo, R. (Eds.), *Marketing semiotics research: Consumption trends, AI, and the economics of experience* (pp. 209–236). IGI Global Scientific Publishing., DOI: 10.4018/979-8-3693-8024-6.ch007

AnySilicon. (n.d.). High bandwidth memory – White paper. https://anysilicon.com/high-bandwidth-memory-white-paper/

Apaydın, N. (2024). Raising awareness on digital pollution: Individual and organizational productivity. *Studies in Social Science Insights*, *4*(3), 88–107. DOI: 10.53035/SOSSCI.99

Arafura Rare Earths Limited. (n.d.). *Nolans rare earths project* [Corporate website]. https://www.arafurarareearths.com/projects/nolans/

Arévalo, P., & Jurado, F. (2024). Impact of Artificial Intelligence on the Planning and Operation of Distributed Energy Systems in Smart Grids. *Energies*, *17*(17), 4501. DOI: 10.3390/en17174501

Arfelli, F., Coralli, I., Cespi, D., Ciacci, L., Fabbri, D., Passarini, F., & Spada, L. (2024). The environmental stake of Bitcoin mining: Present and future challenges. *Applied Sciences (Basel, Switzerland)*, *14*(20), 9597. Advance online publication. DOI: 10.3390/app14209597

Armanios, D. E., & Tucci, C. L. (2025). Building organizational and project readiness for AI: A dialectical approach. *International Journal of Project Management*, *43*(8), 102792. DOI: 10.1016/j.ijproman.2025.102792

Arthur, W. B. (1989). Competing technologies, increasing returns, and lock-in by historical events. *Economic Journal (London)*, *99*(394), 116–131. DOI: 10.2307/2234208

At Meta. (2025). Meta 2025 sustainability report. https://sustainability.atmeta.com/wp-content/uploads/2025/08/Meta_2025-Sustainability-Report_.pdf

Attah, R. U., Garba, B. M. P., Gil-Ozoudeh, I., & Iwuanyanwu, O. (2024a). Strategic frameworks for digital transformation across logistics and energy sectors: Bridging technology with business strategy. *Open Access Research Journal of Science and Technology*, *12*(2), 70. DOI: 10.53022/oarjst.2024.12.2.0142

Attah, R. U., Garba, B. M. P., Gil-Ozoudeh, I., & Iwuanyanwu, O. (2024b). Digital transformation in the energy sector: Comprehensive review of sustainability impacts and economic benefits. *International Journal of Advanced Economics*, *6*(12), 760–776. DOI: 10.51594/ijae.v6i12.1751

Australian Government, Department of Industry, Science and Resources. (2023). Critical minerals strategy 2023–2030. Commonwealth of Australia. https://www.industry.gov.au/publications/critical-minerals-strategy-2023-2030

Ayadi, R., Forouheshfar, Y., & Moghadas, O. (2025). Enhancing system resilience to climate change through artificial intelligence: A systematic literature review. *Frontiers in Climate*, *7*, 1585331. Advance online publication. DOI: 10.3389/fclim.2025.1585331

Baca, M. C., & Greene, J. (2019, September 20). Amazon, Google, other tech employees protest in support of climate action. *The Washington Post*. https://www.washingtonpost.com/technology/2019/09/20/amazon-google-other-tech-employees-protest-support-climate-action/

Baccini, P., & Brunner, P. H. (2012). *Metabolism of the Anthroposphere: Analysis, Evaluation, Design* (2nd ed.). MIT Press. DOI: 10.7551/mitpress/8720.001.0001

Bailey, G., Farinha, J., Mochan, A., & Pólvora, A. (2024). *Eyes on the future: Signals from recent reports on emerging technologies and breakthrough innovations to support European Innovation Council strategic intelligence* (Vol. 1). Publications Office of the European Union., DOI: 10.2760/144136

Bair, J. (2005). Global capitalism and commodity chains: Looking back, going forward. *Competition & Change*, *9*(2), 153–180. DOI: 10.1179/102452905X45382

Baldassarre, B., Buesa, A., Albizzati, P. F., Jakimów, M., Saveyn, H. G. M., & Carrara, S. (2023). *Analysis of circular economy research and innovation (R&I) intensity for critical products in the supply chains of strategic technologies*. Publications Office of the European Union. DOI: 10.2760/582527

Baldassarre, B., & Carrara, S. (2025). Critical raw materials, circular economy, and sustainable development: EU policy reflections for future research and innovation. *Resources, Conservation and Recycling*, *215*, 108060. DOI: 10.1016/j.resconrec.2024.108060

Balsalobre-Lorente, D., Abbas, J., He, C., Pilar, L., & Shah, S. A. R. (2023). Tourism, urbanization and natural resources rents matter for environmental sustainability: The leading role of AI and ICT on sustainable development goals in the digital era. *Resources Policy*, *82*, 103445. DOI: 10.1016/j.resourpol.2023.103445

Banza Lubaba Nkulu, C., Casas, L., Haufroid, V., De Putter, T., Saenen, N. D., Kayembe-Kitenge, T., Musa Obadia, P., Kyanika Wa Mukoma, D., Lunda Ilunga, J.-M., Nawrot, T. S., Luboya Numbi, O., Smolders, E., & Nemery, B. (2018). Sustainability of artisanal mining of cobalt in DR Congo. *Nature Sustainability*, *1*(9), 495–504. DOI: 10.1038/s41893-018-0139-4 PMID: 30288453

Bârã, A., & Oprea, S. (2024). Machine Learning Algorithms for Power System Sign Classification and a Multivariate Stacked LSTM Model for Predicting the Electricity Imbalance Volume. *International Journal of Computational Intelligence Systems*, *17*(1), 80. Advance online publication. DOI: 10.1007/s44196-024-00464-1

Bardi, U. (2014). *Extracted: How the Quest for Mineral Wealth is Plundering the Planet*. Chelsea Green Publishing.

Barnett-Itzhaki, Z., & Tsoury, A. (2025). From awareness to action: A UK-based study on public perceptions of digital pollution. *Sustainability (Basel), 17*(17), 7839. Advance online publication. DOI: 10.3390/su17177839

Bashir, F., Ahmed, S., & Marouf, M. (2020). An Exploratory Study of Students' Learning Performance In Flipped Classroom Using Decision Tree and Regression. *2020 3rd International Conference on Computing, Mathematics and Engineering Technologies (ICoMET)*, 1–6. DOI: 10.1109/iCoMET48670.2020.9074083

Baskaran, G. (2024, January 8). What China's ban on rare earths processing technology exports means. Center for Strategic and International Studies. https://www.csis.org/analysis/what-chinas-ban-rare-earths-processing-technology-exports-means

Baskaran, G., & Schwartz, M. (2024, October 7). From mine to microchip: Addressing critical mineral supply chain risks in semiconductor production. Center for Strategic and International Studies. https://www.csis.org/analysis/mine-microchip

Batmunkh, A. (2022). Carbon footprint of the most popular social media platforms. *Sustainability (Basel), 14*(4), 2195. Advance online publication. DOI: 10.3390/su14042195

Battery Council International. (2025). *AI and data centers: Battery applications in stationary energy storage*. Battery Council International. https://batterycouncil.org/battery-facts-and-applications/essential-applications/stationary-energy-storage/ai-data-centers/

BBC News. (2021). US leads Bitcoin mining as China ban takes effect. https://www.bbc.com/news/technology-58896545

Bekteshi, L. (2025). Education in the Era of AI and Immersive Technologies: A Systematic Review. *Journal of Research in Engineering and Computer Sciences, 3*(1), 1–14. DOI: 10.63002/jrecs.31.779

Berg, R. C., Polo Anaya, E., & Ziemer, H. (2024, May 15). Mineral demands for resilient semiconductor supply chains. Center for Strategic and International Studies. https://www.csis.org/analysis/mineral-demands-resilient-semiconductor-supply-chains

Berkhout, F., & Hertin, J. (2004). De-materialising and re-materialising: Digital technologies and the environment. *Futures, 36*(8), 903–920. DOI: 10.1016/j.futures.2004.01.003

Berners-Lee, M. (2010). *How bad are bananas? The carbon footprint of everything*. Greystone Books.

Bhardwaj, E., Alexander, R., & Becker, C. (2025). Limits to AI growth: The ecological and social consequences of scaling. arXiv preprint arXiv:2501.17980.

BHP. (2025, January). Why AI tools and data centres are driving copper demand. BHP. https://www.bhp.com/news/bhp-insights/2025/01/why-ai-tools-and-data-centres-are-driving-copper-demand

Bibikar, S., Vikalo, H., Wang, Z., & Chen, X. (2022). Federated dynamic sparse training: Computing less, communicating less, yet learning better. *In Proceedings of the AAAI Conference on Artificial Intelligence*, 6080–6088. DOI: 10.1609/aaai.v36i6.20555

Biçer, M. (2022). *Öğretmen adaylarının dijital çevre ve dijital atık kavramlarıyla ilgili algılarının incelenmesi* [Doctoral dissertation, Kastamonu Üniversitesi].

Birch, K., & Muniesa, F. (2020). *Assetization: Turning Things into Assets in Technoscientific Capitalism*. MIT Press. DOI: 10.7551/mitpress/12075.001.0001

Bittle, K., & El-Gayar, O. (2025). Generative AI and academic integrity in higher education: A systematic review and research agenda. *Information (Basel)*, *16*(4), 296. DOI: 10.3390/info16040296

Blakemore, S. J., & Frith, C. (2003). Self-awareness and action. *Current Opinion in Neurobiology*, *13*(2), 219–224. DOI: 10.1016/S0959-4388(03)00043-6 PMID: 12744977

Blühdorn, I. (2024). *Unhaltbarkeit: Auf dem Weg in eine andere Moderne*. Suhrkamp.

Bo, X. (2025). A Review of Detection, Evolution, and Data Reconstruction Strategies for False Data Injection Attacks in Power Cyber-Physical Systems [Review of *A Review of Detection, Evolution, and Data Reconstruction Strategies for False Data Injection Attacks in Power Cyber-Physical Systems*]. *arXiv (Cornell University)*. Cornell University. /arxiv.2501.10441DOI: 10.20944/preprints202501.1098.v1

Bohm, S., & Wirtz, G. (2022). Cloud-Edge orchestration for smart cities: A review of Kubernetes-based orchestration Architectures. *EAI Endorsed Transactions on Smart Cities*, *6*(18), 1–19. DOI: 10.4108/eetsc.v6i18.1197

Bolón-Canedo, V., Morán-Fernández, L., Cancela, B., & Alonso-Betanzos, A. (2024). A review of green artificial intelligence: Towards a more sustainable future. *Neurocomputing*, *599*, 128096. DOI: 10.1016/j.neucom.2024.128096

Bolton, P., & Kacperczyk, M. (2021). Do investors care about carbon risk? *Journal of Financial Economics*, *142*(2), 517–549. DOI: 10.1016/j.jfineco.2021.05.008

Bonilla, S. H., Silva, H. R. O., Terra da Silva, M., Franco Gonçalves, R., & Sacomano, J. B. (2018). Industry 4.0 and sustainability implications: A scenario-based analysis of the impacts and challenges. *Sustainability (Basel), 10*(10), 3740. Advance online publication. DOI: 10.3390/su10103740

Bostandoost, R., Hanafy, W. A., Lechowicz, A., Bashir, N., Shenoy, P., & Hajiesmaili, M. (2024). Data-driven Algorithm Selection for Carbon-Aware Scheduling. *ACM SIGEnergy Energy Informatics Review, 4*(5), 148–153. DOI: 10.1145/3727200.3727222

Branigan, T. (2010, October 25). Chinese moves to limit mineral supplies sparks struggle over rare earths. The Guardian. https://www.theguardian.com/business/2010/oct/25/china-cuts-rare-earths-exports

Braun, V., & Clarke, V. (2022). Thematic analysis: A practical guide. *Sage (Atlanta, Ga.).*

Brevini, B. (2021). *Is AI good for the planet?* Polity Press.

Bridge, G. (2008). Global production networks and the extractive sector: Governing resource-based development. *Journal of Economic Geography, 8*(3), 389–419. DOI: 10.1093/jeg/lbn009

Bughin, J., Seong, J., Manyika, J., Chui, M., & Joshi, R. (2018). Notes from the AI frontier: Modeling the impact of AI on the world economy. *McKinsey Global Institute, 4*(1), 2–61.

Bullard, R. D. (1920). *Dumping in Dixie: Race, Class, and Environmental Quality* (3rd ed.). Westview Press.

Cabrera, C., Bastidas, V., Schooling, J., & Lawrence, N. D. (2025, January 1). The Systems Engineering Approach in Times of Large Language Models. *Proceedings of the ... Annual Hawaii International Conference on System Sciences/Proceedings of the Annual Hawaii International Conference on System Sciences*. DOI: 10.24251/HICSS.2025.666

Cacciatore, A., Berardini, D., Scaraggi, V., Mancini, A., Moccia, S., & Migliorelli, L. (2025). Online Knowledge Distillation and Deep Supervision in HRNet: Green AI for Preterm Infants' Pose Estimation. *ACM Transactions on Computing for Healthcare, 6*(4), 1–19. DOI: 10.1145/3757067

Cao, L. (2023). AI in Finance: Challenges, Techniques, and Opportunities. *ACM Computing Surveys, 55*(3), 1–38. DOI: 10.1145/3502289

Capgemini. (2021). *Sustainable IT*. https://www.capgemini.com/de-de/wp-content/uploads/sites/5/2021/05/Sustainable_IT_Report_2021_Capgemini.pdf

Carbon Credits. (2025, December 11). *Google and NextEra team up to build gigawatt-scale AI data centers powered by clean energy*. https://carboncredits.com/google-and-nextera-team-up-to-build-gigawatt-scale-ai-data-centers-powered-by-clean-energy/

Carlson, L., Grove, S. J., & Kangun, N. (1993). A content analysis of environmental advertising claims: A matrix method approach. *Journal of Advertising*, *22*(3), 27–39. DOI: 10.1080/00913367.1993.10673409

Carrara, S., Bobba, S., Blagoeva, D., Alves Dias, P., Cavalli, A., Georgitzikis, K., Grohol, M., Itul, A., Kuzov, T., Latunussa, C., Lyons, L., Malano, G., Maury, T., Prior Arce, Á., Somers, J., Telsnig, T., Veeh, C., Wittmer, D., Black, C., & Christou, M. (2023). *Supply chain analysis and material demand forecast in strategic technologies and sectors in the EU – A foresight study*. Publications Office of the European Union. DOI: 10.2760/386650

Cascio, E. L., Girardin, L., Ma, Z., & Maréchal, F. (2021). How Smart is the Grid? *Frontiers in Energy Research*, *9*, 637447. Advance online publication. DOI: 10.3389/fenrg.2021.637447

Castilloux, R. (2019, April 25). CHARTS: Rare earth export restrictions, price spikes and the risks of demand destruction. MINING.COM. https://www.mining.com/featured-article/charts-rare-earth-export-restrictions-price-spikes-and-the-risks-of-demand-destruction/

Cavus, M. (2025). Advancing Power Systems with Renewable Energy and Intelligent Technologies: A Comprehensive Review on Grid Transformation and Integration [Review of *Advancing Power Systems with Renewable Energy and Intelligent Technologies: A Comprehensive Review on Grid Transformation and Integration*]. *Electronics, 14*(6), 1159. Multidisciplinary Digital Publishing Institute. DOI: 10.3390/electronics14061159

Cellard, L. (2022). Algorithms as figures: Towards a post-digital ethnography of algorithmic contexts. *New Media & Society*, *24*(4), 982–1000. DOI: 10.1177/14614448221079032

Cellard, L., & Marquet, C. (2023). Frictions sous-marines. Écologies et politiques des fragilités de l'interconnexion numérique. *Revue d'Anthropologie des Connaissances*, *17*(4). Advance online publication. DOI: 10.4000/rac.31070

Cendekiawan, K. A., & Wardhani, F. A. (2025). The Role of Information Technology and Product Innovation on the Competitive Advantage of Companies at Jember East Java. *Dirya: Journal of Economic Management*, *2*(1), 40–47.

Chamanara, S., Ghaffarizadeh, S. A., & Madani, K. (2023). The environmental footprint of Bitcoin mining across the globe: Call for urgent action. *Earth's Future, 11*(10), Article e2023EF003871. https://doi.org/DOI: 10.1029/2023EF003871

Chen, Y., Yan, W., Zhang, H., Liu, Y., Jiang, Z., & Zhang, X. (2022, August 14). A Data-Driven Design Approach for Carbon Emission Prediction of Machining. *Volume 2: 42nd Computers and Information in Engineering Conference (CIE)*. DOI: 10.1115/DETC2022-90465

Chen, C., Shahbaz, P., & Haq, S. U. (2025). Transforming students' green behavior through environmental education: The impact of institutional practices and policies. *Frontiers in Psychology, 15*, 1499781. Advance online publication. DOI: 10.3389/fpsyg.2024.1499781 PMID: 39881698

Chen, X., Jaffari, A. A., Muzaffar, A., & Rahman, N. (2026). Integrating explainable AI and corporate social responsibility for sustainable practices. *Technological Forecasting and Social Change, 223*, 124444. DOI: 10.1016/j.techfore.2025.124444

Chen, Y.-S. (2010). The drivers of green brand equity: Green brand image, green satisfaction, and green trust. *Journal of Business Ethics, 93*(2), 307–319. DOI: 10.1007/s10551-009-0223-9

Chouksey, A., Rajan, A. K., Gurjar, V., Tiwari, R., & Mishra, P. K. (2026). The green paradox: The climate, environmental, and sustainability implications of artificial intelligence. *Global Environmental Change Advances, 6*, 100029. DOI: 10.1016/j.gecadv.2025.100029

Chung, J.-W., Gu, Y., Jang, I., Meng, L., Bansal, N., & Chowdhury, M. (2024). Reducing energy bloat in large model training. In *Proceedings of the ACM SIGOPS 30th Symposium on Operating Systems Principles* (pp. 144–159). https://doi.org/DOI: 10.1145/3694715.3695970

Clemm, C., Stobbe, L., Wimalawarne, K., & Druschke, J. (2024). Towards Green AI: Current Status and Future Research. *2024 Electronics Goes Green 2024+ (EGG)*, 1–11. DOI: 10.23919/EGG62010.2024.10631247

Cohen, L., Manion, L., & Morrison, K. (2007). *Research methods in education* (6th ed.). Routledge. DOI: 10.4324/9780203029053

Competition and Markets Authority. (2021). *Green claims code: Making environmental claims*. CMA.

Corrado, C., Hulten, C., & Sichel, D. (2009). Intangible capital and U.S. economic growth. *Review of Income and Wealth, 55*(3), 661–685. DOI: 10.1111/j.1475-4991.2009.00343.x

Costanza, R., d'Arge, R., de Groot, R., Farber, S., Grasso, M., Hannon, B., Limburg, K., Naeem, S., O'Neill, R. V., Paruelo, J., Raskin, R. G., Sutton, P., & van den Belt, M. (1997). The value of the world's ecosystem services and natural capital. *Nature, 387*(6630), 253–260. DOI: 10.1038/387253a0

Couldry, N., & Mejias, U. A. (2019). *The Costs of Connection: How Data Is Colonizing Human Life and Appropriating It for Capitalism*. Stanford University Press.

Crawford, K. (2021). *Atlas of AI: Power, politics, and the planetary costs of artificial intelligence*. Yale University Press.

Creswell, J. W. (2012). *Educational research: Planning, conducting, and evaluating quantitative and qualitative research* (4th ed.). Pearson.

Creswell, J. W., & Plano Clark, V. L. (2018). *Designing and conducting mixed methods research* (3rd ed.). SAGE.

Crusoe. (2025, May 21). *Crusoe, Blue Owl Capital, and Primary Digital Infrastructure enter second phase of $15 billion joint venture to fund AI data center in Abilene, Texas*. https://www.crusoe.ai/resources/newsroom/crusoe-blue-owl-capital-and-primary-digital-infrastructure-enter-joint-venture

Daly, H. E. (1996). *Beyond Growth: The Economics of Sustainable Development*. Beacon Press.

Damayanti, D., Suyeno, S., & Putra, L. R. (2023). Evaluation of Malang City Government Policy on Traffic Engineering in Overcoming Congestion (Study at the Malang City Transportation Agency). *Public Response, 17*(6), 17–27.

Dangelico, R. M., & Vocalelli, D. (2017). 'Green Marketing': An analysis of definitions, strategy steps, and tools through a systematic review of the literature. *Journal of Cleaner Production, 165*, 1263–1279. DOI: 10.1016/j.jclepro.2017.07.184

Dasari, K., Raj, V. H., Nijhawan, G., Kalra, R., Pahwa, S., & Abdul-Zahra, D. S. (2024). Integration of AI in Distributed Energy Resource Management for Enhanced Load Balancing and Grid Stability. *E3S Web of Conferences, 505*, 3008. DOI: 10.1051/e3sconf/202450503008

Das, B. (2024). Digital pollution: A warning to our planet. *Asian Journal of Management, 15*(2), 211–216. DOI: 10.52711/2321-5763.2024.00033

Davey, C. (2023, March 28). The environmental impacts of cobalt mining in Congo. Earth.Org. https://earth.org/cobalt-mining-in-congo/

David, P. A. (1985). Clio and the economics of QWERTY. *The American Economic Review, 75*(2), 332–337.

Davidson, H. (2025, October 9). China steps up control of rare-earth exports citing 'national security' concerns. The Guardian. https://www.theguardian.com/world/2025/oct/09/china-steps-up-control-rare-earth-exports-national-security-concerns

De Vries, A. (2023). The growing energy footprint of artificial intelligence. *Joule*, *7*(10), 2191–2194. DOI: 10.1016/j.joule.2023.09.004

De Vries, A., Gallersdörfer, U., Klaaßen, L., & Stoll, C. (2022). Revisiting Bitcoin's carbon footprint. *Joule*, *6*(3), 498–502. DOI: 10.1016/j.joule.2022.02.005

Delmas, M. A., & Burbano, V. C. (2011). The drivers of greenwashing. *California Management Review*, *54*(1), 64–87. DOI: 10.1525/cmr.2011.54.1.64

Department for Business and Trade. (2025, November 24). Critical minerals strategy: Vision 2035. UK Government. https://assets.publishing.service.gov.uk/media/69244c80367485ea116a5675/uk-critical-minerals-strategy-vision-2035.pdf

Dhar, P. (2020). The carbon impact of artificial intelligence. *Nature Machine Intelligence*, *2*(8), 423–425. DOI: 10.1038/s42256-020-0219-9

Diantoro, K., Supriyanti, D., Ardi, A., Sanjaya, Y. P. A., & Watini, S. (2023). Implications of Distributed Energy Development in Blockchain-Based Institutional Environment. [ATT]. *Aptisi Transactions On Technopreneurship*, *5*(2sp), 209–220. DOI: 10.34306/att.v5i2sp.343

Díaz-García, V., Montero-Navarro, A., Rodríguez-Sánchez, J.-L., & Gallego-Losada, R. (2022). Digitalization and digital transformation in higher education: A bibliometric analysis. *Frontiers in Psychology*, *13*, 1081595. Advance online publication. DOI: 10.3389/fpsyg.2022.1081595 PMID: 36533011

Divya, S., Panda, S., Hajra, S., Jeyaraj, R., Paul, A., Park, S. H., Kim, H. J., & Oh, T. H. (2023). Smart data processing for energy harvesting systems using artificial intelligence. *Nano Energy*, *106*, 108084. DOI: 10.1016/j.nanoen.2022.108084

Dobbe, R., & Whittaker, M. (2019). AI and climate change: How they're connected, and what we can do about it. *AI Now Institute*. https://ainowinstitute.org/publication/ai-and-climate-change-how-theyre-connected-and-what-we-can-do-about-it

Dodge, J., Prewitt, T., & Tachet des Combes, R.. (2022). Measuring the carbon intensity of AI in cloud instances. In *Proceedings of the 2022 ACM Conference on Fairness, Accountability, and Transparency* (pp. 1877–1894). Association for Computing Machinery. https://doi.org/DOI: 10.1145/3531146.3533234

Dong, H., Munir, A., Tout, H., & Ganjali, Y. (2021). Next-Generation Data Center Network Enabled by Machine Learning: Review, Challenges, and Opportunities. *IEEE Access : Practical Innovations, Open Solutions*, *9*, 136459–136475. DOI: 10.1109/ACCESS.2021.3117763

Dong, J., Ning, P., Zhao, H., & Song, C. (2025). Decentralized peer-to-peer energy trading: A blockchain-enabled pricing paradigm. *Journal of King Saud University. Computer and Information Sciences*, *37*(1-2), 10. Advance online publication. DOI: 10.1007/s44443-025-00025-2

Dosi, G. (1982). Technological paradigms and technological trajectories. *Research Policy*, *11*(3), 147–162. DOI: 10.1016/0048-7333(82)90016-6

DoubleVerify. (2020). Global online content consumption doubles in 2020, research shows. https://doubleverify.com/newsroom/global-online-contentconsumption-doubles-in-2020-research-shows/

Dulloo, R. (2026b). Sustainable adoption and continued use of voice-based AI: Examining the environmental impact through the lens of the Technology Acceptance Model. In *Congress on Control, Robotics, and Mechatronics* (Lecture Notes in Networks and Systems, Vol. 1850, pp. 45–57). Springer. DOI: 10.1007/978-981-96-9771-7_4

Dulloo, R., Biradar, K., & Srijaa, M. (2026d). Understanding Consumer Behavior in Sustainable Aviation: An Integrated Theoretical Model Examining Environmental Values, Technology Acceptance, and Travel Context. In *World Conference on Information Systems for Business Management* (Lecture Notes in Networks and Systems,vol 1744, pp. 375-385). Springer DOI: 10.1007/978-3-032-12996-3_36

Dulloo, R., Singh, A., & Kiran, V. (2026). AI-driven sustainable marketing in the metaverse and consumer purchase intentions: A TPB analysis. In *World Conference on Information Systems for Business Management* (Lecture Notes in Networks and Systems, Vol. 1742, pp. 376–386). Springer DOI: 10.1007/978-3-032-12987-1_35

Dulloo, R. (2026a). Harnessing artificial intelligence for global portfolio management: Enhancing agile practices in diverse enterprise environments. In *Harnessing artificial intelligence to ensure diverse global teams* (pp. 191–228). Academic Press., DOI: 10.1016/B978-0-443-34135-9.00015-5

Dulloo, R., Appolloni, A., & Mathiyazhagan, K. (2025a). *AI-Powered Communication: Strategic Framework for Technological Innovation, Brand Perception, and Sustainable Marketing. Strategic Blueprints for AI-Driven Marketing in the Digital Era*. IGI Global Scientific Publishing., DOI: 10.4018/979-8-3373-3897-2.ch001

Dulloo, R., Silic, M., & Appolloni, A. (2025b). *Key enablers of Marketing 6.0 in the metaverse: how technological, human-centric, sustainable, and seamless approaches drive performance outcomes. Leveraging AI-Powered Marketing in the Experience-Driven Economy*. IGI Global Scientific Publishing. DOI: 10.4018/979-8-3693-9561-5.ch016

Dunlap, R. E., Van Liere, K. D., Mertig, A. G., & Jones, R. E. (2000). Measuring endorsement of the New Ecological Paradigm: A revised NEP scale. *The Journal of Social Issues*, *56*(3), 425–442. DOI: 10.1111/0022-4537.00176

Dutt, A., & Karuna, G. (2024). Machine learning approaches for fault detection in renewable microgrids. *MATEC Web of Conferences, 392*, 1192. DOI: 10.1051/matecconf/202439201192

Dutt, A., Rani, M. N. S., Bisht, M. S., Chandna, M., & Singla, A. (2024). Machine Learning Approaches for Fault Detection in Renewable Microgrids. *E3S Web of Conferences, 511*, 1030. DOI: 10.1051/e3sconf/202451101030

Ebers, M., Hoch, V. R. S., Rosenkranz, F., Ruschemeier, H., & Steinrötter, B. (2021). The European Commission's Proposal for an Artificial Intelligence Act—A Critical Assessment by Members of the Robotics and AI Law Society (RAILS). *J, 4*(4), 589–603. DOI: 10.3390/j4040043

Ebert, K., Alder, N., Herbrich, R., & Hacker, P. (2026). AI, climate, and regulation: From data centers to the AI Act. *Computer Law & Security Review*, *61*, 106326. DOI: 10.1016/j.clsr.2026.106326

Edler, J., Blind, K., Kroll, H., & Schubert, T. (2023). *Strategic autonomy and sovereignty in the innovation and economic domains*. Publications Office of the European Union.

Eduljee, G. H., Verbeke, F., Marvin, H. J. P., & van der Berg, M. (2024). Per- and polyfluoroalkyl substances and emerging contaminants in semiconductor fabrication: Occupational and community exposure dimensions. *Environmental Health Perspectives*, *132*(8), 087001. PMID: 39106155

Edwards, P. N. (2010). *A vast machine: Computer models, climate data, and the politics of global warming*. MIT Press.

Ensmenger, N. (2018). The environmental history of computing. *Technology and Culture*, *59*(4, Supplement), S7–S33. DOI: 10.1353/tech.2018.0148 PMID: 30595595

Epstein, G. A. (2005). *Financialization and the World Economy*. Edward Elgar. DOI: 10.4337/9781781008263

Erdem, D. (2024). *Synchribble: Designing to increase digital carbon footprint awareness* [Unpublished manuscript].

Esho, E. O., Akinyelu, A. A., & Dinis, M. A. P. (2026). Sustainable generative AI and quantum computing: Review assessment on the environmental impact of generative AI and quantum technologies. *Frontiers in Sustainability*, *7*, 1726832. Advance online publication. DOI: 10.3389/frsus.2026.1726832

European Commission, Joint Research Centre. (2025). *Strategic technologies and industrial policies for competitiveness and sustainability: CONCORDi 2025 policy brief(JRC144677)*. Publications Office of the European Union., DOI: 10.2760/477729

European Commission. (2023). *Proposal for a Directive on substantiation and communication of explicit environmental claims*. European Commission.

European Commission. (2023). Proposal for a regulation of the European Parliament and of the Council establishing a framework for ensuring a secure and sustainable supply of critical raw materials (Critical Raw Materials Act) (COM(2023) 160 final). EUR-Lex. https://eur-lex.europa.eu/legal-content/EN/TXT/HTML/?uri=CELEX:52023PC0160

European Commission. (2024, August 8). Artificial Intelligence Act: Call for tenders to measure and foster energy efficient and low emission artificial intelligence in the EU. *Digital Strategy*. https://digital-strategy.ec.europa.eu/en/funding/artificial-intelligence-act-call-tenders-measure-and-foster-energy-efficient-and-low-emission

European Parliament and Council of the European Union. (2024). Regulation (EU) 2024/1252 establishing a framework for ensuring a secure and sustainable supply of critical raw materials. Official Journal of the European Union.

Extractive Industries Transparency Initiative. (2025). EITI 2025 progress report: 57 implementing countries and the agenda for critical minerals transparency. EITI International Secretariat. https://eiti.org

Fairclough, N. (1992). *Discourse and social change*. Polity Press.

Fairclough, N. (2003). *Analysing discourse: Textual analysis for social research*. Routledge. DOI: 10.4324/9780203697078

Faizan, M., Brenner, T., Foerster, F., Wittwer, C., & Koch, B. (2019). Decentralized bottom-up energy trading using Ethereum as a platform. *The Journal of Energy Markets*, *12*(2), 19–48. DOI: 10.21314/JEM.2019.193

Falk, J., Schaltegger, S., & Stern, T. (2025). Comparative life cycle assessment of AI and traditional computing infrastructure. *Journal of Industrial Ecology*. Advance online publication.

Fatima, A., Shaheen, A., Ahmed, S., Fazal, B., Ahmad, F., Liew, T. W., & Ahmed, Z. (2024). Exploring the use of gamification in human-centered agile-based requirements engineering. *Frontiers of Computer Science*, *6*, 1442081. Advance online publication. DOI: 10.3389/fcomp.2024.1442081

Ferreboeuf, H., Berthoud, F., Bihouix, P., Fabre, P., Kaplan, D., & Lefèvre, L. (2019). *Lean ICT: Towards digital sobriety*. The Shift Project. https://theshiftproject.org/en/article/lean-ict-our-new-report/

Foroohar, R. (2018, December 18). Year in a word: Techlash. *Financial Times*. https://www.ft.com/content/76578fba-fca1-11e8-ac00-57a2a826423e

Forti, V., Baldé, C. P., Kuehr, R., & Bel, G. (2020). The global e-waste monitor 2020: Quantities, flows and the circular economy potential. United Nations University. https://ewastemonitor.info/wp-content/uploads/2020/11/GEM_2020_def_july1_low.pdf

Fraenkel, J. R., Wallen, N. E., & Hyun, H. H. (2012). *How to design and evaluate research in education* (8th ed.). McGraw-Hill.

Frimpong, V. (2025). *What counts as Green AI? Mapping efficiency, sustainability, and critical-ecological strands in a fragmented discourse*. Qeios. DOI: 10.32388/8FB2P4.2

Fusch, P. I., & Ness, L. R. (2015). Are we there yet? Data saturation in qualitative research. *The Qualitative Report*, *20*(9), 1408–1416. DOI: 10.46743/2160-3715/2015.2281

Geels, F. W. (2002). Technological transitions as evolutionary reconfiguration processes: A multi-level perspective and a case-study. *Research Policy*, *31*(8–9), 1257–1274. DOI: 10.1016/S0048-7333(02)00062-8

Georgitzikis, K., Kavvadias, K., Kushnir, D., Gumiel, J. C., Mancini, L., Eynard, U., Bobba, S., Pavel, C., Carrara, S., Maury, T., Peffet, C., Deboyer, S., & Mathieux, F. (2025). *Substitution and reduction of critical and strategic raw materials in clean energy technologies (EUR 40297 EN)*. Publications Office of the European Union., DOI: 10.2760/930966

Gereffi, G. (1994). The organization of buyer-driven global commodity chains: How U.S. retailers shape overseas production networks. In Gereffi, G., & Korzeniewicz, M. (Eds.), *Commodity Chains and Global Capitalism* (pp. 95–122). Praeger.

Gereffi, G., Humphrey, J., & Sturgeon, T. (2005). The governance of global value chains. *Review of International Political Economy, 12*(1), 78–104. DOI: 10.1080/09692290500049805

Gökçek, M., Bozdağ, E., & Demirbag, H. (2019). Carbon footprint and digitalization: Environmental impacts. *Journal of Environmental Engineering, 3*(1), 78–92.

Goldman Sachs. (2024). *Generational growth: AI, data centers and the coming US power demand surge*. Goldman Sachs Research.

Gopstein, A., Nguyen, C., O'Fallon, C. M., Hastings, N., & Wollman, D. A. (2021). *NIST framework and roadmap for smart grid interoperability standards, release 4.0*. DOI: 10.6028/NIST.SP.1108r4

Government of Canada. Natural Resources Canada. (2022). Canadian critical minerals strategy. https://www.canada.ca/en/campaign/critical-minerals-in-canada/canadian-critical-minerals-strategy.html

Goyal, G. (2025). AI-Driven Fault Detection and Diagnosis in Smart Grids for Enhanced Power System Reliability. *Journal of Information Systems Engineering & Management, 10*(42s), 845–859. DOI: 10.52783/jisem.v10i42s.8203

Graedel, T. E., Barr, R., Chandler, C., Chase, T., Choi, J., Christoffersen, L., Friedlander, E., Henly, C., Jun, C., Nassar, N. T., Schechner, D., Warren, S., Yang, M., & Zhu, C. (2012). Methodology of metal criticality determination. *Environmental Science & Technology, 46*(2), 1063–1070. DOI: 10.1021/es203534z PMID: 22191617

Graedel, T. E., Harper, E. M., Nassar, N. T., Nuss, P., & Reck, B. K. (2015). Criticality of metals and metalloids. *Proceedings of the National Academy of Sciences of the United States of America, 112*(14), 4257–4262. DOI: 10.1073/pnas.1500415112 PMID: 25831527

Graedel, T. E., & Reck, B. K. (2016). Six years of criticality assessments: What have we learned so far? *Journal of Industrial Ecology, 20*(4), 692–699. DOI: 10.1111/jiec.12305

Green Digest. (n.d.). NVIDIA environmental and social impact. https://www.greendigest.co/p/nvidia-environmental-and-social-impact

Greenpeace International. (n.d.). The environmental cost behind NVIDIA's rise as the world's biggest AI chipmaker: Who is paying the price? https://www.greenpeace.org/international/story/79910/the-environmental-cost-behind-nvidias-rise-as-worlds-biggest-ai-chipmaker-who-is-paying-the-price/

Gregg, M., & Strengers, Y. (2024). Getting beyond net zero dashboards in the information technology sector. *Energy Research & Social Science, 108*, 103397. Advance online publication. DOI: 10.1016/j.erss.2023.103397

Grohol, M., & Veeh, C. (2023). Study on the critical raw materials for the EU 2023: Final report. European Commission.

Gupta, U., Kim, Y. G., Lee, S., Tse, J., Lee, H.-H. S., Wei, G.-Y., Brooks, D., & Wu, C.-J. (2020). *Chasing Carbon: The Elusive Environmental Footprint of Computing*. https://arxiv.org/abs/2011.02839

Gupta, U., Kim, Y. G., Lee, S., Tse, J., Lee, H.-H. S., Wei, G.-Y., Brooks, D., & Wu, C.-J. (2022). Chasing carbon: The elusive environmental footprint of computing. *IEEE Micro, 42*(4), 37–47. DOI: 10.1109/MM.2022.3163226

Haberman, B., Lev, E., & Langley, D. (2003). Action research as a tool for promoting teacher awareness of students' conceptual understanding. *SIGCSE Bulletin, 35*(3), 144–148. DOI: 10.1145/961290.961552

Hacker, P. (2024). Sustainable AI regulation. *Common Market Law Review, 61*(2), 345–386. DOI: 10.54648/COLA2024025

Hafez, S., Alkhedher, M., Ramadan, M., Gad, A., Alhalabi, M., Yaghi, M., Jama, M., & Ghazal, M. (2025). Advancements in grid resilience: Recent innovations in AI-driven solutions. *Results in Engineering, 26*, 105042. DOI: 10.1016/j.rineng.2025.105042

Halpern, O. (2021). Planetary intelligence. In Roberge, J., & Castelle, M. (Eds.), *The cultural life of machine learning* (pp. 227–256). Palgrave Macmillan., DOI: 10.1007/978-3-030-56286-1_8

Hamid, M., & Ganne, A. (2023). ARTIFICIAL INTELLIGENCE IN ENERGY MARKETS AND POWER SYSTEMS. *International Research Journal of Modernization in Engineering Technology and Science*. Advance online publication. DOI: 10.56726/IRJMETS35943

Hamid, M., Zeshan, F., Ahmad, A., Munawar, S., Aimeur, E., Ahmed, S., Abu Elsoud, M., & Yousif, M. (2020). An intelligent decision support system for effective handling of IT projects. *Journal of Intelligent & Fuzzy Systems, 38*(3), 2635–2647. DOI: 10.3233/JIFS-179550

Hartmann, P., & Apaolaza-Ibáñez, V. (2012). Consumer attitude and purchase intention toward green energy brands: The roles of psychological benefits and environmental concern. *Journal of Business Research, 65*(9), 1254–1263. DOI: 10.1016/j.jbusres.2011.11.001

Henderson, P., Hu, J., Romoff, J., Brunskill, E., Jurafsky, D., & Pineau, J. (2020). Towards the systematic reporting of the energy and carbon footprints of machine learning. *Journal of Machine Learning Research, 21*, 1–43.

Hess, J. C. (2024, June 20). Chip production's ecological footprint: Mapping climate and environmental impact. Interface. https://www.interface-eu.org/publications/chip-productions-ecological-footprint

Hilty, L. M., & Aebischer, B. (Eds.). (2015). *ICT innovations for sustainability* (Vol. 310). Springer., DOI: 10.1007/978-3-319-09228-7_1

Hintemann, R., & Hinterholzer, S. (2019). Energy consumption of data centers worldwide. In *Proceedings of the 6th International Conference on ICT for Sustainability (ICT4S)*.

Honari, K., Zhou, X., Rouhani, S., Dick, S., Liang, H., Li, Y., & Miller, J. L. (2022). *A Scalable Blockchain-based Smart Contract Model for Decentralized Voltage Stability Using Sharding Technique*. 124. DOI: 10.1109/Blockchain55522.2022.00026

Hooli, M. (2025). The Carbon Footprint of AI: Can Smart Tech Be Truly Sustainable? *Sarcouncil Journal of Engineering and Computer Sciences, 4*(8), 629–635.

Hornborg, A. (2011). *Global Ecology and Unequal Exchange: Fetishism in a Zero-Sum World*. Routledge.

Hrithika, L. (2025). SUSTAINABLE AI: STRATEGIES TO REDUCE THE ENVIRONMENTAL FOOTPRINTS OF AI. *International Journal of Advances in Business and Management Research, 03*(SUP-02), 216–224. DOI: 10.62674/ijabmr.2025.v3isupp2.0021

Hua, H., Li, Y., Wang, T., Dong, N., Li, W., & Cao, J. (2023). Edge Computing with Artificial Intelligence: A Machine Learning Perspective. *ACM Computing Surveys, 55*(9), 1–35. DOI: 10.1145/3555802

Hughes, T. P. (1987). The evolution of large technological systems. In Bijker, W. E., Hughes, T. P., & Pinch, T. (Eds.), *The Social Construction of Technological Systems* (pp. 51–82). MIT Press.

Hussaine, S., & Aldeen , N. (2026). Exploring the Role of Quantum Computing in Advancing Green Computing Technologies. *Journal of Sustainable Smart Systems in Education & Environment, 1*(01), 56–68. DOI: 10.66823/jsssee.7

Hyperscale, A. F. L. (2024). Optical fibre innovation: Driving seamless flow in hyperscale data centres [Corporate technical brief]. AFL Hyperscale. https://www.aflhyperscale.com/articles/fiber-optic-innovation-driving-seamless-flow-in-hyperscale-data-centers/

IEA. (2023). Data Centres and Data Transmission Networks. Https://Www.Iea.Org/Energy-System/Buildings/Data-Centres-and-Data-Transmission-Networks

IEA. (2023). *Electricity 2024: Analysis and forecast to 2026*. International Energy Agency.

IMARC Group. (2024). *Tantalum capacitors market: Global industry trends, share, size, growth, opportunity and forecast 2024–2032*. IMARC Services Private Limited. https://www.imarcgroup.com/tantalum-capacitors-market

Info, Q. (2025). Meta carbon AI. https://www.infoq.com/news/2025/10/meta-carbon-ai/

Initiative for Responsible Mining Assurance. (2025). IRMA Standard for Responsible Mining IRMA-STD-001 (Version 2.0) [Standard and audit protocol]. IRMA. https://responsiblemining.net

International Council on Clean Transportation. (2024, December). *Electrifying road transport with less mining: A global and regional battery material outlook* (White Paper). ICCT. https://theicct.org/publication/ev-battery-materials-demand-supply-dec24/

International Energy Agency (IEA). (2023). Data Centres and Data Transmission Networks. https://www.iea.org/energy-system/buildings/data-centres-and-data-transmission-networks

International Energy Agency. (2021). The role of critical minerals in clean energy transitions. https://www.iea.org/reports/the-role-of-critical-minerals-in-clean-energy-transitions

International Energy Agency. (2023). *Critical minerals market review 2023*. IEA Publications.

International Energy Agency. (2024). Global critical minerals outlook 2024. IEA Publications. https://www.iea.org/reports/global-critical-minerals-outlook-2024

International Energy Agency. (2025). Energy and AI. IEA. https://www.iea.org/reports/energy-and-ai

International Energy Agency. (2025). Global critical minerals outlook 2025. IEA Publications. https://www.iea.org/reports/global-critical-minerals-outlook-2025

International Resource Panel (IRP). (2019). Global Resources Outlook 2019: Natural Resources for the Future We Want. United Nations Environment Programme. https://digitallibrary.un.org/record/4047319?v=pdf

International Seabed Authority. (2025). Status of negotiations on the regulations for exploitation of mineral resources in the Area. International Seabed Authority Secretariat, Kingston, Jamaica. https://www.isa.org.jm

International, N. S. F. (2015). *Environmental leadership standards for servers (NSF/ANSI 426) (Draft document)*. NSF International.

IRDO. (2025). Poster 63 [Conference poster]. https://www.irdo.si/irdo2025/posters/63.pdf

Irhab, MD, & Nugroho, A. (2024). Application of decision tree regressor in business intelligence development for predicting tourist movement in east java. JSR: Robotic Information System Network, 8(2), 126–135.

Isac, N., Javed, A., Radulescu, M., Cismasu, I. D. L., Yousaf, Z., & Serbu, R. S. (2025). Is greenwashing impacting on green brand trust and purchase intentions? Mediating role of environmental knowledge. *Environment, Development and Sustainability*, *27*(9), 21329–21346. DOI: 10.1007/s10668-023-04352-0

Jamil, S., Qureshi, R., & Ahmed, S. (2020). Analyzing Effects of Resource Distribution on Performance of Fog Infrastructure. *2020 Global Conference on Wireless and Optical Technologies (GCWOT)*, 1–5. DOI: 10.1109/GCWOT49901.2020.9391604

Jancovic, M., & Keilbach, J. (2023). Streaming against the environment: Digital infrastructures, video compression, and the environmental footprint of video streaming. In Es, K., & Verhoeff, N. (Eds.), *Situating data: Inquiries in algorithmic culture* (pp. 85–102). Amsterdam University Press. DOI: 10.2307/jj.362382.9

Jeanquartier, F., Jean-Quartier, C., Rieder, P., Misirlić, V., Pasero, C., Hohensinner, R., Müller, H., & Holzinger, A. (2026). Assessing the carbon footprint of language models: Towards sustainability in AI. *Resources, Conservation and Recycling*, *226*, 108670. DOI: 10.1016/j.resconrec.2025.108670

Jeffry, L., Ong, M. Y., Nomanbhay, S., Mofijur, M., Mubashir, M., & Show, P. L. (2021). Greenhouse gases utilization: A review. *Fuel*, *301*, 121017. Advance online publication. DOI: 10.1016/j.fuel.2021.121017

Jerez, B., Garcés, I., & Torres, R. (2021). Lithium extractivism and water injustices in the Salar de Atacama, Chile: The colonial shadow of green electromobility. *Political Geography*, *87*, 102382. DOI: 10.1016/j.polgeo.2021.102382

Jevons, W. S. (1865). *The Coal Question: An Inquiry Concerning the Progress of the Nation, and the Probable Exhaustion of Our Coal Mines* (2nd ed.). Macmillan., https://oll.libertyfund.org/titles/jevons-the-coal-question

Jha, A., Gupta, G., Rau, S., Meshram, P., & Labhane, N. (2022). Green Computing Approach in ICT Components. *Vidyabharati International Interdisciplinary Research Journal*, 564–569.

Jiang, S., Li, Y., Lu, Q., Hong, Y., Guan, D., Xiong, Y., & Wang, S. (2021). Policy assessments for the carbon emission flows and sustainability of Bitcoin blockchain operation in China. *Nature Communications*, *12*(1), 1938. Advance online publication. DOI: 10.1038/s41467-021-22256-3 PMID: 33824331

Jiménez, A., & Oleson, J. C. (2022). The crimes of digital capitalism. *Mitchell Hamline Law Review*, *48*(4), 971–1018.

Jobin, A., Ienca, M., & Vayena, E. (2019). The global landscape of AI ethics guidelines. *Nature Machine Intelligence*, *1*(9), 389–399. DOI: 10.1038/s42256-019-0088-2

Johnson, G. E., Sebastian-Cardenas, D. J., Balamurugan, S. P., Harun, N. F., Mukherjee, M., Blonsky, M., Markel, T., & Johnson, B. J. (2022). A Unified Testing Platform to Mature Blockchain Applications for Grid Emulation Environments: Preprint. *OSTI OAI (U.S. Department of Energy Office of Scientific and Technical Information)*. https://www.osti.gov/biblio/1889263

Jones, N. (2018). How to stop data centres from gobbling up the world's electricity. *Nature*, *561*(7722), 163–166. DOI: 10.1038/d41586-018-06610-y PMID: 30209383

Jouppi, N. P., & Swing, A. (2023). A Machine Learning Supercomputer with an Optically Reconfigurable Interconnect and Embeddings Support. *2023 IEEE Hot Chips 35 Symposium (HCS)*, 1–24. DOI: 10.1109/HCS59251.2023.10254691

Kaack, L. H., Donti, P. L., Strubell, E., Kamiya, G., Creutzig, F., & Rolnick, D. (2022). Aligning AI with climate change mitigation. *Nature Climate Change*, *12*(6), 518–527. DOI: 10.1038/s41558-022-01377-7

Kaplan, J., McCandlish, S., Henighan, T., Brown, T. B., Chess, B., & Child, R.. (2020). Scaling laws for neural language models. *arXiv*. https://arxiv.org/abs/2001.08361

Kapp, K. W. (1950). *The Social Costs of Private Enterprise*. Harvard University Press.

Karasar, N. (2013). *Bilimsel araştırma yöntemi* (25th ed.). Nobel Yayın Dağıtım.

Kaur, S., Kumar, R., Singh, K., & Huang, Y. (2024). Leveraging Artificial Intelligence for Enhanced Sustainable Energy Management. *Journal of Sustainability for Energy*, *3*(1), 1–20. DOI: 10.56578/jse030101

Keller, K. L. (1993). Conceptualizing, measuring, and managing customer-based brand equity. *Journal of Marketing*, *57*(1), 1–22. DOI: 10.1177/002224299305700101

Keller, K. L. (2001). Building customer-based brand equity: A blueprint for creating strong brands. *Marketing Management*, *10*(2), 15–19.

Kim, J., Kim, J., & Yang, H. (2022). Loneliness and the use of social media to follow celebrities: A moderating role of social media literacy. *The Social Science Journal*, *59*(1), 116–130.

Kim, M.-K., Yoo, T.-A., & Chung, J.-B. (2026). Toward sustainable Generative AI: A scoping review of carbon footprint and environmental impacts across training and inference stages. *IEEE Access : Practical Innovations, Open Solutions*, *14*, 18881–18896. Advance online publication. DOI: 10.1109/ACCESS.2026.3659894

Kitchin, R. (2014). The Data Revolution: Big Data, Open Data, Data Infrastructures and Their Consequences. *Sage (Atlanta, Ga.)*. Advance online publication. DOI: 10.4135/9781473909472

Köhler, S., & Pizzol, M. (2019). Life cycle assessment of Bitcoin mining. *Environmental Science & Technology*, *53*(23), 13598–13606. DOI: 10.1021/acs.est.9b05687 PMID: 31746188

Konat, G., Salihoğlu, E., & Han, A. (2026). Financial investments in AI-based technologies and carbon footprint in selected advanced industrial economies. *Carbon Balance and Management*, *21*(1), 31. DOI: 10.1186/s13021-025-00383-4 PMID: 41483362

Konuk, F. A., Rahman, S. U., & Salo, J. (2015). Antecedents of green behavioral intentions: A cross-country study of Turkey, Finland and Pakistan. *International Journal of Consumer Studies*, *39*(6), 586–596. DOI: 10.1111/ijcs.12209

Koohestani, R., de Bekker, P., Koç, B., & Izadi, M. (2026). Benchmarking AI Models in Software Engineering: A Review, Search Tool, and Unified Approach for Elevating Benchmark Quality. *IEEE Transactions on Software Engineering*, *52*(2), 651–674. DOI: 10.1109/TSE.2025.3644183

Kopka, A., & Grashof, N. (2022). Artificial intelligence: Catalyst or barrier on the path to sustainability? *Technological Forecasting and Social Change*, *175*, 121318. DOI: 10.1016/j.techfore.2021.121318

Krippner, G. R. (2005). The financialization of the American economy. *Socio-economic Review*, *3*(2), 173–208. DOI: 10.1093/SER/mwi008

Kumar, P., & Ghodeswar, B. M. (2015). Factors affecting consumers' green product purchase decisions. *Marketing Intelligence & Planning*, *33*(3), 330–347. DOI: 10.1108/MIP-03-2014-0068

Kunkel, S., & Matthess, M. (2020). Digital transformation and environmental sustainability in industry: Putting expectations in Asian and African policies into perspective. *Environmental Science & Policy*, *112*, 318–329. DOI: 10.1016/j.envsci.2020.06.022

Kunnathully, K., Jinesh, M., Satyanarayana, P., Esther, V., Pathak, G. P., & Manjunath, T. C. (2026). Building human rights by design: Ethical AI development for religious freedom. In Chakraborty, S. (Ed.), *Intersections of AI and the freedom of religion or belief* (pp. 81–104). IGI Global Scientific Publishing., DOI: 10.4018/979-8-3373-3830-9.ch004

Kurniawan, T., Sihombing, A. M., & Berliane, A. (2023). Konstruksi Politik Hukum Pidana Terhadap Delik Perzinaan Dalam Rancangan Kitab Undang-Undang Hukum Pidana. *Binamulia Hukum*, *12*(1), 11–24. DOI: 10.37893/jbh.v12i1.445

Kuzior, A., Sira, M., & Brożek, P. (2022). Using Blockchain and Artificial Intelligence in Energy Management as a Tool to Achieve Energy Efficiency. *Virtual Economics*, *5*(3), 69–90. DOI: 10.34021/ve.2022.05.03(4)

Lachuer, J., & Ben Jabeur, S. (2022). Explainable artificial intelligence modeling for corporate social responsibility and financial performance. *Journal of Asset Management*, *23*(7), 619–630. DOI: 10.1057/s41260-022-00291-z

Lally, N., Kay, K., & Thatcher, J. (2022). Computational parasites and hydropower: A political ecology of Bitcoin mining on the Columbia River. *Environment and Planning. E, Nature and Space*, *5*(1), 18–38. DOI: 10.1177/2514848619867608

Land, P. P. C. (n.d.). Meta confronts sustainability tensions amid massive AI infrastructure expansion. https://ppc.land/meta-confronts-sustainability-tensions-amid-massive-ai-infrastructure-expansion

Lannelongue, L., Grealey, J., & Inouye, M. (2021). Green Algorithms: Quantifying the Carbon Footprint of Computation. *Advanced Science (Weinheim, Baden-Württemberg, Germany)*, *8*(12), 2100707. Advance online publication. DOI: 10.1002/advs.202100707 PMID: 34194954

Laurens, S., & Salauddin, J. (2024). Implementation of the Smart City Concept in Improving the Quality of Urban Public Services. JIIP - Scientific Journal of Educational Sciences, 7(11), 13130-13137. doi:10.54371/jiip.v7i11.6291

Lawshe, C. H. (1975). A quantitative approach to content validity. *Personnel Psychology, 28*(4), 563–575. DOI: 10.1111/j.1744-6570.1975.tb01393.x

Lee, C., Jin, J., Kim, T., Kim, H., & Park, E. (2024). OWQ: Outlier-Aware Weight Quantization for Efficient Fine-Tuning and Inference of Large Language Models. *Proceedings of the AAAI Conference on Artificial Intelligence, 38*(12), 13355–13364. DOI: 10.1609/aaai.v38i12.29237

Lepawsky, J. (2018). *Reassembling rubbish: Worlding electronic waste*. MIT Press. DOI: 10.7551/mitpress/11111.001.0001

Lessner, P. (2025, March 12). *Q&A: AI drives fresh Ta capacitor demand* [Interview by Argus Media]. Argus Media. https://www.argusmedia.com/en/news-and-insights/latest-market-news/2737718-q-a-ai-drives-fresh-ta-capacitor-demand

Levenda, A. M., & Mahmoudi, D. (2019). Silicon Forest and server farms: The (urban) nature of digital capitalism in the Pacific Northwest. *Culture Machine*, •••, 18.

Li, J., Fu, Q., & Ruan, P. (2024). Research on Algorithm Driven Intelligent Management and Control Technology for Future Power Grid. *EAI Endorsed Transactions on Energy Web, 11*. Advance online publication. DOI: 10.4108/ew.5824

Li, L., Jamieson, K., DeSalvo, G., Rostamizadeh, A., & Talwalkar, A. (2018). Hyperband: A novel bandit-based approach to hyperparameter optimization. *Journal of Machine Learning Research, 18*(185), 1–52.

Lima, F. M., Lovon-Canchumani, G. A., & Teixeira, M. A. (2023). Life cycle assessment in the mining sector: A systematic review. *Resources Policy, 78*, 102889. DOI: 10.1016/j.resourpol.2022.102889

Li, P., Yang, J., Islam, M. A., & Ren, S. (2025). *Making AI Less "Thirsty": Uncovering and Addressing the Secret Water Footprint of AI Models*. https://arxiv.org/abs/2304.03271

Liu, M., Yuan, Z., & Ping, W. (2025). Artificial intelligence and green total factor energy efficiency: Evidence from non-linear models. *Applied Economics, 1*. Advance online publication. DOI: 10.1080/00036846.2025.2480218

Liu, V., & Yin, Y. (2024). Green AI: Exploring carbon footprints, mitigation strategies, and trade offs in large language model training. *Discover Artificial Intelligence, 4*(1), 49. DOI: 10.1007/s44163-024-00149-w

Li, Y., Wen, Y., Tao, D., & Guan, K. (2020). Transforming Cooling Optimization for Green Data Center via Deep Reinforcement Learning. *IEEE Transactions on Cybernetics, 50*(5), 2002–2013. DOI: 10.1109/TCYB.2019.2927410 PMID: 31352360

Lopes, R. P., Mesquita, C., de la Cruz del Río-Rama, M., & Álvarez-García, J. (2018). Developing sustainability awareness in higher education. In *Strategies and best practices in social innovation* (pp. 131–152). Springer. DOI: 10.1007/978-3-319-89857-5_9

Lowder, T., & Xu, K. (2020). *The Evolving U.S. Distribution System: Technologies, Architectures, and Regulations for Realizing a Transactive Energy Marketplace*. DOI: 10.2172/1659880

Luccioni, A. S., Strubell, E., & Crawford, K. (2025). From efficiency gains to rebound effects: The problem of Jevons' paradox in AI's polarized environmental debate. In *Proceedings of the 2025 ACM Conference on Fairness, Accountability, and Transparency (FAccT '25)*. DOI: 10.1145/3715275.3732007

Luo, H., Huang, T.-Y., Wu, X., & Zhao, F. (2025). Life cycle assessment for primary gallium production at industrial-scale. *The International Journal of Life Cycle Assessment, 30*(7), 1545–1559. DOI: 10.1007/s11367-025-02492-1

Luo, T., Wong, W.-F., Goh, R. S. M., Do, A. T., Chen, Z., Li, H., Jiang, W., & Yau, W. (2023). Achieving Green AI with Energy-Efficient Deep Learning Using Neuromorphic Computing. *Communications of the ACM, 66*(7), 52–57. DOI: 10.1145/3588591

Lv, Y. (2023). Transitioning to sustainable energy: Opportunities, challenges, and the potential of blockchain technology. *Frontiers in Energy Research, 11*, 1258044. Advance online publication. DOI: 10.3389/fenrg.2023.1258044

Lyon, T. P., & Maxwell, J. W. (2011). Greenwash: Corporate environmental disclosure under threat of audit. *Journal of Economics & Management Strategy, 20*(1), 3–41. DOI: 10.1111/j.1530-9134.2010.00282.x

Madanchian, M., & Taherdoost, H. (2025). Ethical theories, governance models, and strategic frameworks for responsible AI adoption and organizational success. *Frontiers in Artificial Intelligence, 8*, 1619029. Advance online publication. DOI: 10.3389/frai.2025.1619029 PMID: 40741285

Maharani, ES, Fatima, S., & Pratiwi, A. (2024). Malang City Government Strategy in Improving the Quality of Public Services: Critical Analysis and Prospects for Improvement. Idarotuna: Journal of Administrative Science, 5(1), 71-84.

Mahendra, G. S., Wali, M., Idwan, H., Listartha, I. M. E., Yuliastuti, G. E., Sasongko, D., & Jude, G. A. (2022). *Keamanan komputer*. Galiono Digdaya.

Makridakis, S. (2017). The forthcoming Artificial Intelligence (AI) revolution: Its impact on society and firms. *Futures*, *90*, 46–60. DOI: 10.1016/j.futures.2017.03.006

Mancheri, N. A., Sprecher, B., Bailey, G., Ge, J., & Tukker, A. (2019). Effect of Chinese policies on rare earth supply chain resilience. *Resources, Conservation and Recycling*, *142*, 101–112. DOI: 10.1016/j.resconrec.2018.11.017

Marks, L. U., Makonin, S., Rodriguez-Silva, A., & Przedpełski, R. (2021). *Tackling the carbon footprint of streaming media* [SSHRC Knowledge Synthesis Grant Final Report].

Márquez-Ramos, L. (2021). Does digitalization in higher education help to bridge the gap between academia and industry? An application to COVID-19. *Industry and Higher Education*, *35*(6), 630–637. DOI: 10.1177/0950422221989190

Matuštík, J., & Kočí, V. (2021). What is a footprint? A conceptual analysis of environmental footprint indicators. *Journal of Cleaner Production*, *285*, 124833. DOI: 10.1016/j.jclepro.2020.124833

McKinsey & Company. (2023). *As organizations rapidly deploy GenAI tools, survey respondents expect significant effects on their industries and workforces The state of AI in 2023: GenAI's breakout year*.

Meadows, D. H. (2008). *Thinking in Systems: A Primer*. Chelsea Green Publishing.

Meta Platforms, Inc. (2024). *2024 sustainability report*. Meta. https://sustainability.atmeta.com/wp-content/uploads/2024/08/Meta-2024-Sustainability-Report.pdf

Microsoft. (2019a, April 15). We're increasing our carbon fee as we double down on sustainability. https://blogs.microsoft.com/on-the-issues/2019/04/15/were-increasing-our-carbon-fee-as-we-double-down-on-sustainability/

Microsoft. (2025, September 18). *Made in Wisconsin: The world's most powerful AI datacenter*. Microsoft On the Issues. https://blogs.microsoft.com/on-the-issues/2025/09/18/made-in-wisconsin-the-worlds-most-powerful-ai-datacenter/

Microsoft. (2025a, May 29). Environmental sustainability report. https://blogs.microsoft.com/on-the-issues/2025/05/29/environmental-sustainability-report/

Microsoft. (2025b, September 16). Microsoft invests $30 billion in UK AI future. https://blogs.microsoft.com/on-the-issues/2025/09/16/microsoft-30-billion-uk-ai-future/

Miles, M. B., & Huberman, A. M. (1994). *Qualitative data analysis: An expanded sourcebook* (2nd ed.). Sage.

MINING.COM. (2024, July 23). *AI data centers to worsen copper shortage – BNEF*. MINING.COM. https://www.mining.com/ai-data-centers-to-worsen-copper-shortage-bnef/

Ministry of Economy, Trade and Industry. (2020). International resource strategy [ ]. Government of Japan. https://www.meti.go.jp/press/2020/03/20200330001/20200330001.html

Ministry of Economy, Trade and Industry. (2023). Policy for measures to ensure stable supply of critical minerals. Government of Japan. https://www.meti.go.jp/policy/economy/economic_security/metal/index.html

Ministry of Trade, Industry and Energy. (2023). Strategy for securing reliable critical minerals supply. Government of the Republic of Korea.

Mirindi, D., Khang, A., & Mirindi, F. (2025). *Artificial Intelligence (AI) and Automation for Driving Green Transportation Systems: A Comprehensive Review* (pp. 1–19). DOI: 10.1007/978-3-031-72617-0_1

Mitu, N. E., & Mitu, G. T. (2024). The hidden cost of AI: Carbon footprint and mitigation strategies. *Revista de Ştiinţe Politice*, *84*, 9–16.

MLQ.ai. (n.d.). Sam Altman counters AI environmental criticisms by comparing to human energy needs. https://mlq.ai/news/sam-altman-counters-ai-environmental-criticisms-by-comparing-to-human-energy-needs/

Monstadt, J., & Saltzman, K. (2025). HOW DATA CENTERS HAVE COME TO MATTER : Governing the Spatial and Environmental Footprint of the 'Digital Gateway to Europe.'. *International Journal of Urban and Regional Research*, *49*(4), 757–778. DOI: 10.1111/1468-2427.13316

Montakhabi, M., Madhusudan, A., Mustafa, M., Vanhaverbeke, W., Almirall, E., & van der Graaf, S. (2023). Leveraging blockchain for energy transition in urban contexts. *Big Data & Society*, *10*(2), 20539517231205503. Advance online publication. DOI: 10.1177/20539517231205503

Mordor Intelligence. (2025). *Tantalum capacitors market*. https://www.mordorintelligence.com/industry-reports/tantalum-capacitors-market

Muhie, S. H. (2022). Novel approaches and practices to sustainable agriculture. *Journal of Agriculture and Food Research*, *10*, 100446. DOI: 10.1016/j.jafr.2022.100446

Muimba-Kankolongo, A., Banza Lubaba Nkulu, C., Mwitwa, J., Kampemba, F. M., Mulele Nabuyanda, M., Haufroid, V., Smolders, E., & Nemery, B. (2022). Impacts of trace metals pollution of water, food crops, and ambient air on population health in Zambia and the DR Congo. *Journal of Environmental and Public Health, 4515115*(1), 4515115. Advance online publication. DOI: 10.1155/2022/4515115 PMID: 35844936

Mukherjee, M., Sebastian-Cardenas, D. J., Balamurugan, S. P., Johnson, G. E., Gourisetti, S. N. G., Credle, S., Cutler, D., Bhattarai, B., & Johnson, B. J. (2022). A Blockchain Based Co-Simulation Framework for Integrating DERs to Wholesale Electricity Markets: Preprint. *OSTI OAI (U.S. Department of Energy Office of Scientific and Technical Information).* https://www.osti.gov/biblio/1889264

Mustafa, A., Kazmi, M., Khan, H. R., Qazi, S. A., & Lodi, S. H. (2022). Towards a Carbon Neutral and Sustainable Campus: Case Study of NED University of Engineering and Technology. *Sustainability (Basel), 14*(2), 794. DOI: 10.3390/su14020794

Mustafaoğlu, F. M., & Alkan, F. (2025). Prediction of middle school students' recycling behaviors with machine learning algorithms. *Science Education International, 36*(2), 209–218. DOI: 10.33828/sei.v36.i2.8

Naeem, R., Bajwa, A. M., Sattar, H., & Naeem, B. (2023). Effect of social media usage on digital carbon footprint: Analyzing awareness level of social media users of Punjab. https://doi.org/DOI: 10.21203/rs.3.rs-3225346/v1

Nassar, N. T., & Fortier, S. M. (2021). *Methodology and technical input for the 2021 review and revision of the U.S. Critical Minerals List* (Open-File Report 2021-1045). U.S. Geological Survey. DOI: 10.3133/ofr20211045

Nassar, N. T., Shojaeddini, E., Alonso, E., Jaskula, B., & Tolcin, A. (2024). Quantifying potential effects of China's gallium and germanium export restrictions on the U.S. economy (Open-File Report 2024–1057). U.S. Geological Survey. DOI: 10.3133/ofr20241057

National Energy System Operator. (2024, November 5). *Clean Power 2030: Advice on achieving clean power by 2030.* https://www.neso.energy/publications/clean-power-2030

National Research Council. (2008). *Minerals, critical minerals, and the U.S. economy.* National Academies Press., DOI: 10.17226/12034

Negroponte, N. (1995). *Being Digital.* Knopf.

Newell, S. J., Goldsmith, R. E., & Banzhaf, E. J. (1998). The effect of misleading environmental claims on consumer perceptions of advertisement. *Journal of Marketing Theory and Practice*, *6*(2), 48–60. DOI: 10.1080/10696679.1998.11501795

Nguyen, T. C. (2025). AI as a solution for tackling the sustainable energy challenge in developing countries. *Energy Sustainable Development. Energy for Sustainable Development*, *87*, 101761. DOI: 10.1016/j.esd.2025.101761

Nhi, D. T. T., Chuloy, M., & Glomann, L. (2024). Environmental impact of video streaming from users' perspectives. *Human Side of Services Engineering, 143.*

Nikolinakos, N. Th. (2023). *A European Approach to Excellence and Trust: The 2020 White Paper on Artificial Intelligence* (pp. 211–280). DOI: 10.1007/978-3-031-27953-9_5

Nobrega, C., & Varon, J. (2020). Big tech goes green(washing): Feminist lenses to unveil new tools in the master's houses. In *Global Information Society Watch 2020: Technology, the environment and a sustainable world: Responses from the Global South* (pp. 26–33). Association for Progressive Communications.

Nolden. (2021). Blockchain Contradictions in Energy Service and Climate Markets. In *CRC Press eBooks* (p. 1). Informa. DOI: 10.1201/9781003208365-1

Noor, Q., Ilyas, A., Javaid, Z., & Khan, H. (2023). Framing a knowledge domain visualization on green human resource management: A bibliometric analysis from 2008-2022. *Pakistan Journal of Humanities and Social Sciences*, *11*(4), 4200–4212. DOI: 10.52131/pjhss.2023.1104.0684

Nost, E., & Colven, E. (2022). Earth for AI: A political ecology of data-driven climate initiatives. *Geoforum*, *130*, 23–34. DOI: 10.1016/j.geoforum.2022.01.016

NVIDIA. (2024, July 12). NVIDIA GB200 NVL72 delivers trillion-parameter LLM training and real-time inference. NVIDIA Technical Blog. https://developer.nvidia.com/blog/nvidia-gb200-nvl72-delivers-trillion-parameter-llm-training-and-real-time-inference/

NVIDIA. (2025). *GB200 NVL72: Rack-scale Blackwell platform specifications* [Product page]. NVIDIA Corporation. https://www.nvidia.com/en-us/data-center/gb200-nvl72/

Nyilasy, G., Gangadharbatla, H., & Paladino, A. (2014). Perceived greenwashing: The interactive effects of green advertising and corporate environmental performance on consumer reactions. *Journal of Business Ethics*, *125*(4), 693–707. DOI: 10.1007/s10551-013-1944-3

Obringer, R., Rachunok, B., Maia-Silva, D., Arbabzadeh, M., Nateghi, R., & Madani, K. (2021). The overlooked environmental footprint of increasing Internet use. *Resources, Conservation and Recycling*, *167*, 105389. Advance online publication. DOI: 10.1016/j.resconrec.2020.105389

Obstfeld, M., Shambaugh, J. C., & Taylor, A. M. (2005). The trilemma in history: Tradeoffs among exchange rates, monetary policies, and capital mobility. *The Review of Economics and Statistics*, *87*(3), 423–438. DOI: 10.1162/0034653054638300

Office of the United States Trade Representative. (2014, March 26). United States wins victory in rare earths dispute with China: WTO report finds China's export restraints breach WTO rules. https://ustr.gov/about-us/policy-offices/press-office/press-releases/2014/March/US-wins-victory-in-rare-earths-dispute-with-China

Oliveira, A. P., Carraquico, T., & Martinez-Perez, C. (2026). Beyond efficiency: A systematic review of energy consumption and carbon footprint across the AI Lifecycle. *Sustainability (Basel)*, *18*(3), 1359. DOI: 10.3390/su18031359

Omar, N. N., & Saleh, M. A.Nuhayd Nadi OmarMahsen Abdulkarim Saleh. (2024). Power to the People: Advancing Resilient and Sustainable Decentralized Energy Distribution Systems. *Proceedings of London International Conferences*, *11*(11), 125–139. DOI: 10.31039/plic.2024.11.251

Oo, K. T., Jonah, K. A. Z. O. R. A., & Thin, M. M. Z. (2023). A systematic review of the pros and cons of digital pollution and its impact on the environment. *Journal of Sustainability and Environmental Management*, *2*(1), 61–73. DOI: 10.3126/josem.v2i1.53118

Open, A. I. (2025, September 23). *OpenAI, Oracle, and SoftBank expand Stargate with five new AI data center sites*. https://openai.com/index/five-new-stargate-sites/

Organisation for Economic Co-operation and Development. (2011). OECD due diligence guidance for responsible supply chains of minerals from conflict-affected and high-risk areas. OECD Publishing. https://www.oecd.org/content/dam/oecd/en/publications/reports/2011/07/oecd-due-diligence-guidance-for-responsible-supply-chains-of-minerals-from-conflict-affected-and-high-risk-areas_g1g13355/9789264111110-en.pdf

Organisation for Economic Co-operation and Development. (2023). Critical minerals market review 2023. OECD Publishing. https://www.oecd.org/en/publications/critical-minerals-market-review-2023_9cdf8f39-en.html

Ortar, N., Taylor, A. R., Velkova, J., ... (2023). Powering 'smart' futures: Data centres and the energy politics of digitalisation. In *Energy futures: Anthropocene challenges, emerging technologies and everyday life* (pp. 125–168). De Gruyter.

Owen, J. R., Kemp, D., Lechner, A. M., Harris, J., Zhang, R., & Lèbre, É. (2023). Energy transition minerals and their intersection with land-connected peoples. *Nature Sustainability*, *6*(2), 203–211. DOI: 10.1038/s41893-022-00994-6

Packard, V. (1960). *The Waste Makers*. David McKay Company.

Pandey, D., Agrawal, M., & Pandey, J. S. (2011). Carbon footprint: Current methods of estimation. *Environmental Monitoring and Assessment*, *178*(1–4), 135–160. DOI: 10.1007/s10661-010-1678-y PMID: 20848311

Paré, G., & Kitsiou, S. (2017). Methods for literature reviews. In *Handbook of eHealth evaluation: An evidence-based approach.* University of Victoria., https://www.ncbi.nlm.nih.gov/books/NBK481583/

Parguel, B., Benoît-Moreau, F., & Larceneux, F. (2011). How sustainability ratings might deter 'greenwashing': A closer look at ethical corporate communication. *Journal of Business Ethics*, *102*(1), 15–28. DOI: 10.1007/s10551-011-0901-2

Pasek, A. (2019). Managing carbon and data flows: Fungible forms of mediation in the cloud. *Culture Machine*, *18*, 1–15.

Patterson, D., Gonzalez, J., Le, Q., Liang, C., Munguia, L.-M., Rothchild, D., So, D., Texier, M., & Dean, J. (2021). Carbon and the broad question of AI's environmental impact. arXiv preprint arXiv:2104.10350.

Patterson, D., Gonzalez, J., Le, Q., Liang, C., Munguia, L., & Rothchild, D.. (2021). Carbon emissions and large neural network training. *arXiv*. https://arxiv.org/abs/2104.10350

Patterson, D., Gonzalez, J., Le, Q., Liang, C., Munguia, L., Rothchild, D., So, D., Texier, M., & Dean, J. (2021). Carbon emissions and large neural network training. *ArXiv*, *2104*(10350), 1–22.

Paul, S. G., Saha, A., Arefin, M. S., Bhuiyan, T., Biswas, A. A., Reza, A. W., Alotaibi, N. M., Alyami, S. A., & Moni, M. A. (2023). A Comprehensive Review of Green Computing: Past, Present, and Future Research. *IEEE Access : Practical Innovations, Open Solutions*, *11*, 87445–87494. DOI: 10.1109/ACCESS.2023.3304332

Pe'er, S., Goldman, D., & Yavetz, B. (2007). Environmental literacy in teacher training: Attitudes, knowledge, and environmental behavior of beginning students. *The Journal of Environmental Education*, *39*(1), 45–59. DOI: 10.3200/JOEE.39.1.45-59

Peattie, K., & Charter, M. (2003). Green marketing. In Baker, M. (Ed.), *The marketing book* (5th ed., pp. 726–755). Butterworth-Heinemann.

Pell, R., Wall, F., Yan, X., Li, J., & Zeng, X. (2019). Life cycle assessment in the minerals industry: Current practice, harmonization efforts, and potential improvement through the integration with process simulation. *Journal of Cleaner Production*, *232*, 155–173. DOI: 10.1016/j.jclepro.2019.05.264

Pereira, A. C., & Romero, F. (2017). A review of the meanings and the implications of the Industry 4.0 concept. *Procedia Manufacturing*, *13*, 1206–1214. DOI: 10.1016/j.promfg.2017.09.032

Perez, C. (2002). *Technological Revolutions and Financial Capital: The Dynamics of Bubbles and Golden Ages*. Edward Elgar. DOI: 10.4337/9781781005323

Perkins, D. N., Drisse, M. N. B., Nxele, T., & Sly, P. D. (2014). E-waste: A global hazard. *Annals of Global Health*, *80*(4), 286–295. DOI: 10.1016/j.aogh.2014.10.001 PMID: 25459330

Peters, M. A. (2017). Technological unemployment: Educating for the fourth industrial revolution. *Educational Philosophy and Theory*, *49*(1), 1–6. DOI: 10.1080/00131857.2016.1177412

Pew Charitable Trusts. (2025, January 15). The world currently lacks the ability to govern deep-sea mining. The Pew Charitable Trusts. https://www.pew.org/en/about/news-room/opinion/2025/01/15/the-world-currently-lacks-the-ability-to-govern-deep-sea-mining

Peykani, P., Emrouznejad, A., Ghanidel, S., Javadi-Sisi, I., & Mirjalili, S. (2026). Green Artificial Intelligence: A Comprehensive Review of Metrics, Tools, Challenges, Trends, and Future Prospects. *Archives of Computational Methods in Engineering*. Advance online publication. DOI: 10.1007/s11831-026-10546-2

Pham, B. T., & Huynh, T. H. (2025). Greenwashing and brand avoidance: Examining green scepticism, green perceived risk, and green consumer confusion. *Journal of Services Marketing*, •••, 1–12.

Piotr. (2024). Innovative Solutions for the Future Development of the Energy Sector. *EUROPEAN RESEARCH STUDIES JOURNAL*, 297. DOI: 10.35808/ersj/3435

Plantin, J.-C., Lagoze, C., Edwards, P. N., & Sandvig, C. (2018). Infrastructure studies meet platform studies in the age of Google and Facebook. *New Media & Society*, *20*(1), 293–310. DOI: 10.1177/1461444816661553

Polaris Market Research. (2022). *Thermal interface materials market share, size, trends, industry analysis report: 2022–2030*. Polaris Market Research. https://www.polarismarketresearch.com/industry-analysis/thermal-interface-materials-market

Polimeni, J. M., Mayumi, K., Giampietro, M., & Alcott, B. (2008). *The Jevons Paradox and the Myth of Resource Efficiency Improvements*. Earthscan.

Porawagamage, G. D., Dharmapala, K., Chaves, J. S., Villegas, D. C., & Rajapakse, A. (2024). A review of machine learning applications in power system protection and emergency control: opportunities, challenges, and future directions [Review of *A review of machine learning applications in power system protection and emergency control: opportunities, challenges, and future directions*]. *Frontiers in Smart Grids, 3*. Frontiers Media. DOI: 10.3389/frsgr.2024.1371153

Qi, C., Shen, S., Li, R., Zhao, Z., Liu, Q., Liang, J., & Zhang, H. (2021). An efficient pruning scheme of deep neural networks for Internet of Things applications. *EURASIP Journal on Advances in Signal Processing*, *2021*(1), 31. DOI: 10.1186/s13634-021-00744-4

Rabbani, H., & Yuliani, D. (2025). HEXAHELIX COLLABORATION STRATEGY BASED ON EMPOWERMENT OF THE SERIKANDI BIRU COMMUNITY IN UTILIZING GARUT REGENCY RESOURCES. *DEVELOPMENT: Journal of Community Engagement*, *4*(3), 422–439. DOI: 10.46773/djce.v4i3.2564

Rachmawati, R., Susilowati, A. P. E., Radika, P., Sofie, T., & Aditya, M. Y. (2024). Sustainable Regional Development Through the Development of Smart Village and Smart Branding: A Case Study of Bangsri Village, Blora Regency. *Intelligent Systems Journal*, *7*(2), 199–214. DOI: 10.37396/jsc.v7i2.441

Rafat, K., Islam, S., Mahfug, A., Al, , & Hossain, M. (2023). Mitigating carbon footprint for knowledge distillation based deep learning model compression. *PLoS One*, *18*(5), e0285668. DOI: 10.1371/journal.pone.0285668 PMID: 37186614

Rahadiansyah, M. (2023). *PENGARUH KUALITAS PRODUK, HARGA, DAN LOKASI TERHADAP KEPUTUSAN PEMBELIAN PADA RESTORAN POPOLAMAMA DI TANGERANG* (Doctoral dissertation, Universitas Satya Negara Indonesia).

Rana, S. (2025). AI-DRIVEN FAULT DETECTION AND PREDICTIVE MAINTENANCE IN ELECTRICAL POWER SYSTEMS: A SYSTEMATIC REVIEW OF DATA-DRIVEN APPROACHES, DIGITAL TWINS, AND SELF-HEALING GRIDS [Review of AI-DRIVEN FAULT DETECTION AND PREDICTIVE MAINTENANCE IN ELECTRICAL POWER SYSTEMS: A SYSTEMATIC REVIEW OF DATA-DRIVEN APPROACHES, DIGITAL TWINS, AND SELF-HEALING GRIDS]. *American Journal of Advanced Technology and Engineering Solutions, 1*(1), 258–289. DOI: 10.63125/4p25x993

Raper, R., Boeddinghaus, J., Coeckelbergh, M., Gross, W., Campigotto, P., & Lincoln, C. N. (2022). Sustainability Budgets: A Practical Management and Governance Method for Achieving Goal 13 of the Sustainable Development Goals for AI Development. *Sustainability (Basel), 14*(7), 4019. DOI: 10.3390/su14074019

Rasool, S., & Jacobsen, R. H. (2024). Perspective Chapter: Leveraging Self-Sovereign Identity to Introduce ReCert – A Foundational Framework for Decentralized BCTE Solutions. In *IntechOpen eBooks*. IntechOpen. DOI: 10.5772/intechopen.1007156

Rawat, A., Arora, M., Bhadula, G., & Sharma, S. (2023). A Blockchain Based Energy Management System for Microgrids. *International Journal for Research in Applied Science and Engineering Technology, 11*(4), 1313–1319. DOI: 10.22214/ijraset.2023.50314

Reddi, V. J., Cheng, C., Kanter, D., Mattson, P., Schmuelling, G., & Wu, C.-J. (2021). The Vision Behind MLPerf: Understanding AI Inference Performance. *IEEE Micro, 41*(3), 10–18. DOI: 10.1109/MM.2021.3066343

Reuters. (2025, October 15). *Meta commits $1.5 billion for AI data center in Texas.* https://www.reuters.com/business/meta-commits-15-billion-ai-data-center-texas-2025-10-15/

Richie, C., Hinrichs-Krapels, S., Dobbe, R., French, P., Wei, J. C. J., Diehl, J. C., & Kong, R. (2025). Environmental impacts of artificial intelligence in health care: Considerations and recommendations. *Health and Technology, 15*(6), 1087–1093. DOI: 10.1007/s12553-025-01003-4

Ritchie, H., Rosado, P., & Roser, M. (2020). Greenhouse gas emissions. *Our World in Data.* https://ourworldindata.org/greenhouse-gas-emissions

Rockström, J., Steffen, W., Noone, K., Persson, Å., Chapin, F. S.III, Lambin, E., Lenton, T. M., Scheffer, M., Folke, C., Schellnhuber, H. J., Nykvist, B., de Wit, C. A., Hughes, T., van der Leeuw, S., Rodhe, H., Sörlin, S., Snyder, P. K., Costanza, R., Svedin, U., & Foley, J. A. (2009). A safe operating space for humanity. *Nature, 461*(7263), 472–475. DOI: 10.1038/461472a PMID: 19779433

Roczen, N., Kaiser, F. G., Bogner, F. X., & Wilson, M. (2014). A competence model for environmental education. *Environment and Behavior*, *46*(8), 972–992. DOI: 10.1177/0013916513492416

Rohde, F., Nasruddin, Z., Kotova, E., & Ammon, S. (2026). Navigating justice in AI lifecycles: Ethical perspectives on infrastructural prerequisites and their environmental impacts. *AI and Ethics*, *6*(1), 156. DOI: 10.1007/s43681-026-01008-3

Rolnick, D., Donti, P. L., Kaack, L. H., Kochanski, K., Lacoste, A., & Sankaran, K.. (2019). Tackling climate change with machine learning. *arXiv*. https://arxiv.org/abs/1906.05433

Rolnick, D., Donti, P. L., Kaack, L. H., Kochanski, K., Lacoste, A., Sankaran, K., Ross, A. S., Milojevic-Dupont, N., Jaques, N., Waldman-Brown, A., Luccioni, A., Maharaj, T., Sherwin, E. D., Mukkavilli, S. K., Kording, K. P., Gomes, C., Ng, A. Y., Hassabis, D., Platt, J. C., & Bengio, Y. (2019). *Tackling Climate Change with Machine Learning*. https://arxiv.org/abs/1906.05433

Roos, G. (2025b, September). *Australia–EU critical minerals partnerships: Navigating supply security and climate commitments* (Road to Belém Briefing #1). Institute for Sustainability, Energy and Resources & Institute for International Trade, The University of Adelaide.

Roos, G. (2025c, November). Stimulating demand for green premium products in OECD countries: Policy tools, theoretical perspectives, and empirical evidence from the iron–steel value chain. Discussion paper prepared for the Fortum Nordic Energy Debate, 25 November 2025, Finlandia Hall, Helsinki, Finland.

Roos, G. (2025a, May). *Increasing unevenness in trade-related sustainability policy: The ASEAN perspective (Policy Brief No. 29)*. Institute for International Trade, The University of Adelaide. https://iit.adelaide.edu.au

Rust, R. T. (2020). The future of marketing. *International Journal of Research in Marketing*, *37*(1), 15–26. DOI: 10.1016/j.ijresmar.2019.08.002

Saba, G., Ahmed, S., Sania, H., Khan, T. A., & Mohamad Nasir, H. B. (2026). An empirical comparison of AI assisted software refactoring tools. *Scientific Reports*, *16*(1), 20971. Advance online publication. DOI: 10.1038/s41598-026-49590-0 PMID: 42032054

Safari, A., Gharehbagh, H. K., Nazari-Heris, M., & Oshnoei, A. (2023). DeepResTrade: A peer-to-peer LSTM-decision tree-based price prediction and blockchain-enhanced trading system for renewable energy decentralized markets. *Frontiers in Energy Research, 11*, 1275686. Advance online publication. DOI: 10.3389/fenrg.2023.1275686

Sağlam-kaya, Y. (2019). Öğretmen adaylarının teknopedagojik eğitim yeterliklerinin çeşitli değişkenler ve öğretmen öz yeterlikleri bağlamında incelenmesi. *Journal of Theoretical Educational Science, 12*(1), 185–204. DOI: 10.30831/akukeg.420909

Salesforce. (n.d.-a). Sustainable AI. https://www.salesforce.com/news/stories/sustainable-ai/

Salesforce. (n.d.-b). Salesforce AI sustainability outlook. https://www.salesforce.com/en-us/wp-content/uploads/sites/4/documents/company/sustainability/salesforce-ai-sustainability-outlook.pdf

SalesforceBen. (n.d.). Salesforce tackles AI's environmental issues with 6-point plan. https://www.salesforceben.com/salesforce-tackle-ais-environmental-issues-with-6-point-plan/

Sánchez-Maroño, N., Rodríguez-Arias, A., Dumitru, A., Lema-Blanco, I., Guijarro-Berdiñas, B., & Alonso-Betanzos, A. (2022). How Agent-based modeling can help to foster sustainability projects. *Procedia Computer Science, 207*, 2546–2555. DOI: 10.1016/j.procs.2022.09.313

Scala, F., Flesca, S., & Pontieri, L. (2025). *Play it Straight: An Intelligent Data Pruning Technique for Green-AI* (pp. 69–85). DOI: 10.1007/978-3-031-78977-9_5

Schlosberg, D. (2007). *Defining Environmental Justice: Theories, Movements, and Nature*. Oxford University Press. DOI: 10.1093/acprof:oso/9780199286294.001.0001

Schneider Electric. (2025, May 20). AI-tolerant UPSs: The first line of defense in data center resilience. Schneider Electric Blog. https://blog.se.com/datacenter/2025/05/20/ai-tolerant-ups-first-line-defense-data-center-resilience/

Schrijvers, D., Hool, A., Blengini, G. A., Chen, W. Q., Dewulf, J., Eggert, R., van Ellen, L., Gauss, R., Goddin, J., Habib, K., Hagelüken, C., Hirohata, A., Hofmann-Amtenbrink, M., Kosmol, J., Le Gleuher, M., Grohol, M., Ku, A., Lee, M.-H., Liu, G., & Wäger, P. (2020). A review of methods and data to determine raw material criticality. *Resources, Conservation and Recycling, 155*, 104617. DOI: 10.1016/j.resconrec.2019.104617

Schwab, K. (2016). *The fourth industrial revolution*. World Economic Forum.

Schwartz, R., Dodge, J., Smith, N. A., & Etzioni, O. (2020). Green AI. *Communications of the ACM*, *63*(12), 54–63. DOI: 10.1145/3381831

Securities and Exchange Board of India. (2021). *Business Responsibility and Sustainability Report (BRSR)*. SEBI.

SemiAnalysis. (2024, July 17). *GB200 hardware architecture: Component supply chain and bill of materials*. SemiAnalysis. https://semianalysis.com/2024/07/17/gb200-hardware-architecture-and-component/

SemiAnalysis. (2025, September 16). *xAI's Colossus 2 – First gigawatt datacenter in the world, unique RL methodology, capital raise*. https://newsletter.semianalysis.com/p/xais-colossus-2-first-gigawatt-datacenter

Sgueglia, A., Sorbo, A. D., Visaggio, C. A., & Canfora, G. (2022). A systematic literature review of IoT time series anomaly detection solutions. *Future Generation Computer Systems*, *134*, 170–186. DOI: 10.1016/j.future.2022.04.005

Shan, S., Yang, S., Becerra, V. M., Deng, J., & Li, H. (2023). A Case Study of Existing Peer-to-Peer Energy Trading Platforms: Calling for Integrated Platform Features. *Sustainability (Basel)*, *15*(23), 16284. DOI: 10.3390/su152316284

Sharma, P., & Dash, B. (2023). *The digital carbon footprint: Threat to an environmentally sustainable future*. SSRN. https://doi.org/ (replace with actual if available) DOI: 10.2139/ssrn.XXXX

Shehabi, A., Smith, S. J., Hubbard, A., Newkirk, A., Lei, N., Siddik, M. A. B., Holecek, B., Koomey, J., Masanet, E., & Sartor, D. (2024). *2024 United States data center energy usage report*. Lawrence Berkeley National Laboratory., DOI: 10.71468/P1WC7Q

Sigalos, M. (2025, October 29). *Amazon opens $11 billion AI data center Project Rainier in Indiana*. CNBC. https://www.cnbc.com/2025/10/29/amazon-opens-11-billion-ai-data-center-project-rainier-in-indiana.html

Silicon, A. N. G. L. E. (2025, October 23). *Crusoe lands $1.3B to accelerate buildout of large-scale AI data centers*. https://siliconangle.com/2025/10/23/crusoe-lands-1-3b-accelerate-buildout-large-scale-ai-data-centers/

Singh, A. (2012). Environmental education for sustainable development. *Jyotirmay, 18*.

Singh, A. R., Kumar, R. S., Bajaj, M., Khadse, C. B., & Зайцев, Є. (2024). Machine learning-based energy management and power forecasting in grid-connected microgrids with multiple distributed energy sources. *Scientific Reports, 14*(1). Advance online publication. DOI: 10.1038/s41598-024-70336-3 PMID: 39160194

Sonderegger, T., Berger, M., Alvarenga, R., Bach, V., Cimprich, A., Dewulf, J., Frischknecht, R., Guinée, J., Helbig, C., Huppertz, T., Jolliet, O., Motoshita, M., Northey, S., Peña, C. A., Rugani, B., Sahnoune, A., Schrijvers, D., Schulze, R., Sonnemann, G., & Young, S. B. (2020). Mineral resources in life cycle impact assessment—Part I: A critical review of existing methods. *The International Journal of Life Cycle Assessment, 25*(4), 784–797. DOI: 10.1007/s11367-020-01736-6

Sovacool, B. K., Ali, S. H., Bazilian, M., Radley, B., Nemery, B., Okatz, J., & Mulvaney, D. (2020). Sustainable minerals and metals for a low-carbon future. *Science, 367*(6473), 30–33. DOI: 10.1126/science.aaz6003 PMID: 31896708

Sprecher, B., Daigo, I., Murakami, S., Kleijn, R., Vos, M., & Kramer, G. J. (2015). Framework for resilience in material supply chains, with a case study from the 2010 rare earth crisis. *Environmental Science & Technology, 49*(11), 6740–6750. DOI: 10.1021/acs.est.5b00206 PMID: 25965803

Srnicek, N. (2017). *Platform Capitalism*. Polity Press.

Stadlthanner, K. A., Andreu, L., Font, X., Ribeiro, M. A., & Currás-Pérez, R. (2022). How environmental gain messages affect cause involvement, attitude and behavioural intentions: The moderating effects of CSR scepticism and biospheric values. *Corporate Communications, 27*(4), 781–799. DOI: 10.1108/CCIJ-11-2021-0125

Stanford. (2025). *AI Index Report 2025*.

Stavins, R. N. (2003). Experience with market-based environmental policy instruments. In Mäler, K.-G., & Vincent, J. R. (Eds.), *Handbook of Environmental Economics* (pp. 355–435). Elsevier.

Strader, T., Huang, Y.-H., & Tu, Y.-J. (2024). Artificial intelligence and environmental sustainability: Review and research directions. *Journal of International Technology and Information Management, 32*(4).

Strubell, E., Ganesh, A., & McCallum, A. (2019). *Energy and Policy Considerations for Deep Learning in NLP*. https://arxiv.org/abs/1906.02243 DOI: 10.18653/v1/P19-1355

Strubell, E., Ganesh, A., & McCallum, A. (2020). Energy and policy considerations for modern deep learning research. *Proceedings of the AAAI Conference on Artificial Intelligence, 34*(09), 13693–13696. DOI: 10.1609/aaai.v34i09.7123

Sukmana, A. H. (2022). Pengaruh strategi pemasaran pasca pandemi covid-19 terhadap minat nasabah menabung di Bank BSI Mataram. *Jurnal Perbankan Syariah, 1*(2), 26–35. DOI: 10.20414/jps.v1i2.6324

Szabo, S., & Webster, J. (2021). Perceived greenwashing: The effects of green marketing on environmental and product perceptions. *Journal of Business Ethics, 171*(4), 719–739. DOI: 10.1007/s10551-020-04461-0

Tabassum, S., Babu, A. R. V., & Dheer, D. K. (2024). A comprehensive exploration of IoT-enabled smart grid systems: Power quality issues, solutions, and challenges. *Science and Technology for Energy Transition, 79*, 62. DOI: 10.2516/stet/2024056

Taffel, S., Bedford, L., & Mann, M.. (2022). Ecocide isn't ethical: Political ecology and capitalist AI ethics. In Phan, T., Goldenfein, J., & Kuch, D. (Eds.), *Economics of virtue: The circulation of 'ethics' in AI* (pp. 59–86). Institute of Network Cultures.

Talens Peiró, L., & Ardente, F. (2015). *Analysis of material efficiency requirements of enterprise servers*. Publications Office of the European Union., DOI: 10.2788/409022

Tanaka, K. (2026). Carbon-aware energy scheduling for renewable-dominant smart grid infrastructure. *Natural Journal of Renewable Energy Systems and Innovation, 2*(3), 43–50.

The Economic Times. (2025, October 9). *TCS announces 1-GW data centre plan with $6.5 billion investment.* https://m.economictimes.com/tech/information-tech/tcs-announces-1-gw-data-centre-plan-with-6-5-billion-investment/articleshow/124429399.cms

The Greens/EFA in the European Parliament. (2023). Breakthrough for regulating AI and banning discriminatory technologies. https://www.greens-efa.eu/en/article/press/breakthrough-for-regulating-ai-and-banning-discriminatory-technologies

The Guardian. (2025, June 27). Google emissions and AI electricity demand derail green efforts. https://www.theguardian.com/technology/2025/jun/27/google-emissions-ai-electricity-demand-derail-efforts-green

The Oregon Group. (2025). *Rare earths: Critical minerals and energy intelligence.* The Oregon Group. https://theoregongroup.com/rare-earths/

The Shift Project. (2019). *Climate crisis: The unsustainable use of online video—The practical case study of online video.* https://theshiftproject.org/en/article/unsustainable-use-online-video/

Toderas, M. (2025). Artificial intelligence for sustainability: A systematic review and critical analysis. *Sustainability (Basel), 17*(17), 8049. DOI: 10.3390/su17178049

Tomar, P. S., & Grover, V. (2024). Transforming the Energy Sector: Addressing Key Challenges through Generative AI, Digital Twins, AI, Data Science and Analysis. *EAI Endorsed Transactions on Energy Web*, *10*. Advance online publication. DOI: 10.4108/ew.4825

Tomlinson, B., Black, R. W., Patterson, D. J., & Torrance, A. W. (2024). The carbon emissions of writing and illustrating are lower for AI than for humans. *Scientific Reports*, *14*(1), 54271. Advance online publication. DOI: 10.1038/s41598-024-54271-x PMID: 38355820

Trading Economics. (n.d.). Neodymium. Retrieved December 15, 2025, from https://tradingeconomics.com/commodity/neodymium

Trafigura. (2024). *AI could add 1 million tons to copper demand by 2030*. https://www.trafigura.com/news-and-insights/in-the-news/2024/ai-could-add-1-million-tons-to-copper-demand-by-2030-says-trafigura/

U.S. Congress. (2022a). CHIPS and Science Act (Pub. L. No. 117–167). https://www.congress.gov/117/plaws/publ167/PLAW-117publ167.pdf

U.S. Congress. (2022b). Inflation Reduction Act of 2022 (Pub. L. No. 117–169). https://www.congress.gov/117/plaws/publ169/PLAW-117publ169.pdf

U.S. Department of State. (2022). Minerals Security Partnership [Fact sheet]. https://www.state.gov/minerals-security-partnership

U.S. Department of State. (2025, December 11). Pax Silica. https://www.state.gov/pax-silica/

U.S. Geological Survey. (2024). Mineral commodity summaries 2024. DOI: 10.3133/mcs2024

United Nations Human Rights Council. (2011). *Guiding Principles on Business and Human Rights: Implementing the United Nations 'Protect, Respect and Remedy' Framework (HR/PUB/11/04)*. Office of the United Nations High Commissioner for Human Rights.

United States Geological Survey. (2025). *Mineral commodity summaries 2025*. U.S. Geological Survey., DOI: 10.3133/mcs2025

University of Cambridge. (2022). *Cambridge Blockchain Network Sustainability Index (CBECI)*. https://ccaf.io/cbnsi/cbeci

Uzondu, N. C., & Lele, D. D.Nwankwo Charles UzonduDominic Dummene Lele. (2024). Comprehensive analysis of integrating smart grids with renewable energy sources: Technological advancements, economic impacts, and policy frameworks. *Engineering Science & Technology Journal*, *5*(7), 2334–2363. DOI: 10.51594/estj.v5i7.1347

Van Wynsberghe, A. (2021). Sustainable AI: AI for sustainability and the sustainability of AI. *AI and Ethics*, *1*(3), 213–218. DOI: 10.1007/s43681-021-00043-6

Vantage Data Centers. (2025, October 22). *OpenAI, Oracle and Vantage Data Centers announce Stargate data center site in Wisconsin*. https://vantage-dc.com/news/openai-oracle-and-vantage-data-centers-announce-stargate-data-center-site-in-wisconsin/

Velkova, J. (2021). Thermopolitics of data: Cloud infrastructures and energy futures. *Cultural Studies*, *35*(4–5), 663–683. DOI: 10.1080/09502386.2021.1895243

Vera, M. L., Torres, W. R., Galli, C. I., Chagnes, A., & Flexer, V. (2023). Environmental impact of direct lithium extraction from brines. *Nature Reviews. Earth & Environment*, *4*(2), 149–165. DOI: 10.1038/s43017-022-00387-5

Versino, C., Heine, P., & Du, T. (2025). *Strategic trade atlas 2019–2023: Country- and commodity-based views*. Publications Office of the European Union., DOI: 10.2760/7177906

Vertiv Holdings. (n.d.). *Data centre power density trends and infrastructure requirements*. Vertiv Holdings Co.

Vinuesa, R., Azizpour, H., Leite, I., Balaam, M., Dignum, V., Domisch, S., Felländer, A., Langhans, S. D., Tegmark, M., & Fuso Nerini, F. (2020). The role of artificial intelligence in achieving the Sustainable Development Goals. *Nature Communications*, *11*(1), 233. DOI: 10.1038/s41467-019-14108-y PMID: 31932590

Vivoda, V. (2023). Friend-shoring and critical minerals: Exploring the role of the Minerals Security Partnership. *Energy Research & Social Science*, *100*, 103085. DOI: 10.1016/j.erss.2023.103085

Vivoda, V., & Matthews, R. (2024). 'Friend-shoring' as a panacea to Western critical mineral supply chain vulnerabilities. *Mineral Economics*, *37*(3), 463–476. DOI: 10.1007/s13563-023-00402-1

Wang, B., Rau, P.-L. P., & Yuan, T. (2022). Measuring user competence in using artificial intelligence: Validity and reliability of artificial intelligence literacy scale. *Behaviour & Information Technology*, *42*(9), 1324–1337. DOI: 10.1080/0144929X.2022.2072768

Wan, Y., Qu, Y., Gao, L., & Xiang, Y. (2022). Privacy-preserving blockchain-enabled federated learning for B5G-Driven edge computing. *Computer Networks, 204*, 108671. DOI: 10.1016/j.comnet.2021.108671

Wark, S., & Soncul, Y. (2022). A media-theoretical armamentarium: From the pharmakon to pharmacologies of media. *Media Theory*, *6*(2), 1–24. DOI: 10.70064/mt.v6i2.879

Waseem, U. (2025, April 11). *AI infrastructure engineering: Building enterprise-scale machine learning systems*. Wevolver. https://www.wevolver.com/article/ai-infrastructure

Wathuge, A., & Sedera, D. (2021). An awareness-based model to minimize the environmental damage of internet usage: A longitudinal study. arXiv. https://arxiv.org/abs/2111.04453

Wen, H., Lee, C.-C., & Song, Z. (2021). Digitalization and environment: How does ICT affect enterprise environmental performance? *Environmental Science and Pollution Research International*, *28*(39), 54826–54841. DOI: 10.1007/s11356-021-14474-5 PMID: 34014481

Widodo, N. (2016). Development of e-Government in Regional Government in the Context of Realizing a Smart City (Study in the Regional Government of Malang City). *Scientific Journal of Public Administration*, *2*(4), 227–235. DOI: 10.21776/ub.jiap.2016.002.04.15

Wiedmann, T. O., Schandl, H., Lenzen, M., Moran, D., Suh, S., West, J., & Kanemoto, K. (2015). The material footprint of nations. *Proceedings of the National Academy of Sciences of the United States of America*, *112*(20), 6271–6276. DOI: 10.1073/pnas.1220362110 PMID: 24003158

Williams, E. D., Ayres, R. U., & Heller, M. (2002). The 1.7 kilogram microchip: Energy and material use in the production of semiconductor devices. *Environmental Science & Technology*, *36*(24), 5504–5510. DOI: 10.1021/es025643o PMID: 12521182

Winata, R., & Friantini, R. N. (2019). Pengaruh motivasi belajar terhadap prestasi belajar matematika siswa kelas VIII SMPN 1 Kuala Behe. [Jurnal Ilmiah Pendidikan Matematika]. *JIPM*, *7*(2), 85–92. DOI: 10.25273/jipm.v7i2.3663

Winfrey, O., & Jung, C. G. (2022). Self-awareness and skill development. In *Leadership: Leaders, followers, environments* (p. 47).

World Bank. (2020). Minerals for climate action: The mineral intensity of the clean energy transition. https://documents.worldbank.org/en/publication/documents-reports/documentdetail/099052423172525564

World Economic Forum. (2025). AI data centres and critical minerals: Infrastructure requirements for the AI transition. World Economic Forum.

World Energy Council. (2023). World Energy Trilemma Index 2023: Frameworks for resilient energy transitions. World Energy Council, London. https://www.worldenergy.org/transition-toolkit/world-energy-trilemma-framework

Wu, C., Raghavendra, R., Gupta, U., Acun, B., Ardalani, N., Maeng, K., Chang, G., Aga, F., Huang, J., Bai, C., Gschwind, M., Gupta, A., Ott, M., Melnikov, A., Candido, S., Brooks, D., Chauhan, G., Lee, B., Lee, H., & Hazelwood, K. (2022). Sustainable AI: Environmental implications, challenges and opportunities. *Proceedings of the Machine Learning Systems*, 1–19.

Wyatt, S. (2008). Technological determinism is dead; long live technological determinism. In Hackett, E., Amsterdamska, O., & Lynch, M. (Eds.), *Handbook of science and technology studies* (pp. 165–180). MIT Press.

Xiao, G., Lin, J., Seznec, M., Wu, H., Demouth, J., & Han, S. (2023). SmoothQuant: Accurate and efficient post-training quantization for large language models. *Proceedings of Machine Learning Research*, *202*, 38087–38099.

Xu, Y., Ahokangas, P., Yrjölä, S., & Koivumäki, T. (2019). The fifth archetype of electricity market: The blockchain marketplace. *Wireless Networks*, *27*(6), 4247–4263. DOI: 10.1007/s11276-019-02065-9

Xu, Z., Sun, D.-W., Zeng, X.-A., Liu, D., & Pu, H. (2015). Research developments in methods to reduce the carbon footprint of the food system: A review. *Critical Reviews in Food Science and Nutrition*, *55*(9), 1270–1286. DOI: 10.1080/10408398.2013.821593 PMID: 24689789

Yadav, S. K., Banerjee, A., Jhariya, M. K., Meena, R. S., Raj, A., Khan, N., & Sheoran, S.. (2022). Environmental education for sustainable development. In *Natural resources conservation and advances for sustainability* (pp. 415–431). Elsevier. DOI: 10.1016/B978-0-12-822976-7.00010-7

Yanarateş, E. (2022). Dijital çevre ve dijital atık. In *Dijitalleşme* (p. 159).

Yang, Q., & Wang, H. (2021).. . *Exploring Blockchain for The Coordination of Distributed Energy Resources*, *1*, 1–6. Advance online publication. DOI: 10.1109/CISS50987.2021.9400211

Yin, X., Wu, G., Wei, J., Shen, Y., Qi, H., & Yin, B. (2022). Deep Learning on Traffic Prediction: Methods, Analysis, and Future Directions. *IEEE Transactions on Intelligent Transportation Systems, 23*(6), 4927–4943. DOI: 10.1109/TITS.2021.3054840

Yoon, Y., Gürhan-Canli, Z., & Schwarz, N. (2006). The effect of corporate social responsibility (CSR) activities on companies with bad reputations. *Journal of Consumer Psychology, 16*(4), 377–390. DOI: 10.1207/s15327663jcp1604_9

Zhao, L., Liu, S., Xin, T., Tan, J., Wang, X., Li, Y., Bian, Z., Chen, Y., Kong, F., Bian, J., Qian, C., & Zhang, Z. (2026). AI agent in healthcare: Applications, evaluations, and future directions. *Npj Artificial Intelligence, 2*(1), 31. DOI: 10.1038/s44387-026-00076-4

Zhao, P., Li, S., Cao, Z., Hu, P. J., Gu, C., Yan, X., Huo, D., Luo, T., & Wang, Z. (2023). Socially Governed Energy Hub Trading Enabled by Blockchain-Based Transactions. *IEEE Transactions on Computational Social Systems, 1*. Advance online publication. DOI: 10.1109/TCSS.2023.3308608

Zhong, B., Song, Y., Feng, G. C., Shi, J., Zhu, Y., Xie, L., Zhou, W. A., Yu, S., Lu, Y., Qin, Y., & Xiong, Z. (2025). AI imaginaries shape technological identity and digital futures. *Computers in Human Behavior, 169*, 108682. DOI: 10.1016/j.chb.2025.108682

Zimba, A., Phiri, K. O., Mulenga, M., & Mukupa, G. (2025). A systematic literature review of blockchain technology and energy efficiency based on consensus mechanisms, architectural innovations, and sustainable solutions. *Discover Analytics, 3*(1), 14. Advance online publication. DOI: 10.1007/s44257-025-00041-6

Zuboff, S. (2019). *The Age of Surveillance Capitalism*. PublicAffairs.

Zuboff, S. (2019). *The age of surveillance capitalism: The fight for a human future at the new frontier of power*. PublicAffairs.

About the Contributors

Momina Shaheen is an academic and researcher in computer science with over 8 years of experience in teaching and research. She currently serves as a Lecturer in Computing and Programme Leader at the University of Roehampton London. Her work focuses on edge computing, the Internet of Things, federated learning, and cybersecurity, with applications across smart cities, healthcare, finance, and education. Shaheen holds a B.Sc. in Information Technology, an M.Eng. in Software Engineering, and defended her Ph.D. in Computer Science, focusing on improving deep learning performance in federated machine learning. Her academic contributions include over 46 peer-reviewed publications in Q1 journals, multiple book chapters, and editorial work, earning her an h-index of 9 and over 400 citations. She has taught a wide range of undergraduate and postgraduate modules, including algorithms, software development, artificial intelligence, and data visualization. In addition to her teaching, she leads international collaborations and actively contributes to curriculum design, grant writing, and supervision of research projects. She also served as a reviewer for prominent journals such as Nature (under Early Career Reviewer Program), MDPI Electronics, IEEE Access, IET Information Security, Springer Scientific Reports, PLOS ONE, and ACM Transactions, and has chaired international conferences. Shaheen is committed to fostering innovation in computing education, and intelligent systems.

***

Faizan Ahmad is an Associate Professor (Senior Lecturer) at the University of East London. An internationally recognized researcher, he holds the UK Global Talent designation (endorsed by The Royal Society) and is a Fellow of the Higher Education Academy. His research focuses on Human-Centered AI (HCAI) and intelligent immersive learning technologies, specifically the design of adaptive, personalized learning solutions within Virtual Reality (VR), Augmented Reality

(AR), and gamified environments. His methodology integrates multimodal learning analytics and computational modeling to bridge cognitive, emotional, and behavioral insights. With nearly a decade of international academic experience, Dr. Ahmad has published over 40 research articles, predominantly in top-tier, Q1 journals including Virtual Reality and the International Journal of Human-Computer Interaction. His scholarly work has accumulated over 1,550 citations with a cumulative impact factor exceeding 110. As a Principal and Co-Investigator, he has secured nearly £1 million in research funding from agencies such as the NSFC, CAS, and HEC. Dr. Ahmad serves on the editorial boards of Nature Portfolio's Scientific Reports and Springer Nature's Humanities & Social Sciences Communications, and has edited ten volumes for major publishers including Cambridge University Press, Springer, and Elsevier.

Sohaib Ahmed is working as a Senior Lecturer in the School of Computer Science, Faculty of Innovation and Technology, Taylor's University, Malaysia. He received PhD in Information Technology (IT) from Massey University, New Zealand. In addition, he received his Postgraduate Diploma (PGD) from Massey University, New Zealand with distinction. Earlier, he graduated as BS in Computer Engineering from Sir Syed University of Engineering & Technology, Karachi, Pakistan. He has around 18 years of teaching and research experience. He has published more than 40 peer-reviewed articles in high-quality journals and conferences. He voluntarily provides editorial services to various high-impact factor journals. He participated as an organizing committee member at international conferences. His research interests include AI in education, educational technology, AI in software engineering, gamification, and agile software development. He holds certifications in Certified Scrum Master (CSM) and Gemini Certified Faculty.

Fatma Alkan is works as professor in Hacettepe University Faculty of Education, Chemistry Teaching Program. She has published many articles in national and international peer-reviewed journals. She works as a field editor in peer-reviewed journals.

Rhytheema Dulloo, Ph.D. in Marketing, is working as a Professor with Mittal School of Business, Lovely Professional Univeristy, India. She is a leading researcher at the forefront of understanding consumer behavior in the digital age. Her doctoral dissertation, "Consumers' online shopping behavior through use of mobile apps in the e-tailing industry," delves into the complex world of mobile commerce and its impact on consumer decision-making. This work has laid the foundation for her continued exploration of the consumer landscape, shaped by her keen interest in AI, marketing, consumer behavior, and retail management. Dr. Dulloo's scholarly contributions have achieved significant recognition in the global academic community. She has published 3 books with IGI Global, USA, which are indexed by Scopus—one of the world's most prestigious academic databases. From AI-powered marketing strategies to bridging global educational divides, these book publications represent years of collaborative research and knowledge sharing. Dr. Dulloo's research prowess was recently recognized with the "Best Research Project" award for her study on "Voice Based Artificial Intelligence Technologies: Study on User Acceptance in the Era of Digital Disruption" and "Driving Sustainable Consumption through Immersive Technologies: Exploring Consumer Motivations for VR/AR in Indian Digital Marketing." These accolades demonstrate her profound understanding of the transformative power of AI and its potential to reshape the marketing landscape. Dr. Dulloo's thirst for knowledge extends beyond the classroom. Her participation in the Singapore immersion Study Tour and Interdisciplinary workshop at Management Development Institute Singapore, further broadened her perspective and exposed her to cutting-edge insights on marketing strategies in the global marketplace. Driven by an insatiable curiosity and a commitment to continuous learning, Dr. Dulloo actively invests in her professional development. Her academic journey extends beyond her Ph.D., boasting completed courses from prestigious institutions like the University of London (Research Methodology), University of Virginia (Introduction to Personal Branding), University of Illinois at Urbana-Champaign (Marketing Management) to name a few. This diverse academic exposure has equipped her with a comprehensive understanding of marketing fundamentals and the latest trends shaping the industry.

Barno Matchanova is working at Department of National Idea and Philosophy, Urgench State Pedagogical Institute, Uzbekistan.

Eylül Özülüş is currently pursuing a master's degree in the Chemistry Education Program within the Department of Mathematics and Science Education at Hacettepe University in Ankara, Turkey. She completed her undergraduate education as the top student in the Chemistry Education Department at the Faculty of Education, Hacettepe University. Her research interests include chemistry education, teacher education, organic chemistry, and digitalization in education.

P. Selvakumar currently serves as an Associate Professor in the Department of Science and Humanities at Nehru Institute of Technology, Coimbatore, Tamil Nadu, India. With a teaching career spanning over a decade, he previously held positions as an Assistant Professor in the Department of Science and Humanities at various engineering colleges in Coimbatore from 2011 to 2023. Dr. P. Selvakumar's primary area of academic and research interest is phytochemistry, where he has made substantial contributions. His research credentials include 30 journal publications, 14 patents, and 25 conference papers. He has actively promoted faculty and student development by organizing 5 Faculty Development Programs (FDPs) and completing 3 NPTEL courses along with 60 Coursera certifications. His dedication to student success is evident in his role as a committed mentor, consistently guiding and supporting his students academically and personally. Recognized for his excellence in teaching, Dr. P. Selvakumar has received numerous accolades, including Best Teacher Awards. He is also an active member of MISTE (The Indian Society for Technical Education) and continues to play a vital role in both acade

Priya Rani is a Research Scholar in the area of finance, with a focused research interest in Environmental, Social, and Governance (ESG) frameworks, corporate finance and sustainability.

Göran Roos is a globally recognised expert in intellectual capital management, manufacturing competitiveness, industry policy, and innovation systems. With around 19,000 Google citations and more than 300 publications, he has established himself as a leading authority on strategic management, economic complexity, and sustainable industrial development. Academic Credentials and Appointments Professor Roos holds a PhD from the University of South Australia (2019), an MBA from INSEAD (1989), and an MSc in Engineering Physics from Chalmers University of Technology (1982). His current academic positions and fellowships include: • Adjunct Professor at the Institute for Sustainability, Energy and Resources (ISER), The University of Adelaide (2023-2025) • Fellow of the Australian Academy of Technological Sciences and Engineering (ATSE) since 2014 • Fellow of the Royal Swedish Academy of Engineering Sciences (IVA) since 2015 Areas of Expertise Intellectual Capital and Knowledge Management Professor Roos is a world-renowned expert in intellectual capital, having developed theories and implemented practical solutions that have shaped the

Ayşem Seda(Önen) Yücel is currently working as a Professor in the Department of Mathematics and Science Education at Hacettepe University, Ankara, Turkey. Prof. Dr. (Önen) Yücel holds a B.Sc. and M.Sc. degree from the Faculty of Education at Hacettepe University, and a Ph.D. in Chemistry from the Faculty of Science at Ankara University. Her research interests include organic chemistry, chemistry education, and teacher education. She has over 25 years of teaching experience in chemistry education. Prof. Dr. (Önen) Yücel has published numerous papers in prestigious journals and has participated in various conferences.

Index

A

B

C

D

E

F

G

W

www.ingramcontent.com/pod-product-compliance
Lightning Source LLC
LaVergne TN
LVHW061526290926
849162LV00004B/687